Proceedings in Life Sciences

Localization and Orientation in Biology and Engineering

Edited by
D. Varjú and H.-U. Schnitzler

With 196 Figures

Springer-Verlag
Berlin Heidelberg New York Tokyo 1984

Prof. D. Varjú
Lehrstuhl für Biokybernetik
Universität Tübingen
Auf der Morgenstelle 28
7400 Tübingen

Prof. H.-U. Schnitzler
Lehrstuhl für Tierphysiologie
Universität Tübingen
Auf der Morgenstelle 28
7400 Tübingen

ISBN 3-540-12741-0 Springer-Verlag Berlin Heidelberg New York Tokyo
ISBN 0-387-12741-0 Springer-Verlag New York Heidelberg Berlin Tokyo

Library of Congress Cataloging in Publication Data. Deutsche Gesellschaft für Kybernetik. Kongress (8th: 1983: University of Tübingen) Localization and orientation in biology and engineering. (Proceedings in life sciences). 1. Orientation (Physiology)–Congresses. 2. Biological control systems–Congresses. 3. Bioengineering–Congresses. 4. Guidance systems (Flight)–Congresses. I. Varjú, Dezsö, 1932–. II. Schnitzler, H.-U., 1939–. III. Title. IV. Series. QP443.D48 1983 001,53'2 83-20172

Typesetting, printing and bookbinding: Brühlsche Universitätsdruckerei, Giessen
2131/3130-543210

At its last meeting on March 22, 1983, the council of member organizations of the German Society of Cybernetics has elected

Professor Dr. Ing. h.c. Winfried Oppelt

as its Honorary President. On this occasion the German Society of Cybernetics is devoting these proceedings of its last conference to Professor Oppelt.

Professor Oppelt was born on June 5, 1912, in Hanau. On finishing school he studied Technical Physics at the Technical University in Darmstadt. He began his professional career in 1934 at the German Research Institution of Aeronautics in Berlin. In 1937 he became a member of the Anschütz Corporation in Kiel and in 1942 a member of the Aeronautics Corporation at Hakenfelde near Berlin. In 1945 (after the war) Professor Oppelt worked at the Wöhlerinstitut in Braunschweig and later (1949 to 1950) became head of a research and development laboratory of the Hartmann and Braun Corporation in Frankfurt. In 1956 he was appointed Professor at the Technical University in Darmstadt, a remarkable honour, as this was the first chair in Germany in Automatic Feedback and Control Technique.

Professor Oppelt has received numerous honours and awards. In 1965 he was elected Doctor honoris causa by the Technical University in Munich; 1971 he received the Grashof award of the VDI, and in 1980 the Ring of honour of the VDE.

Since its foundation, Professor Oppelt has been an active member of the German Society of Cybernetics. We gratefully acknowledge his valuable advice and his suggestions that have contributed tremendously to the value of the scientific activities of our Society.

For the Presidency of the   W. von SEELEN
German Society of Cybernetics

Preface

The German Society of Cybernetics organizes international conferences on selected interdisciplinary topics in regular 3-year intervals. The aim of these meetings is to bring together scientists who work in quite different disciplines, but are confronted with related problems and use the same or similar approaches. The topic of the 1983 conference which was held on March 23–25 at the University of Tübingen came from a typical field of research in which engineers, biologists, and physicists share a common interest.

We do not want to discuss here in detail the common principles which are used by nature and by engineers to solve the problems associated with localization and orientation, since the reader will find enough examples in this volume. The question, however, whether the participants of such meetings can really profit from each other, deserves some further consideration.

First, there is the difficulty of finding a common language. This still seems to be a problem, although in some fields the language of engineers and biologists has become very similar over the years, an impression we also gained during the conference. Most of the authors made a great effort to use a vocabulary which is understandable to people outside their own field of research, but, admittedly, not all succeeded.

Second, there is the question whether the chance to exchange ideas in such an interdisciplinary meeting is mutual. About two decades ago the representants of the "bionics" movement believed that the best solution to some engineering problems could be found by looking at how animals cope with similar tasks. They initiated biological research in the hope that the aquired knowledge might be applicable to engineering problems. This has so far not been very successful. Apparently, not all the principles nature invented during evolution are also optimal (and economical) solutions for technological problems. Furthermore, many useful techniques had been discovered and applied by physicists and engineers long before biologists recognized that animals also make use of them. It appears, therefore, that knowledge of technology more often helps biologists to understand *their* systems, or at least to ask the right questions, than vice versa. Also, some of the new developments in technology offer new and powerful tools for data acquisition and processing in biological experiments.

We believe, however, that the conference has been – and these proceedings will be – beneficial to both biologists and engineers. The latter might learn how nature developed systems which make use of information from several sensory modalities to achieve an "optimal" performance. Even if this knowledge will rarely be of immediate use, we hope that the representatives of the other disciplines will at least be entertained by learning how animals manage to find their ways about. After all, their performance is often astonishing considering that for data processing some of them use brains occupying less space than a pinhead.

The congress has been sponsored by the Deutsche Forschungsgemeinschaft, the Ministerium für Wissenschaft und Kunst Baden-Württemberg, by the Companies Daimler Benz AG, IBM Germany, Karl Zeiss AG, EAI Aachen, Dynamit Nobel AG, and generously supported by the president and administration of the University Tübingen. We gratefully acknowledge all the financial and administrative help.

Tübingen, January 1984   D. VARJÚ

                                                            H.-U. SCHNITZLER

Contents

Chapter II Orientation and Path Control

**Chapter III Localization and Identification of Targets by Active
Systems**

Chapter IV Navigation, Bird Migration and Homing

Contributors

You will find the addresses at the beginning of the respective contribution

Chapter I Localization, Identification and Tracking of Signal Sources

The Problem of Orientation in Photogrammetry from the Phototheodolite to Earth Resources Satellites

F. ACKERMANN[1]

1 The Geometrical Concept of Photogrammetry

The term photogrammetry has been in use for about 100 years, its two constituents, "image" and "measurement", representing the two essential components which characterize the discipline. Until recently these concepts were restricted to photographic images and to purely geometrical measurements. Photogrammetry is in principle an indirect measuring method as it refers to the image as a basic tool. The image is used in two ways: First, it is the source of information for the interpretation process, i.e. for recognition and identification of objects, and secondly is the basis for measuring and deriving geometrical properties of objects, including their horizontal and vertical positioning.

The first photogrammetric utilization of photographs took place shortly after the invention of photography. The French Colonel Laussedat tried, around 1850, to derive from photographs, by graphical construction, a cadastral plan of the village of Versailles. In Germany, the architect Meydenbauer introduced the method in 1858 for architectural surveys.

These initial examples led to today's range of application of photogrammetry. The majority of applications are in the field of surveying and mapping. On the other hand, there is a great variety of special applications in science and engineering, as everything which can be photographed can be the object of photogrammetric measurements.

The theory of photogrammetry has, until recently, concentrated entirely on the geometrical aspects of image mensuration. The aspect of image contents and image interpretation did not receive any methodical attention for a long time. It was considered trivial, especially in connection with stereoscopic viewing of stereo-images, which has been a standard photogrammetric technique for the three-dimensional recognition and mensuration of objects since about 1900.

The development of photogrammetry started with terrestrial photogrammetry. From about 1920, the systematic and industrial development of air survey methods took place, with automatic cameras, precise measuring and plotting instruments, and specialization in application of surveying and mapping. During the last 20 years electronic data processing has developed photogrammetry to a new level of performance representing the present status of application.

In practice, aerial photographs are taken from flight altitudes between 500 m and 12 km, conventional photo scales of wide-angle or super-wide-angle photography

---

[1]   Institut für Photogrammetrie, Keplerstraße 11, 7000 Stuttgart, FRG

Localization and Orientation in Biology and Engineering
ed. by Varjú/Schnitzler
© Springer-Verlag Berlin Heidelberg 1984

range from about 1:3,000 to smaller than 1:100,000. The accuracy of photogrammetric positioning is in the order of 5 μm or better in the photo scale. The vertical measuring accuracy is in the order of 1/10,000 to 1/20,000 of the flight altitude.

2 Image Orientation and Object Determination

The mathematical theory of photogrammetry considers an image with regard to its geometrical properties as a perspective projection of the three-dimensional object. The object may be extended and covered by a set of partially overlapping photographs. The restitution reverts the perspective projections, leading to the geometrical reconstruction of the object. The theory systematically considers the relations between object and images. For the reconstruction of the object, three separate steps can be distinguished:

— reconstruction of the perspective bundle of rays, related to the image points of an image. The position of the perspective centre in relation to the image known as interior orientation (and knowledge about image distortion) is required.
— external orientation of the bundle of rays with regard to the reference coordinates system, described by six independent parameters, referring to position (X_0, Y_0, Z_0) and angular attitude (ω, ϕ, κ).
— reconstruction of the three-dimensional object requires at least two partly overlapping orientated bundles of rays. The object points are determined by intersection of related rays.

Generally the three steps are to be solved simultaneously. Normally the parameters of interior orientation are considered to be known from previous camera calibration.

In the case of terrestrial photogrammetry, the parameters of external orientation used to be treated as known, as they could be measured directly by geodetic resection or be fixed by level bubbles. For that purpose, phototheodolites, a combination of camera and theodolite on a tripod, were designed. However, today's numerical methods allow the orientation parameters to be treated as unknowns.

For aerial photographs the direct measurement of the external orientation elements is not possible or precise enough, so they are introduced as unknowns. They are indirectly determined with the aid of control points, the coordinates of which are known, referring to the external reference system.

The 12 known coordinates of 4 control points (in general location) are barely sufficient in determining the 12 unknown external orientation parameters of pairs of photographs which is the basic unit of photogrammetric restitution of spatial objects.

Prior to approximately 1960, the numerical solution of 12 non-linear equations was extremely tedious. The orientation methods applied in connection with analog, optical or mechanical projection instruments for pairs of photographs therefore used non-numerical analog procedures (systematic trial and error with intermediate remeasurement). The orientation was split into two phases. First, in the "relative orientation," only the intersection conditions of the two bundles of rays were taken into account, leading to a "model" of the object. The second phase took care of the remaining seven orientation parameters required for the orientation and scaling of the model, (and the tied-in bundles of rays) with regard to the external coordinate system. This procedure

Fig. 1. The multi-station problem of photogrammetric orientation and point determination

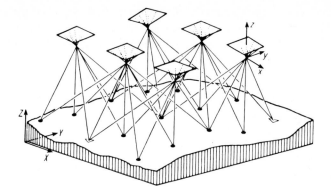

reduces the required minimum number to two planimetric and three vertical control points. However, for reasons of symmetry and accuracy, at least four control points of either type were usually used.

The solution can fail theoretically in individual cases, which, however, are easily circumvented in practice.

Since approx. 1960, computers and numerical procedures permit at the same time the simultaneous orientation of all photographs of an extended photo-coverage (consisting of many photographs) and the determination of all object points involved.

It is still an indirect determination of orientation parameters by means of some known control points. As all tie-relations, given by the conditions of intersection of rays, are also taken into account, the required number of control points can be kept to a minimum. Economic application is mapping of large areas which contain only wide-spaced or incomplete geodetic networks, still the case in many countries.

The aerial photographs are in fact taken sequentially. Nevertheless the geometry of the joint set is considered without the time-component and solved simultaneously, which is permissible as long as the object remains unchanged in time. The conventional photogrammetric survey of moving or changing objects requires simultaneous photography.

The numerical solution of the multi-station case makes use of the general perspective relationships between object points and image points:

$$x_{ij} + v_{ij}^x + \Delta x_{ij} = x_{oj} + c_j \frac{a_{11}^j(X_i - X_{oj}) + a_{21}^j(Y_i - Y_{oj}) + a_{31}^j(Z_i - Z_{oj})}{a_{13}^j(X_i - X_{oj}) + a_{23}^j(Y_i - Y_{oj}) + a_{33}^j(Z_i - Z_{oj})}$$

$$y_{ij} + v_{ij}^y + \Delta y_{ij} = y_{oj} + c_j \frac{a_{12}^j(X_i - X_{oj}) + a_{22}^j(Y_i - Y_{oj}) + a_{32}^j(Z_i - Z_{oj})}{a_{13}^j(X_i - X_{oj}) + a_{23}^j(Y_i - Y_{oj}) + a_{33}^j(Z_i - Z_{oj})}$$

x_{ij}, y_{ij} measured plane cartesian coordinates of image point ij (image of object point i in photograph j)

v_{ij}^x, v_{ij}^y least-squares corrections of image coordinates

x_{oj}, y_{oj}, c_j parameters of interior orientation, mostly known from camera calibration, ($c_j \approx$ focal distance of camera)

X_i, Y_i, Z_i cartesian coordinates of object point i, unknown except for control points

X_{oj}, Y_{oj}, Z_{oj} unknown parameters of exterior orientation, coordinates of perspective centre j

$(a_{11} \ldots a_{33})_j$ unknown orientation parameters, elements of the orthogonal rotation matrix which are non-linear functions of, for instance three independent rotations ω, ϕ, κ which describe the angular attitude of photograph j

$\Delta x_{ij}, \Delta y_{ij}$ correction terms with additional unknown parameters (of the interior orientation) with which unknown systematic image deformation may be corrected.

The total system is set up by letting the indices i and j run through the full range of terrain points i and photographs j. The non-singularity of the system is usually ascertained by geometrical considerations (overlap, location, and number of points). Formal mathematical concepts (eigenvalues and condition of the equation system) are normally inapplicable because of the size of the systems.

The numerical solution of the multi-station case has to overcome a number of difficulties:

— The systems of equations can be very large, up to 10^4 unknowns or more (for instance 1,000 photographs and thousands of object points).
— The equations are non-linear. Approximate values for linearisation have to be obtained automatically from the given data. They are initially imprecise therefore requiring several iterations.
— The least squares normal equations, or the partially reduced normal equations after algebraic elimination of the group of unknown object coordinates, have to be given an optimal structure in order to reduce the computing time. Special algorithms have been developed (minimum bandwidth, progressive partitioning).
— The equation systems are still to large to keep all non-zero coefficients in the core memory even of large computers. Efficient data shifting between internal and external memory has to be organized.
— The possibly large number of given observations (10^4 or more) require efficient data management. In particular the tie-connections have to be established automatically, by elaborate search routines.
— Large sets of observational data always contain some errors. At present algorithms are developed for automatic identification and elimination of gross data errors by applying robust adjustment methods and "data snooping" tests.
— For the evaluation of results, statistical tests and accuracy indicators are also desired.

During the last decade, operational computer programs have been developed for large main frame computers as well as for mini-computers, and are now widely and most successfully used in aerial survey practice. Also, the most favourable accuracy and reliability features of the multi-station solution have been thoroughly investigated.

The multi-station solution described still determines the orientation parameters indirectly, raising the question whether they could not be measured directly. In fact, attempts were made already during the 1930's to directly measure at least some of the orientation paramters of aerial photographs during the photo flight, for instance by barometer, solar periscope or horizon camera. However, such auxiliary orientation data, as they are called, are still not accurate enough for direct use, but can be most helpful if introduced as additional observations into a joint least squares adjustment, together with all photogrammetric data.

At present only differential barometric measurements by instruments called stato-scopes are used, in combination with vertical terrain profiles, measured from the air-craft by radar or laser. Their vertical accuracy is better than 1 m (0.1 mbar), consti-tuting highly effective measurements which allow considerable reduction of the re-quired vertical control points.

In the near future, additional camera orientation data derived from inertial navigation systems will be utilized, giving absolute or relative positions for the air stations as well as tilt information. Unfortunately inertial systems are still quite expensive and not generally used in air survey missions.

Attention is further paid to the American GPS (Global Positioning System) satellites. Eventual-ly there will be 24 geostationary satellites in polar orbits which will allow continuous positioning anywhere on the earth and in the air.

None of the navigation systems is accurate enough to allow direct determination of camera orientation parameters in real time. However, used as additional data in the subsequent joint least squares adjustment they will contribute to the solution of the orientation parameters most effec-tively as in the combination with photogrammetric measurements some of the errors of naviga-tional data (such as drift) can be compensated.

The described methods of photogrammetric orientation and object determination characterize the present status of photogrammetry and are regularly and widely applied in practice. The concept is still entirely geometrical, off-line, and passive, containing no dynamic features, and is therefore hardly interesting from the cybernetic point of view. This situation is beginning to change in connection with the development of remote sensing. Before coming to this, attention is briefly directed towards digital terrain models and image correlation, two developments which evolved in photo-grammetry but which reach beyond its classical concept. They are the result of in-creased application of electronic and digital techniques:

Digital terrain models (DTM), especially in the special form of digital elevation models (DEM), describe a (terrain) surface by a set of points, preferably arranged in a regular grid. They can replace the conventional description of topography by contour maps, or they may be used to derive contour maps. The development of DTM's in photogrammetry began simply, referring to conventional stereoscopic measurement of heights from which the grid points were calculated by more or less elaborate inter-polation procedures, gaining however, a new dimension by the large number of points involved (and in connection with image correlation, see below). A DTM extending over the area of 1 map sheet of a 1:50,000 scale topographic map may contain about 1 million grid points. The Gestalt Photo Mapper, a Canadian instrument for producing differentially rectified orthophotomaps, which operates with image correlation, would deal in this case with about 20 million grid points per map sheet. In addition, digital terrain models are becoming independent products in their own right in multiple use in engineering and science, and are being issued for large regions or entire countries.

Image correlation was introduced in photogrammetry as far back as 1958. It was particularly intended to automatize parallax measurements between stereo-images, thus replacing human stereoscopic vision and stereoscopic measurements. Initially the signals obtained by electronic scanning of local image areas by cathode ray tubes were correlated by electronic means. Now one- or two-dimensional arrays of semi-conductor sensor elements (CCD's) are used and the processing has advanced to digital techniques.

Image correlation can be applied off-line. However, it is here particularly interesting to consider the dynamic mode of operation in real time. Attached to photogrammetric plotting instruments image correlators can scan vertical terrain profiles (or pursue contour lines, as was the case in the first "stereomat"-instrument). They also can assist human operators who trace and plot planimetric features by keeping the three-dimensional measuring mark automatically on the apparent terrain surface.

A particularly advanced design, the Gestalt Photo Mapper mentioned above, uses the parallax discrimination faculty of image correlation for automatic relative orientation of photo-pairs. It also produces differentially rectified "orthophotos" which can be equated to transforming a perspective projection of the terrain into an orthogonal projection. (There are still some difficulties by parallax discontinuities, as in the case of trees or buildings). As a side product dense digital elevation models are issued (with about 800,000 points per stereo-overlap).

The essential point here is that such instruments operate in real time, and that speed is essential for economic reasons.

3 Extension of Photogrammetry by Remote Sensing

During the last 20 years techniques of remote sensing have evolved, and can be seen as a vast and general extension of photogrammetry. There are different directions of extension to be distinguished:

— While airplanes continue to be widely used as platform carriers for cameras and remote sensing equipment, *satellites* have also been in use for more than a decade. Well-known are the earth resources inventory satellites LANDSAT 1–4 (900 km altitude), SEASAT or the French satellite SPOT to be launched in 1984.
— A more fundamental extension is obtained by the development of sensors covering a much wider range of the electromagnetic spectrum of radiation than conventional photography. Three major steps can be distinguished; the first being the extension of photography into the near infrared. Combined with colour photography it led to infra-colour (or false colour) photographs. A parallel development is known as multispectral imagery in which several images (from cameras or scanners, see below) are obtained at the same time, each covering only a narrow band of wavelengths. The second extension with regard to wavelengths concerns the (non-photographic) recording of thermal radiation at wavelengths of 10–14 μm. Ground and water have a natural radiation in that range which coincides with an atmospheric window. The third extension is represented by radar with wavelengths in the cm range. The side-looking radar, especially can provide small-scale radar imagery of large areas, a well-known example being the radar maps of the complete Amazonas region.
The sensor development not only constitutes an extension with regard to the range of wavelengths, but also provides different kinds of information beyond geometry, in particular physical information about wavelengths λ and possibly polarization of passive or reflected active radiation.

— Most remote sensors obtain imagery by scanning techniques, as geometrical information is not the primary objective any more. This implies that we deal with more general image geometry and image distortions as the geometry of perspective projection is no longer maintained. Geometrical quality is partly restored recently by the current development of linear or two-dimensional array cameras, which provide directly digital output.

— Scanners and array cameras provide directly digital imagery, essentially similar to digitizing of photograhic images. As a result the methods of digital image processing are of fundamental importance in photogrammetry and remote sensing, as they reveal the full range of possibilities related to image enhancement, image analysis, image classification, and pattern recognition.

The extent to which the technology of remote sensing has widened and superseded the conventional range of photogrammetric techniques, tasks and possibilities, is clear. It is also evident, however, that the development has been directed mainly towards the field of image interpretation, leaving geometrical considerations largely aside. Therefore, typical applications of remote sensing, relate in the first case to the "soft" geosciences and to regional or global environment monitoring (geology, forestry inventories, vegetation, land use reviews, natural resources inventories). This will not be pursued here any further.

On the other hand, geometrical quality is gaining increasing importance in remote sensing, perhaps not as a primary aim but at least as a necessary precondition for elaborate image information processing. We have therefore, with image geometry and digital image processing, two main fields where strictly photogrammetric considerations and remote sensing techniques meet very closely. In particular the classical photogrammetric problems of sensor orientation and of geometrical object reconstruction reappear in remote sensing.

Images from scanners generally have no fixed image geometry. The instant scan angle is a known geometrical quantity, but the orientation of a sensor is continuously changing. Each picture element (= pixel) is therefore associated with its own set of orientation parameters which are, in principle, a function of time. This general orientation problem is eased, however, in the case of satellites as they move very steadily. Array cameras have common orientation for all elements of a fixed array.

For the solution of the orientation problem of scanner images three cases may be distinguished: (1) In the simplest case the complete image is oriented onto a reference system, represented by some control points, only by a simple transformation. This is permissible if the accuracy requirements are low. (2) The general formulation of the orientation problem would, in principle, apply perspective relationships as described for each pixel. With control points and some interpolation principle with regard to the changes of orientation a solution can in time be obtained. (3) With the help of image correlation scanner images can be matched with aerial photographs of known geometry and known orientation, even if their resolution may be different. A special case of that procedure is given if one scanner image is matched with another scanner image without any attempt to obtain absolute orientation data.

The geometrical reconstruction of an object is not the standard task of remote sensing images. However, the problem exists with regard to differential rectification of images, as two-dimensional image information is mostly to be processed and displayed with regard to a geometrical reference.

A standard objective is therefore the geometrical restoration and *rectification*, including differential rectification of scan-images. The solution involves three points: (1) The orientation problem, as mentioned above, (2) correction for earth curvature and for the effects of relief on the geometry of the image, and (3) consideration of the assumedly known geometry of the sensor.

Correction for earth curvature and distortion into a derived map projection presents no particular difficulty as the parameters are known. The differential rectification for relief displacement makes use of digital terrain models.

Particularly precise rectification of one scan image onto another is required for change detection from images taken at different times. Change detection may refer to image contents or to geometry and the technical solution is assisted by or may rely solely on image correlation.

In other applications of remote sensing vertical information must be obtained about the object rather than assuming it as known. In this case the principle of stereo-overlap is used as known from conventional photogrammetry. Particularly favourable solutions can be expected from array cameras. The necessary information can be obtained by linear arrays in convergent cameras or by special arrangement of linear arrays within one camera (three linear arrays arranged forward, centre, and aft). Matching of identical detail is again obtained by image correlation techniques. There are also other experiments which attempt to obtain vertical information from radar stereo-coverage.

4 The Photogrammetric Orientation Problem and Navigation

It is tempting, at the end of this review, to touch briefly on the problem of navigation and its inverse relation to photogrammetry. Navigation is clearly a dynamic orientation process which operates in real time, and is closely related to photogrammetry, as it is also based on given terrain information and on imaging sensor information obtained in flight (or from any moving vehicle). The common technical processes relate to digital terrain models and digital or optical image processing (image correlation and feature extraction).

The first example refers to robots equipped with stereoscopic vision, seemingly improbable, but the present development is in fact very closely related to photogrammetry. A robot may be equipped with two or more video cameras in fixed relation, which scan the surrounding space. On-line image correlation would produce an oriented digital terrain model of the environment in real time, and this information would be used to steer and control free-moving robots, capable of recognizing and avoiding obstacles.

The next step would be digital recognition of certain objects and of their orientation, enabling robots to grasp and handle objects or tools. It is evident, that the demands on real time performance are extremely high in this case.

Air navigation relies to a great extent on absolute positioning in regional or global coordinate systems (by inertial systems, or satellite navigation systems), while visual navigation maintains its importance in many cases. At present, great efforts are being made to automate this kind of navigation. One possible method is to scan the terrain by laser or radar and to identify by digital image analysis certain features such as rivers, shore lines, roads, railroads, outstanding vegetational features or related smaller objects. Navigational information would then be obtained by comparison (by digital image matching) with stored terrain data of the same kind.

An example worth noting is the utilization of relief information for navigational purposes. Instead of image comparison, in this case, a directly measured digital terrain model is compared with a stored one. Such methods are currently employed in military developments, however there is no doubt that many general applications are feasible and will emerge.

5 Conclusion

This paper has attempted to review the problems and the development of solutions concerning the orientation of images and, related with it, the derivation of object information from images. The review started with the conventional concepts of photogrammetry which were restricted to photographic images and purely geometrical considerations, then moved on via auxiliary orientation data and digital terrain models to the vast extension opened up by the variety of new sensors and the generalized concept of images. The techniques of image processing were shown to be of central importance, even when restricted to mainly geometrical aspects. The development makes full use of the potential of modern technology and has reached a level of performance which was by no means anticipated 20 years ago. Nevertheless we are just beginning to exploit the vast possibilities implied.

This development is a prime example of how a specialized and narrow discipline can emerge into much more general concepts and applications, and how it can influence and combine with other scientific and technical developments.

Digital Array Signal Processing Methods Applied for Advanced Direction Finding Operations

S. BLOCH[1]

1 Introduction

The main problem affecting the operation of direction finders relates to signal reflections. This problem is best known as multipath interference. The surface of the earth, for example, gives rise to significant multipath interferences. Waves reflected by the ground may be regarded as if they were emitted by an image of the signal source, located at a negative elevation. Ground interferences become most severe at low elevation angles as the magnitude of the reflected wave increases and the signal source and its image move close together. Here separation of the direct and reflected signal becomes increasingly difficult. This results in enhanced wave distortions and degraded direction finder (DF) performance. Ground multipath for this reason presents one of the main problems for microwave landing systems where signals transmitted by aircraft at low elevation (about $3°$) need to be precisely evaluated.

The accuracy of DF systems is further influenced by lateral reflections, caused by large topographical and man-made structures. These multipath interferences affect the accuracy of DF in the azimuthal plane. Depending on site conditions, a superposition of many lateral reflections may become very severe. In such cases their composite magnitude may become comparable to, or even larger than, that of the direct signal.

Spatial filtering methods may be used in order to separate the direct signal from the unwanted multipath information. Such methods require the use of wide antenna apertures. Two different approaches, both providing high angular resolution, are feasible: (a) large baseline interferometry; (b) narrow-beam forming techniques. This paper will focus on digital beam forming methods. It should, however, be mentioned that digital techniques may similarly be applied in order to enhance the performance of large baseline interferometers.

2 Discussion

Narrow-beam antennas are characterized by a high sensitivity to signals arriving from a confined spatial sector, referred to as the beamwidth. They are at the same time

---

[1]   Standard Elektrik Lorenz AG (ITT), 7000 Stuttgart 40, FRG

Localization and Orientation in Biology and Engineering
ed. by Varjú/Schnitzler
© Springer-Verlag Berlin Heidelberg 1984

insensitive to signals incident from other directions; these signals become attenuated prior to reaching the receiver front end. Thus, when a narrow-beam antenna is pointed at a signal source the relative amount of interfering signals is effectively reduced. The higher the antenna directivity the better the multipath suppression that can be achieved. Radars make wide use of high resolution directive beams. They utilize large parabolic antennas or array configurations in order to reduce the interference (clutter), and provide accurate angular target information. Unfortunately solutions used in the radar field cannot be readily adapted for DF's. Radars use dedicated high-power transmitters for the illumination of their targets. Their receivers are, therefore, provided with a priori information pertaining to the angle of incidence and (approx.) time of arrival of the incoming signals. Unlike radars, DF's depend on the energy emitted by external sources. In most cases no a priori information is available. Thus, when the DF is expected to detect and evaluate very short signal bursts of unknown origin a simultaneous observation of the entire space is required. The system has to react instantaneously whenever a signal is being emitted. The apparently contradicting requirement for high angular resolution on the one hand, and a continuous observation of the entire space on the other cannot be met by mechanically rotating parabolic antennas or even by electronically steered phased arrays here more sophisticated methods are needed. An appropriate beam-forming method will be described in this paper which is based on a combined operation of antenna arrays and fast digital signal processors. This method provides a high degree of DF accuracy coupled with fast operation allowing the processing of thousands of signal bursts per second. Radio signals $u(t)$, to be evaluated by a DF, may be presented by

$$u(t) = REAL[U(t)] = REAL[A(t)exp\{j(2\pi ft + \phi)\}exp(j\frac{2\pi}{\lambda}x)] ,$$

where $A(t)$ = signal envelope as a function of time (V), f = carrier frequency (Hz), λ = carrier wavelength (m), x = length of propagation path (m), ϕ = phase shift with respect to an arbitrary reference (rads).

The time function $U(t)$ describes the signal received as a rotating phasor of a length $A(t)$. The breakdown of $U(t)$ in two separate exponential terms underlines that the phase condition is a function of both time and propagation distance.

The temporal variations of the amplitude envelope $A(t)$ are usually considerably slower than the periodic changes of the signal carrier. Thus, for fast signal evaluations, lasting only for a few carrier periods, $A(t)$ may be assumed constant [e.g. $A(t) = A$, where A is the envelope's amplitude at the sampling instance].

Assuming the signal is received by an antenna configuration consisting of N omni-directional antenna elements which are grouped together to a so-called linear array (Fig. 1), the complex signals $U_n(t)$ seen by each individual element can then be denoted by

$$U_n(t) = A \ exp\{j(2\pi ft + \phi)\}exp(j\frac{2\pi}{\lambda}x)exp(j\frac{2\pi Dn}{\lambda}\sin\theta) \leqslant n \leqslant N-1 ,$$

where N = number of array elements, D = spacing between adjacent array elements, θ = angle of incidence with respect to broadside, x = path length between signal source and receiving element No. 0.

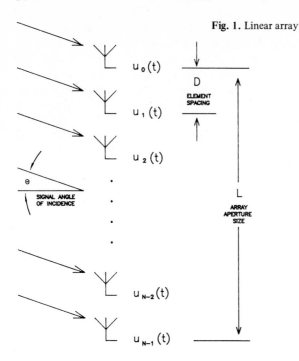

Fig. 1. Linear array

For further analysis it is convenient to relate all array signals to the signal at element No. 0. This gives

$$U_n(t) = U_o(t)\exp(j\frac{2\pi Dn}{\lambda}\sin\theta)\quad 0 \leqslant n \leqslant N-1.$$

In order to generate a directional beam pointed at an angle θ_o, the N complex signals $U_n(t)$ need to be multiplied by the following N complex terms,

$$W_n = \exp(-j\frac{2\pi Dn}{\lambda}\sin\theta_o)\quad 0 \leqslant n \leqslant N-1.$$

Following this multiplication, the N products are summed, thus giving a single time function Y(t).

$$Y(t) = \sum_{n=0}^{N-1} U_o(t)\exp(j\frac{2\pi Dn}{\lambda}\sin\theta)\,W_n$$

$$= U_o(t)\sum_{n=0}^{N-1}\exp j\frac{2\pi Dn}{\lambda}(\sin\theta - \sin\theta_o)$$

$$= U_o(t)\,V(\theta),$$

where θ_o = steering parameter, $V(\theta)$ = directivity factor.

The signal Y(t), containing the information of the entire array, is equivalent to the signal received by a single element (e.g. element No. 0) weighted by an angle-dependent amplification factor (directivity factor). A simple transformation gives

$$|V(\theta)| = \frac{\sin\left\{\frac{\pi ND}{\lambda}(\sin\theta - \sin\theta_o)\right\}}{\sin\left\{\frac{\pi D}{\lambda}(\sin\theta - \sin\theta_o)\right\}} \ .$$

For a selected steering parameter θ_o the system's sensitivity is shifted to an angular sector centred at $\theta = \theta_o$. Thus, the composite signal $Y(t)$ corresponds to the output of a directional antenna pointing at azimuth θ_o. The mathematical beam-steering procedure used here is comparable to the method applied electronically by so-called phased array antennas (which are utilized by advanced tracking radars).

The angular resolution of antenna arrays increases when their aperture becomes larger. In general the relationship between aperture length and antenna beamwidth can be expressed arithmetically by

$$BW = \frac{360}{\pi} \sin^{-1}\left(\frac{\lambda}{ND}\right) \approx \frac{360\lambda}{\pi L} \ ,$$

where BW = antenna beamwidth ($^\circ$), L = array aperture = $(N - 1) \cdot D$ (m).

It is a known property of antenna arrays that their angular resolution becomes excessively degraded at steering angles larger than 60°. For this reason, when omnidirectional coverage is required, orthogonal arrays, each providing coverage in a sector of max. $\pm 60^\circ$, need to be used.

Ambiguity problems, due to aliasing, arise when the spacing between adjacent elements become larger than half the wavelength of the signals incident upon the array. Thus, the number of array elements essentially depends on the aperture size. This means that for high angular resolution a sufficiently large number of elements is required.

The beam-forming method described so far applies in general to all systems using directional antenna beams (e.g. phased array radars). The specific requirements of high performance DF's necessitate a further step leading to the implementation of so-called multiple stacked beam systems. Such systems deploy a large number of interleaving antenna beams, and therefore allow a continuous observation of the entire space. For any signal occurrence there will be at least one antenna beam pointed at the right time in the right direction.

For high performance DF applications wide aperture arrays with a very large number of elements are needed. For the computation of a single beam about N complex multiplications need to be performed (N is the number of elements). It is obvious that an improved angular resolution necessitates the computation of an increased number of interleaving beams. Thus, the mathematical complexity of a multiple stacked beam system increases proportionally to the squared number of elements. Careful design, however, enables a significant reduction in the computational load.

It is, for example, useful to set the number of beams identical to the number of elements N (where N is preferably chosen as an integer power of two, e.g. N = 128).

Conducting the following multiple stacked beam forming operation

$$Y_k(t) = \sum_{n=0}^{N-1} U_n(t) \exp(-j2\pi nk/N) \quad 0 \leqslant k \leqslant N - 1 \ ,$$

provides an expression which is identical to the DFT (Discrete Fourier Transform) of $U_n(t)$. In this case, efficient arithmetical computations, known as FFT algorithms (Fast Fourier Transform), may be used in order to derive the set of N functions $Y_k(t)$.

We can prove by inspection that the N functions denoted $Y_k(t)$ correspond to the N outputs of a multiple stacked beam system with its simultaneous beams directed at

$$\theta_k = \sin^{-1} (\lambda k/DN) \quad 0 \leqslant k \leqslant N - 1 .$$

A closer analysis reveals that the N interleaving lobes denoted by the DFT expression are spread in half beamwidth steps over the entire observation sector. Figure 2 depicts a multiple stacked beam pattern computed for a relatively short (L = 7.5 λ) 16-element array.

Figure 3 shows the basic system configuration of a DF system based on the mathematical principles discussed above. The composite electromagnetic field incident upon the array aperture is sampled by N antenna elements. Each element is equipped with an individual receiver. The received signals are amplified and frequency converted

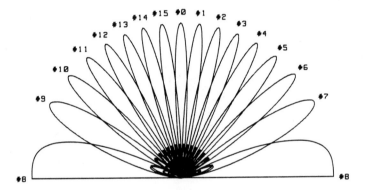

Fig. 2. Multiple stacked beam system. No. of elements N = 16. Aperture size L = 7.5 λ

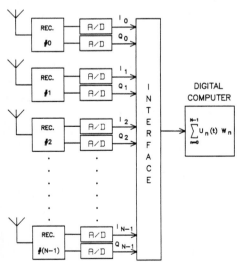

Fig. 3. Digital array signal processing basic system configuration

into base band (this means that the carrier component of the signal is being removed). In order to retain the full phase information, coherent receivers are used. Each of these double channel receivers derives two output signals referred to as the inphase-(I) and quadrature-(Q) components of the received signal u(t). The I and Q information corresponds to the real and imaginary parts of the complex signal $U_n(t)$, respectively. The I and Q baseband signals at the receiver outputs are converted into digital data using analog/digital-converters. From this stage on all further processing is performed digitally.

The digital data at the output of the multi-channel A/D-converter unit carry the full information of the field incident upon the array aperture at the sampling time instant. They actually present a hologram of the incident field. Once stored, this digital information is available for multiple stacked beam calculations. For this task a fast digital computer or dedicated FFT-Processor is used. State-of-the-art processors enable the computation of complex Fourier spectra within a fraction of a millisecond. Such processors allow the evaluation of thousands of signal bursts per second.

The operation of a system using a 60λ aperture and 128 elements is illustrated in Fig. 4. The magnitude of the 128 Fourier coefficients at the output of the DFT processor are shown for two different signal azimuth angles. In the upper plot Fourier coefficient No. 15 is exited, thus indicating a signal incidence from $15°$ azimuth. In the lower plot a signal arriving from $60°$ becomes visible at DFT output No. 50. This means that by simply searching the 128 DFT outputs for the Fourier coefficient with the largest amplitude, received signals can be detected, and their angle of arrival may be determined. As each DFT output corresponds to a highly directive antenna beam (beam width approx. $1°$), the individual outputs are practically unaffected by multi-path interferences and noise.

The implementation of the straight – forward mathematical DF algorithms discussed above presents no problem using todays digital computers. Unfortunately, the hardware realization of the array and of the multiple receiver unit sets a limiting factor. The cost of 128 antenna elements plus coherent receivers is prohibitive. Additionally, half a wavelength spacing between neighbouring omnidirectional array elements gives rise to excessive mutual antenna coupling problems (caused by signals reradiated by receiving array elements, thus perturbing the incoming field at their near proximity). For this reason the number of array elements needs to be reduced and the spacing between them increased. Such antenna configurations are colloquially referred to as "thinned" or "sparse" arrays. Careful design is needed in order to avoid the occurrence of system ambiguities as a result of the thinning procedure. This requires the spacing between elements to be non-uniform. Figure 5 illustrates the changes caused by significantly thinning the afore-mentioned 60λ array from its original 128 elements to merely 14 elements. For the new array configuration, the spacing between elements is sufficiently large to allow a satisfactory decoupling of adjacent receiving channels (no element is separated from its neighbours by less than 3λ). Regardless of the modified array configuration the DFT processing algorithm remains unchanged. The missing 114 signals at the processor input are simply replaced by the complex number 0. The 128 Fourier coefficients at the output of the processor are of course affected by the loss of signal information. The discrete Fourier spectrum at the processor output becomes, as Fig. 5 shows, more disturbed than in the case of

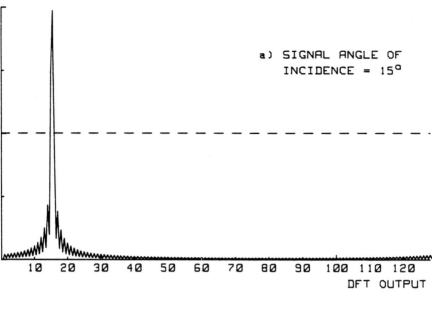

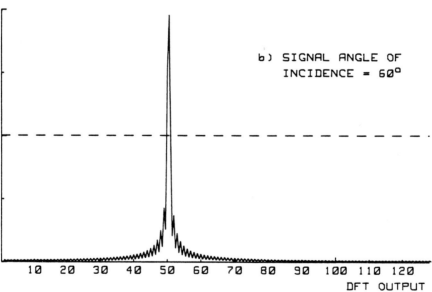

Fig. 4. Linear array configuration. Implementation of a virtual multiple stacked beam system with 128 simultaneous antenna beams. No. of elements 128: element spacing 0.45 λ: aperture size 57 λ

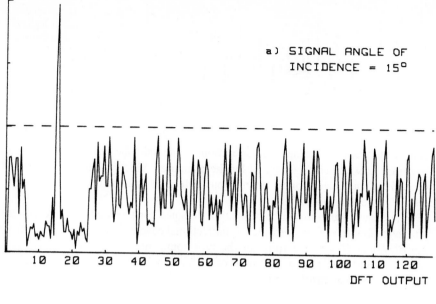

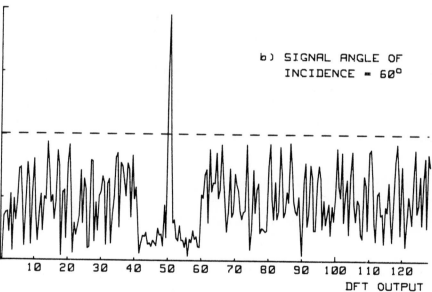

Fig. 5. Thinned linear array configuration. Implementation of a virtual multiple stacked beam system with 128 simultaneous antenna beams. No. of elements 14: element spacing >3.15 λ: aperture size 57 λ

Fig. 6. Orthogonal thinned array antennas for high performance DF operations

a full array, thus indicating a deterioration in the sidelobe characteristics of the 128 virtual stacked beams generated by the computer. Yet, the quality of the DFT output is still sufficient for an unambigous detection and evaluation of the signal received by the antenna.

Figure 6 shows an antenna configuration designed for the needs of precision aircraft navigation. In order to provide omnidirectional coverage, the antenna comprises two orthogonal 60λ sparse arrays of the type commented on above. This DF type provides azimuth accuracy in the order of $0.1°$ (rms) and is capable of coping with relative multipath intensities up to 85%.

3 Conclusions

The performance of antenna arrays may considerably be enhanced by the utilization of fast digital computers. The DF system presented in this paper demonstrates the potential of this state-of-the-art technique. The discussion touched some significant aspects of modern radio navigation (nav. aids), namely beamforming techniques, multiple stacked beam systems as well as sparse array concepts.

Remote Sensing and Imaging with Microwave Sensors

M. VOGEL[1]

1 Introduction

This discussion starts out from the fictitious situation in which a team of scientists and technicians with all theoretical, technical, and computational resources is working in an enclosure that is ideally transparent to microwaves (electromagnetic decimeter, centimeter, and millimeter waves with frequencies from 0.3 to 300 GHz), but opaque otherwise. The shielding properties of clouds and fog are very close to this assumption. What can the team learn about their environment?

2 Passive Sensing

2.1 Field Probing

The team could first try to probe the electric and magnetic wave-fields entering from the outside world by small dipole and loop antennas connected to tunable receivers. By amplitude and relative phase measurements throughout the enclosure with combinations of probes and by tuning through the microwave region they might, for a stationary situation, obtain a fairly complete picture of the microwave radiation fields in the enclosure. From this they could derive a map of the apparent distribution of sources outside, plus some information on their emission spectra.

2.2 Spherical Imaging Array

If the situation were non-stationary, all the data should be taken simultaneously. For this purpose the team could cover a large sphere of diameter D with N outward looking pairs of orthogonally polarized antenna elements, e.g. crossed dipoles. Output signals of $2 \cdot N$ elements would then be available in parallel for amplification and further processing, analog or digital.

One way of analog processing would be to combine the $2 \cdot N$ signals in a passive linear network to obtain $2 \cdot N$ new signals, each coupled to one of $2 \cdot N$ receiving

[1] Deutsche Forschungs- und Versuchsanstalt für Luft- und Raumfahrt, Institut für Hochfrequenztechnik, 8031 Oberpfaffenhofen, FRG

Localization and Orientation in Biology and Engineering
ed. by Varjú/Schnitzler
© Springer-Verlag Berlin Heidelberg 1984

patterns. A very convenient class of receiving patterns is characterized by dominant main lobes (pencil beams) with small side lobes. The lobes of the spherical array should sample the waves coming from all directions within the resolution the array dimension permits.

How many antenna elements are needed on the spherical array? The required number N can be estimated from the sampling theorem, if we consider that subarrays of two elements with the largest possible spacing D provide the highest resolution. The directional response in broadside direction is nearly sinusoidal, the angular period is $2 \cdot \frac{\lambda}{D}$ (λ = wavelength). Two samples are needed per period in angular spacings of $\frac{\lambda}{D}$ in both principal directions, giving a solid angle per sample of $(\frac{\lambda}{D})^2$ and a total number of samples $N = \frac{4 \pi D^2}{\lambda^2}$. The available area for each of N crossed pairs of dipoles on the spherical array of surface area πD^2 is then $(\frac{\lambda}{2})^2$; the elements are spaced by half wavelengths.

2.3 Microwave Camera

If omnidirectional coverage is not required but only a sector of (say) unity solid angle is of interest, a flat, e.g. square, antenna array with area $A = D^2$ and $\frac{4D^2}{\lambda^2}$ crossed dipole pairs will do. Dual polarization signal outputs for only $\frac{A}{\lambda^2}$ pencil beams in the sector are needed (some reduction of the number of antenna elements appears possible).

If separate receivers (radiometers) are connected to all outputs, a suitable display can present a microwave image of the observed sector. This display might well be a three-color display because three parameters are available from each pencil beam: amplitudes for both polarizations and the relative phase between the two signals. These parameters are related to three independent quantities of the four Stokes parameters of the incoming radiation.

If time permits, the number of receivers can be reduced and the image built up successively by beam switching or scanning. Any other antenna type – lens, reflector or horn – of the same aperture area will give similar imaging performance. In place of pencil beams an equivalent number of non-correlating multibeam receiving patterns could be used for image reconstruction. For close range imaging the antenna must be focussed.

In passive sensing range information is only indirectly obtainable from phase front curvature (or for a source with a known emission spectrum seen through a medium with a known dispersion).

The data rate from the channels is given by the frequency bandwidth B of the receivers. However, no useful measurements can be expected at this rate for thermal microwave radiation from the environment. (Man-made microwave emissions and their information content will not be discussed.)

The Rayleigh-Jeans approximation for Planck's black body radiation formula holds quite well in the microwave range. The average single mode power output $\bar{P}$ from an antenna looking at a perfect emitter surface is $\overline{P}$ = kTB, proportional to Boltzmann's constant k, the absolute temperature T of the black body and the frequency bandwidth B. Thermal radiation fluctuates, and the rms uncertainty ΔP_1 for each of the independent power measurements coming at the rate B from the receiver is just equal to the mean power $\overline{P}$. Averaging over many samples in time is necessary to obtain a meaningful value for the mean power and a clear radiometric image. For an ideal (noiseless) radiometer, integrating with a time constant τ, the uncertainty is reduced to $\Delta P = \dfrac{\overline{P}}{\sqrt{B \cdot \tau}}$.

For example $\dfrac{\Delta P}{\overline{P}}$ = 1%: $B \cdot \tau = 10^4$ samples are needed. For B = 10 MHz, this corresponds to τ = 1 ms. An antenna with A = 1 m^2 aperture has, for λ = 1 cm, $\dfrac{A}{\lambda^2} = 10^4$ resolution cells per unity solid angle. One radiometric receiver pair would need 10 s for a complete image (frame). Up to 10^4 receiver pairs could be applied to increase the frame rate of the passive microwave camera. A trade-off between system parameters has to be made for each specific application.

In radiometric practice the Rayleigh-Jeans equation is used to express received power P in terms of an equivalent black body temperature T, the brightness temperature.

2.4 Information Content of Thermal Microwave Radiation

The microwave brightness temperature of the universe is very low. In frequency regions where the atmosphere is transparent for microwaves (atmospheric "windows") the sky appears "cold". In first order the radiation intensity is proportional to the product of temperature and emissivity of the objects in the antenna beam. Emissivity is equal to absorptivitiy and adds up to one with reflectivity. Reflectivity at interfaces depends on electromagnetic material constants ϵ, μ (Fresnel's formulas for smooth interfaces), on surface roughness ($\dfrac{\lambda}{4\pi}$ is the critical scale length, Rayleigh roughness criterion) and on (semi-)transparent covering layers (e.g. vegetation, sand, ice; subsurface features!). Water content has a strong dielectric effect, increasing the reflectivity and internal attenuation. Ice, however, is quite transparent with low reflection. The roughness of water surfaces depends on the wind. Rain absorbs, emits, and scatters at shorter microwaves.

Polarisation effects can be expected at smooth interfaces at the Brewster angle and for 90° scattering in rain or vegetation.

Attenuation and emission in the clear atmosphere is caused by molecular absorption by oxygen and water vapor. This gives access to atmospheric temperature and water vapor content. Frequency dependent penetration depths (range dependent weighting functions) at the wings of O_2 or H_2O spectrum lines allow the measurement of profiles of temperature or humidity from satellites or the ground.

3 Active Sensing

3.1 Degrees of Freedom

Active sensing relies on illumination of the scene with strong microwave sources. The thermal background is considered as noise. Many degrees of freedom exist. Known waveforms can be generated and radiated into known illuminating antenna patterns. The backscattered waves, intercepted by the known receiving antenna, can be compared with the emitted reference waveform. "Questions" may be asked by waveform selection. In a feedback arrangement the waveform selection can be made to depend an received echos. Animal sonar systems are examples of optimized active sensors.

Our fictitious team in the enclosure could convert their passive imager into an active imager by adding a dual polarization microwave illuminator flood-lighting the scene and feeding the reference waveforms to the receiver channels.

3.2 Scattering Matrices

Any waveform can be represented by a superposition of continuous monochromatic waves. For one such spectral component, the comparison of received waves and emitted waves (each has two polarization components) results in a set of four relative amplitudes and four associated relative phases (from each pencil beam the scattering matrix with eight parameters). If reciprocity holds (exchange of receiver and transmitter gives the same result), six parameters are independent. Only five parameters are object descriptors because all phases have a common dependence on the distance between sensor and object. The scattering matrix depends on frequency, aspect and, if changes occur in the scene, on time. If completely known it defines the response of the target to any combination of transmitter and receiver polarizations and contains all information about the object measurable by backscatter experiments. Bistatic observations (illuminator and imager spatially separated) would give two more degrees of freedom.

3.3 Radar Cross Sections

The five object descriptors of the scattering matrix can be measured and expressed in terms of power ratios for at least 5 independent combinations of illuminator and receiver polarizations, or, normalized with respect to distance, in terms of five radar cross sections σ.

3.4 Range-Doppler Mapping

Active systems can measure range R from the frequency dependence of the echophase

$$\phi = \frac{4\pi}{\lambda} \cdot R = \frac{4\pi}{c} \cdot f \cdot R,$$ and range rate $\dot{R}$ (radial velocity) from the time dependence of

the phase (Doppler shift). Power ratios or radar cross sections can be mapped over a range-Doppler shift plane. Targets in the same range cell but differing in radial velocities can be separated.

The frequency bandwidth of the radar waveform determines the resolution in range, and the time duration of the Doppler analysis the resolution in range rate. A wide frequency sweep during a pulse (chirp) can make the range resolution much shorter than pulse length.

3.5 Synthetic Apertures

The availability of a reference phase from the illuminator allows sampling the echo field by a moving receiver (or by a moving illuminator/receiver combination) and processing the stored data in such a way that the final angular resolution corresponds to the synthetic aperture covered by the sampling.

In airborne sideward looking radar the *linear synthetic aperture* (flight velocity times the dwell time in the real antenna beam) increases proportional to range, and the ultimate azimuth resolution becomes equal to half the length L (not shorter than about λ) of the radar antenna, independent of range and wavelength. This is ideal for satellites or planetary probes, but the handling and processing of the very large amounts of data becomes a problem.

Objects moving steadily in the scene will be displaced in the image, and irregular movement will blur the image of the object.

Circular synthetic apertures are generated if an object on a turntable is rotated in front of a non-moving radar. Angular apertures of 1 radian give an azimuth resolution of $\frac{\lambda}{2}$ near the center, independent of range. For full rotations tomographic mapping is feasible without range resolution requirements. This technique is good for the identification of scattering centers on objects.

Two- and three-dimensional synthetic apertures can be used. This leads deeper into the realm of microwave holography.

3.6 Scattering Centers and Echo Fluctuation

Most secondary radiation amplitudes from the scatterer's surface elements cancel at the point of observation because the phase varies rapidly over the surface. Scattering centers, bright spots, occur at points where the phase is stationary. At these points we would, in the optical analog, see mirror images of the light source. The local shape determines the backscatter intensity and its wavelength dependence (λ^{+2} to λ^{-2}). An estimate of the scattering center intensity can be made if all ray paths in a phase tolerance region of $\Delta\phi = 1$ around a maximal or minimal path are considered equiphase while the rest is ignored.

In complicated scenarios, normally several to many scattering centers are in the antenna beam. All contributions add up coherently, and the total amplitude and phase is a matter of chance.

The echo intensity will be very sensitive to changes in aspect, wavelength and object shape and will tend to fluctuate. For controlled variations of wavelength or aspect the fluctuation contains information about the scattering center locations in range or azimuth.

A single radar measurement of a resolution cell containing more than one scattering center is not meaningful under constant conditions. An image will be speckled. Averaging over many decorrelated measurements is essential for a quantitative evaluation. For more than three scattering centers of equal magnitude, the rms power uncertainty for a single measurement is approximately $\Delta P_1 = \overline{P}$ or $\Delta \sigma_1 = \overline{\sigma}$.

3.7 Information Content of Scattered Coherent Microwaves

Radar systems can measure azimuth, elevation, range, and range-rate. Object information is contained in the scattering matrix or the set of radar cross sections as functions of aspect, frequency, and time. What has been said about reflectivity in Sect. 2.4 is also valid for active remote sensing. The information content of the full polarization scattering matrix is still a research topic, and is not discussed here.

4 Concluding Comments

The above discussion tried to convey some understanding of the capabilities of microwave sensors as artificial sense organs.

The state of the art was not discussed, although in the oral presentation examples of imagery, etc. were shown.

Not all capabilities are yet exploited. Further developments in the microwave technology of transmitters, receivers and antennas, especially for multichannel systems, are required. Absolute intensity calibration over a large dynamic range poses a difficult but important challenge. Storage of mass data from high data rate sources, and hardware and software for rapid handling and processing of large quantities of data are key requirements for high resolution imaging systems.

The most essential key to the future success of user related applications of microwave sensing is signature research, aiming at understanding the information transfer in wave-object interaction and at the development of inversion techniques for the extraction of user data from sensor data with low ambiguity. Ambiguity is inherent because normally much more object parameters are involved than are measurable. Conclusions have only a certain probability of truth.

In most cases evidence is extracted from spatial, spectral, and temporal sensor data with reference to contextual knowledge, sometimes in a very intelligent "Sherlock Holmes" technique. High resolution imagery gives many clues to the context. Symbiotic multisensor combinations may reduce the inherent ambiguity.

Empirical thematic calibration by learning is based on the mapping of multiparameter sensor data for known objects into a multidimensional signature space, hoping that cells of this space can be assigned to object classes of interest.

References

No references are given in the above text. Further reading may start out from contents and bibliographies of comprehensive books like

1. Skolnik MI (1970) Editor-in-Chief. Radar handbook. McGraw-Hill Book Company New York
2. Reeves RG (1975) Editor-in-Chief. Manual of remote sensing (2 volumes). American Society of Photogrammetry. Falls Church, Virginia (New edition in preparation for 1983)

The Application of Spread Spectrum Techniques in Satellite Orbit Determination and Navigation Systems

S. STARKER[1]

The principle of a *Spread Spectrum System* is shown in Fig. 1. A signal of a small bandwidth B_1 is spread over a wide frequency band B_2 for transmission. After being received, the signal is despread to the original bandwidth B_1.

The relation between bandwidth B_1, B_2, signal power S_1, S_2, and noise power density N_0 can be described by the well-known equation for the information rate I.

$$I = B_1 \, \log_2 \left(1 + \frac{S_1}{N_0 B_1}\right) = B_2 \, \log_2 \left(1 + \frac{S_2}{N_0 B_2}\right) .$$

From this equation follows

$$S_2 < S_1 \text{ when } B_2 > B_1 \text{ and } I = \text{const and } N_0 = \text{const} .$$

Apart from the advantage of saving signal power, the spreading of bandwidth is an efficient method for interference rejection and jamming resistance. Another advantage is the low density power spectrum of signals which causes negligble disturbancies to other systems and which is difficult to detect by unauthorized users.

There are divers methods for *signal spreading and despreading*: frequency modulation and demodulation with analog ("chirp") or digital signals ("frequency hopping"), or phase modulation with a broadband analog or digital modulation signal. In the latter case despreading of the received signal has to be performed by the narrow band correlation with a replica of the broadband modulation signal.

Special advantages can be obtained, if a binary Pseudo Noise (PN)-Code is used for signal spreading. Thus code division multiplexing can be realized with selective addressing capability and with good multipath resistance. Moreover this kind of spread spectrum signal can be used for high resolution ranging in a very similar way as with pulsed signals. In Fig. 2 the equivalence is shown between a PN-code signal and a pulsed signal by comparison of the autocorrelationfunctions $R(\tau)$. These are very similar when both signals have the same bandwidth and repetition rate, and when the PN-

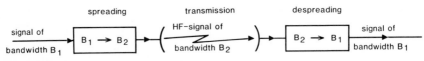

Fig. 1. Principle of a spread spectrum communication system

[1]   DFVLR = Deutsche Forschungs- und Versuchsanstalt für Luft- und Raumfahrt, Institut für Hochfrequenztechnik, 8031 Weßling/Obb., FRG

Localization and Orientation in Biology and Engineering
ed. by Varjú/Schnitzler
© Springer-Verlag Berlin Heidelberg 1984

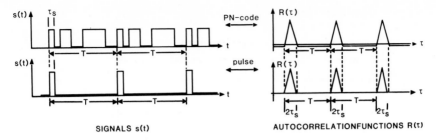

Fig. 2. Equivalence of pulse and PN-code signals with respect to ranging applications

code sequence is not too short. Because of this similarity accuracies of the same order can be expected from both signal types. But there are decisive technical advantages when PN-codes in connection with the Spread Spectrum Techniques are used: low continuous transmitting power can be used instead of high pulsed peak power; the correlation detection of the received signal allows despreading of the broadband signal to the small information bandwidth with very small delay time variations; and finally these PN-code ranging signals can be combined very efficiently with communication links. This last point is of special importance in navigation systems.

In Fig. 3 a functional block diagram of the technical realization of a Spread Spectrum System using PN-codes for ranging applications is given.

For the generation of the PN-code stable oscillators and shift registers are necessary in the transmitter and receiver. In the receiver the generated replica of the PN-code has to be shifted in such a position that the correlation $R(\tau)$ between the received signal and the replica is a maximum. After this code acquisition has been carried out by the Delay Lock Loop (DLL), the reproduced PN-code can be used for despreading the message.

A special state of the shift register in the PN-code generator (the "all-one state") can be used simultaneously as a time mark for ranging and time transfer measurements.

At the Institute for Radio Frequency Technology at DFVLR several experiments are going on with the aim of investigating future applications of this technique. Two-way ranging measurements of distances between ground stations and the geosynchronous satellites OTS-2 and Symphonie B were carried out. In both cases a modified PN-code was used with 250 bit codelength and with a bit rate of 100 kbit/s. The time jitter was $\sigma_t \lesssim 0.5$ ns by averaging over 2.5 ms, this is adequate to a statistical range error $\sigma_r < 10$ cm. The transmitting power during Symphonie measurements was 5 W.

Another experiment which will use and test spread spectrum techniques is the Space Shuttle experiment shown in Fig. 4. This is a navigation and time transfer experiment which will be flown in the first German Spacelab mission D1, scheduled for June 1985. One objective of the experiment is to synchronize clocks in distant ground stations and to control the onboard clocks with an accuracy better than 10 ns. Another objective will be to demonstrate one-way ranging methods applied to Shuttle orbit determination and to position determination of ground receiving stations. As stable onboard oscillators a Cs and a Rb clock will be used. The relativistic effect of these clocks is expected to be -25 μs per day.

TRANSMITTER:

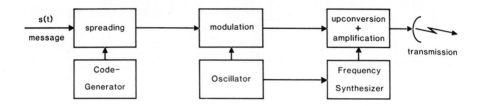

RECEIVER:

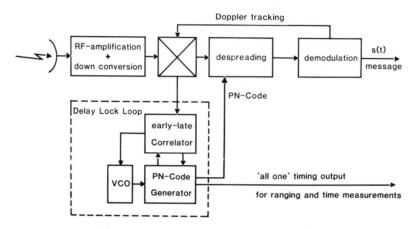

Fig. 3. Functional block-diagram of a spread spectrum system with PN-code for ranging applications

Another experimental activity concerns a satellite navigation system of the United States which is just under construction, called NAVSTAR-Global Positioning System (GPS). This system, as roughly visualized in Fig. 5, could be used by a large number of different users for positioning and time comparisons. PN-codes with bit rates of 1.023 and 10.23 Mbit/s allow positioning accuracies of better than 100 m, or 10 m and time comparisons of better than 50 ns, or 10 ns under special conditions. The satellite signals contain the navigation message and can be received with small spread spectrum receivers having omnidirectional antennas. An experimental receiver of this type was built at DFVLR and preliminary investigations have been carried out. The RMS of single measurements was $\sigma = 54$ ns. By averaging over 300 samples, received during 10 min observation time, the RMS of time comparison measurements could be reduced to $\sigma \approx 3$ ns.

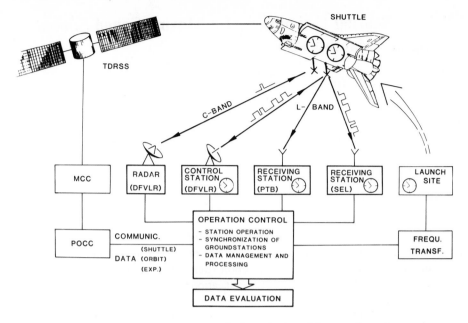

Fig. 4. NAVEX – A Space Shuttle experiment for investigation of time transfer and one way ranging methods using spread spectrum techniques

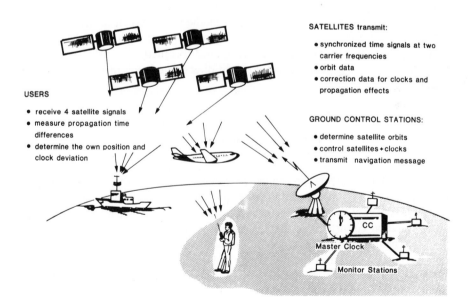

Fig. 5. NAVSTAR Global Positioning System

Passive Bearing Estimation of a Moving Sound Source Using an Adaptive Filter

U. JÄGER[1]

1 Introduction

Those concerned with noise analysis of vehicles often want to know from which point the noise is coming in order to understand its mechanism better and to control or identify it. By knowing the exact position over a given time, one can determine velocity and direction of movement. Applications range from simple detection to vehicle identification and traffic control.

A passive location is accomplished by taking at least two-site bearings, at the intersection of which is the position of the source. Because the sensors (microphones) are located nearby at the side of the road, one has to contend with effects such as diffraction, nonstationariness, reflections, and relatively fast-moving sources. Generally, the input is a broadband stochastic signal with fast-changing and unknown statistics. Passive bearing can be established by measuring the time delay between the signals of two geometrically separated microphones by means of a correlator. An alternative approach has been made here using an iterative procedure which estimates the time delay by means of an adaptive filter. This method was developed originally by the navy for underwater passive bearing purposes, working in the farfield region (Feintuch et al. 1978, 1979). We have successfully applied this method to vehicle detection as well as to estimating position, velocity and direction of movement in the nearfield region.

2 The Adaptive Filter Bearing Estimator

Adaptive signal processing can handle fast varying signals of either unknown or partly unknown statistics, which makes the system of value for everyday applications. Moreover, it is very easy to implant the necessary algorithm into a microprocessor.

The method consists of an adaptive filter lying in one of the microphone channels (Fig. 1). The filter coefficients (weights) are controlled by a LMS (least-mean-square)-algorithm, which updates the weights at each iteration according to $\vec{W}(j+1) = \vec{W}(j) + 2\,\mu e \vec{X}(j)$ in order to minimize the error between the unfiltered reference signal d(j) and the filter output y(j). Hence, for broadband input processes with time delays

[1] Fraunhofer-Institut für Informations- und Datenverarbeitung, Karlsruhe, FRG

Localization and Orientation in Biology and Engineering
ed. by Varjú/Schnitzler
© Springer-Verlag Berlin Heidelberg 1984

Fig. 1. Principle of adaptive
filter time delay estimation

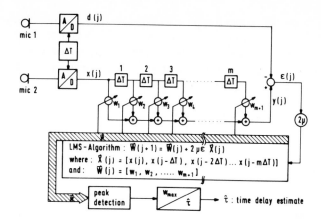

$\tau = n\Delta T$ (n: integer) only one weight (corresponding to the correct delay time) would
be unity and all other weights zero. Thus, the problem of determining the time delay
between the two inputs only requires looking for the maximum value in the weight
vector $\vec{W}$. The maximum's position within the weight vector is then converted into a
time delay estimate. Computer simulations indicated that the peak weight can be
correctly selected much more quickly than the time required for either the mean
square error or the weights to converge, which makes the system suitable for processing time varying input statistics (Read et al. 1981). In case of linearly varying time
delay, the peak can be observed moving through the adaptive filter tapped delay
line, as shown in Fig. 2.

3 Computer Simulations

An adaptive filter bearing estimator with 31 tap bins has been simulated on a digital
computer. Additional interpolation by a factor of 4 (corresponding to a total amount
of 128 vector elements) permitted the detection of delay times lying between the

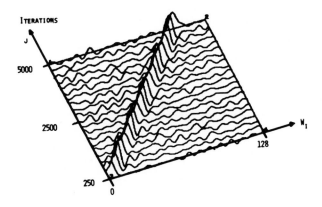

Fig. 2. Weight vector response
to a linear increasing time
delay

taps. The computer-generated signals were lowpass-filtered to simulate the real case, where the signal energy of the vehicles is concentrated below 500 Hz. Simulations were run down to 100 Hz, and the adaption curves examined to obtain several values for μ in order to show that, if μ was properly chosen (Jäger 1982) there need not be a lag beyond the true time delay.

4 Estimating Position, Velocity, and Direction of Movement out of Time Delay Data

A sound source moving along a road produces an "S"-shaped time delay between the signals of two microphones placed in parallel at distance y_i to the moving object, and thus, depending on the specific distance y_i several curves exist (Fig. 3). The boundaries describing the time delay of a source moving at the front and the rear parts of the road respectively, are shown as a series of dashes in Fig. 4, where the limit is given by $|\tau_g| = d/c$ (c: speed of sound). Depending, therefore, upon the chosen value of y_i, there are several distances forming hyperbola across the road all giving the same time delay estimate. However, in most cases, readings taken from the middle of the road are sufficient to give a good estimate. With this velocity estimate and the zero-crossing

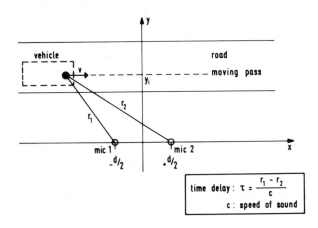

Fig. 3. Sensor configuration

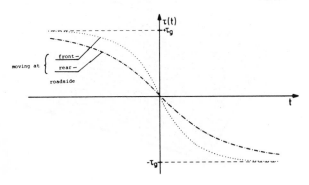

Fig. 4. Typical "S"-shaped time delay limits for a sound source moving along a road

(weight vector peaked at 64th element), determination of position becomes possible. The direction of movement is given by examining the polarity of the time delay curve.

Analysis of the weight vector of actual vehicle-passes is done using a colour display (Fig. 5, grey-scaled to fit requirements of reproduction), which on the left side shows the weight vector with its 128 elements along 512 points in time.

The typical "S"-shape can readily be seen. Evaluating its maximum, the calculated velocity is shown in the lower part of the picture, for the rear (−) middle and front part (0) of the road respectively.

The overlayed bar indicates the specific velocity displayed at the left bottom of the screen. By taking the middle of the road for true parallel distance y_i, a good estimate of velocity is given in most cases. The same is true for higher velocities (Fig. 6).

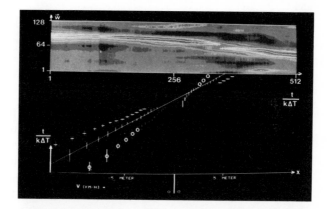

Fig. 5. Analyses of passing vehicles (true speed: 10 km/h, k any integer constant to fit time requirements)

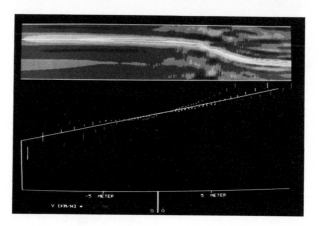

Fig. 6. Analyses of passing vehicles (true speed: 30 km/h)

5 The Maximum Range of Performance and Resolution Capability

In order to analyse positive as well as negative values of time delays, zero will be put in the middle of the weight vector (corresponding to the 64th element) by means of an additional delay in the reference path. Interpolation by a factor L gives $M \cdot L$ resolution cells on each side. Movements are not detected until the time delay goes from one resolution cell to the other, which for the first time occurs at

$$x_g = (ML - 1) \frac{d\tau\, c}{2} \sqrt{1 + \frac{4\,y_i^2}{d^2 - [(ML - 1)d\tau\, c]^2}}$$

where

$$M = \text{int}\,[m/2]$$

and

$$\begin{aligned}
\text{int:}\ &\text{integer of } [...] \\
m:\ &\text{true number of weights (32)} \\
d\tau = \Delta T/L:\ &\text{effective time resolution} \\
\Delta T:\ &\text{sampling time interval} \\
L:\ &\text{interpolation factor.}
\end{aligned}$$

x_g indicates the maximum range of performance. Using for example $\Delta T = 1/1{,}600$ s; $L = 4$; $y_i = 8$ m; $d = 1.5$ m results in $x_{g1} = 26$ m, whereas for the same example, except $y_i = 4$ m gives $x_{g2} = 13$ m. With the formulae given above, the position of the hyperbola across the road could be approximated by a straight line. Hence, the evaluation of an angle of resolution becomes possible, which in this case is about 4 degrees at x_{g1} and about 2 degrees at the zero crossing ($x = 0$) as shown in Fig. 7.

Fig. 7. Maximum range of performance and resolution capability

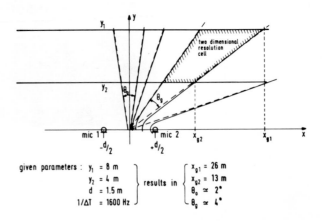

6 Summary

Passive bearing estimation is usually accomplished by determination of the delay times between a set of microphones by means of a correlator. However, an adaptive procedure has been shown to work in the nearfield region where there are fast-changing signal statistics. Thus, detection and control of moving vehicles becomes possible.

Although estimation of the exact position requires a minimum of three microphones, one can reduce the number to two by assuming a known geometry and a constant source velocity over a sufficiently small time interval. Examining this constancy for some proper traces across the road leads to an estimated value of velocity, which, with the zero-crossing in turn leads to the source position, whereas the moving direction is given by the polarity of the "S"-shaped time delay curve. The maximum range of performance and angle of resolution are controlled by the distance between the two microphones and the road, as well as by the sampling frequency. The actual vehicle passes were analyzed to show velocity errors below 10% when taking the middle of the road as the suggested moving trace.

References

Feintuch PL, Reed FA, Bershad NJ (1978) Adaptive tracking system study. Part I. Hughes Aircraft Co., Fullerton AD-A063 392

Feintuch PL, Reed FA, Bershad NJ (1979) Adaptive tracking system study. Part II. Hughes Aircraft Co., Fullerton AD-A078 469

Jäger U (1982) Bestimmung der Einfallsrichtung eines Schallsignals in eine Sensoranordnung mittels adaptiver Filter. Fortschritte der Akustik FASE/DAGA, pp 621–624

Reed FA, Feintuch PL, Bershad NJ (1981) Time delay estimation using the LMS adaptive filter-static behavior. IEEE Trans. on Acoustics. Speech and signal processing, vol ASSP-29, No. 3, pp 561–570

Second-Order Representation of Signals

A.M.H.J. AERTSEN[1], P.I.M. JOHANNESMA, J. BRUIJNS, and P.W.M. KOOPMAN

Dept. of Medical Physics and Biophysics
University of Nijmegen, Nijmegen, The Netherlands

1 Introduction

During the past decades much effort has been devoted to the adequate representation of signals in various fields such as acoustics and (bio-)electrical engineering. Among others, this resulted in a number of second order signal representations, each one with its own particular characteristics and special domain of application. A number of these functionals can be met regularly in the literature dealing with the problems of localization and identification of signal sources, both in technical and biological applications. Especially those second order functionals which have time and/or frequency, in the various possible configurations, as arguments have shown to be of particular interest: e.g. the lagged product function, the ambiguity function (Woodward 1953, Rihaczek 1969), the bispectrum, the Wigner-distribution (Wigner 1932, Claasen and Mecklenbräuker 1980), and the Rihaczek-CoSTID (Johannesma and Aertsen 1983). In the present paper we present a general scheme providing the formal relations between these functionals.

2 Theory

Throughout this paper we will use the analytic signal $\xi(t)$ instead of the real signal $x(t)$, since the former leads to more elegant formulations of important signal characteristics. Based on the analytic signal the lagged product function Π is defined as

$$\Pi(\tau, t; a) = \xi^*[t + (a - \frac{1}{2})\tau]\xi[t + (a + \frac{1}{2})\tau] \qquad -\frac{1}{2} \leqslant a \leqslant \frac{1}{2} \qquad (1)$$

with * denoting complex conjugation.

From the product function the ambiguity function $\Delta(\tau, \nu)$ is derived by Fourier transformation with respect to t. The time-frequency density $\equiv (\omega, t)$ is the Fourier transform with respect to τ while the bispectrum $\Gamma(\omega, \nu)$ is produced by Fourier transformation with respect to both τ and t. This leads to the following relations:

---

[1] Present address: Max-Planck-Institut für Biologische Kybernetik, Tübingen, FRG

Localization and Orientation in Biology and Engineering
ed. by Varjú/Schnitzler
© Springer-Verlag Berlin Heidelberg 1984

— ambiguity function

$$\Delta(\tau, \nu; a) = \int dt\ e^{-i\nu t}\ \Pi(\tau, t; a) \tag{2}$$

— time frequency density or CoSTID

$$\equiv(\omega, t; a) = \int d\tau\ e^{-i\omega t}\ \Pi(\tau, t; a) \tag{3}$$

— bispectrum

$$\Gamma(\omega, \nu; a) = \int d\tau\ e^{-i\omega\tau} \int dt\ e^{-i\nu t}\ \Pi(\tau, t; a) \tag{4}$$

$$= \hat{\xi}^*[\omega - (a + \tfrac{1}{2})\nu]\hat{\xi}[\omega - (a - \tfrac{1}{2})\nu] \tag{5}$$

where $\xi(\omega)$ denotes the Fourier transform of $\xi(t)$, and all integrals have to be taken from $-\infty$ to $+\infty$. A rigorous and detailed discussion on the relations between these functionals and their respective properties, as well as on the influence of the parameter a is outside the scope of the present paper, and will be considered elsewhere (Aertsen et al. in prep.). Figure 1 summarizes the relations between the functionals introduced above. By appropriate choice of the parameter a most second order functionals proposed in the literature can be reached. By taking $a = -\dfrac{1}{2}$ one obtains a quadruplet of functionals with the (complex) Rihaczek-function $\equiv (\omega,\ t;\ -\dfrac{1}{2})$ (Rihaczek 1968; see also companion paper) as a cornerstone, the *Rihaczek quadruplet*; $a = 0$ leads to the *Wigner quadruplet* with the (real) Wigner distribution $\equiv (\omega, t; 0)$ at the corresponding position.

3 Representation

Visualization of a complex second-order functional using standard graphical techniques (3-D, gray-coding) leads to a doublet of pictures: either real and imaginary part, or amplitude and phase. The two can be integrated by using a color mapping of the complex plane (Johannesma et al. 1981).

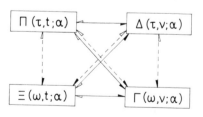

Fig. 1. Generalized diagram of relations between different second order signal representations

Gray-coding representations of the functionals from the two quadruplets have been presented at the conference for various signals (different tone combinations, chirp). From these and other examples the following conclusions can be drawn:

- a shift in frequency and/or time leads to a translation in $\equiv$, a complex modulation in Δ and, in general, a mixture of these in Π and Γ;
- the influence of a is virtually absent in Δ, trivially present in Π and Γ, whereas in $\equiv$ the influence is pronounced and more intricate;
- comparison of the Wigner distribution $\equiv (\omega, t; 0)$ and the Rihaczek-CoSTID $\equiv (\omega, t; \pm \frac{1}{2})$ shows the following (see also Claasen and Mecklenbräuker (1980), Johannesma et al. (1981), and companion paper): (see also Fig. 2):

$\equiv (\omega, t; 0)$	$\equiv (\omega, t; \pm \frac{1}{2})$
– real	– complex
– non-factorable	– factorable
– minimum spread	– larger spread
– cross terms on "diagonal"	– cross terms "off-diagonal"

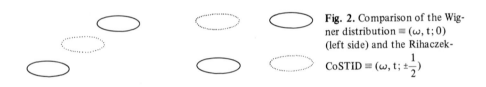

Fig. 2. Comparison of the Wigner distribution $\equiv (\omega, t; 0)$ (left side) and the Rihaczek-CoSTID $\equiv (\omega, t; \pm \frac{1}{2})$

Which functional is optimal in general will depend on the special question of interest. On the whole the CoSTID $\equiv (\omega, t; a)$ appears to be the first natural candidate because of its intuitive base and close relation to auditory perception [e.g. Altes (1980), Aertsen and Johannesma (1981), Hermes et al. (1981)].

Acknowledgements. This investigation forms a part of the research program "Brain and Behaviour: Neural base of localization and orientation".

The research was supported by the Dutch Foundation for the Advancement of Pure Research (ZWO). During this project the first author (AA) was a visiting research fellow with Prof. Braitenberg at the Max-Planck-Institut für Biologische Kybernetik in Tübingen (BRD).

References

Aertsen AMHJ, Johannesma PIM (1981) The spectro-temporal receptive field. A functional characteristic of auditory neurons. Biol Cybern 42:133–143

Altes RA (1980) Detection, estimation, and classification with spectrograms. J Acoust Soc Am 67:1232–1246

Claasen TACM, Mecklenbräuker WFG (1980) The Wigner-distribution – A tool for time-frequency signal analysis. I. Continuous-time signals. II. Discrete-time signals. III. Relations with other time-frequency signal transformation. Philips J Res 35:217–250; 276–300; 372–389

Hermes DJ, Aertsen AMHJ, Johannesma PIM, Eggermont JJ (1981) Spectro-temporal characteristics of single units in the auditory midbrain of the lightly anaesthetised grass frog (Rana temporaria L.) investigated with noise stimuli. Hear Res 5:147–178

Johannesma PIM, Aertsen AMHJ, Cranen L, Erning LJTO van (1981) The phonochrome: a coherent spectro-temporal representation of sound. Hear Res 5:123–145

Johannesma PIM, Aertsen AMHJ (1983) The phonochrome: a coherent spectro-temporal representation of sound. This volume

Rihaczek AW (1968) Signal energy distribution in time and frequency. IEEE Tr on Inf Theory IT-14:369–374

Rihaczek AW (1969) Principles of high resolution radar. McGraw-Hill, New York

Wigner E (1932) On the quantum correction for thermodynamic equilibrium. Phys Rev 40:749–759

Woodward PM (1953) Probability and information theory with applications to radar. Pergamon, London

The Phonochrome –
A Coherent Spectro-Temporal Representation of Sound

P.I.M. JOHANNESMA and A.M.H.J AERTSEN[1]

1 Introduction

In order to find an optimal visual image of sound three general desiderata for an iso-
morphic representation are formulated.

Formal equivalence. The relation of sound and image should be unique. In the
image all information contained in the sound should be preserved including phase
relations. The optical-acoustic map should be one to one.

Operational definition. The representation can be implemented instrumentally on
the base of the sound without additional information regarding context or meaning
and without human interference. Preferentially the representations are made on a
digital computer with conventional displays. Real time operation is a desirable aspect.

Perceptual congruence. Continuity with the tradition of representation of sound
in music (notes on a staff) and in vocalisations (sonogram). Moreover a simple correspondence should exist between perceptual elements with associated distances and
relations in the original sound (audition) and the perceptual elements with associated
distances and relations in the resulting image (visual).

2 Theory and Representation

The mathematical solution starts from the analytic signal $\xi(t)$ associated with the
original sound $x(t)$

$$\xi(t): = x(t) + i\,\tilde{x}(t) \tag{1}$$

where

$$\tilde{x}(t): = \frac{1}{\pi} \int ds \frac{x(s)}{t-s}$$

is the Hilbert transform or quadrature signal of $x(t)$. In the spectral domain

[1] Present address: Max-Planck-Institut für biologische Kybernetik, Tübingen, FRG

$$\tilde{\xi}(\omega) = \begin{cases} 2\,\hat{x}(\omega) & \omega > 0 \\ 0 & \omega < 0 \end{cases} \tag{2}$$

where $\hat{x}(\omega)$ is the spectral transform of $x(t)$ and $\tilde{\xi}(\omega)$ of $\xi(t)$.

The presentation is based upon a second order functional of the analytic signal. Four equivalent functionals exist: (1) product function $\Pi(\tau, t)$; (2) ambiguity function $\Delta(\tau, \nu)$; (3) CoSTID $\Xi(\omega, t)$; (4) bispectrum $\Gamma(\omega, \nu)$. Mathematically these four representations are equivalent, perceptually a spectro-temporal form is desired. Therefore only ambiguity function and CoSTID remain. The similarity with sonogram representing frequency and time, not their differences, eliminates the ambiguity function. As a consequence we propose the Coherent Spectro-Temporal Intensity Density (CoSTID) as the formal base for the visual representation of sound. Mathematically it is defined on the base of the product function

$$\Pi(\tau, t; a) := \xi^*[t + (a - 1/2)\tau]\,\xi[t + (a + 1/2)\tau] \quad -\frac{1}{2} \leqslant a \leqslant \frac{1}{2}. \tag{3}$$

The CoSTID is now the spectral transform with respect to the temporal difference τ

$$\Xi(\omega, t; a) := \int d\tau\, e^{-i\omega\tau}\, \Pi(\tau, t; a). \tag{4}$$

For $a = 0$ the product function Π is Hermitic in τ and the CoSTID is a real valued function, not necessarily positive, of frequency and time.

In physical optics the CoSTID with $a = 0$ has been introduced by Wigner (1932) and used by Bastiaans (1979) and Wolf (1982). In acoustics this function is used by Claasen et al. (1980) and Flandrin and Escudie (1980). For $a = \pm\, 1/2$ the product function Π is not Hermitic and the CoSTID becomes a complex function. However, it can now be written as a product of spectral and temporal aspects with an inter-mediating spectro-temporal (de)modulation.

$$\Xi(\omega, t) := \Xi(\omega, t; -1/2) \tag{5}$$

$$(4) \rightarrow \qquad = \int d\tau\, e^{-i\omega\tau}\, \Pi(\tau, t; -1/2)$$

$$(3) \rightarrow \qquad = \int d\tau\, e^{-i\omega\tau}\, \xi^*(t - \tau)\, \xi(t)$$

$$= \tilde{\xi}^*(\omega)\, e^{-i\omega t}\, \xi(t). \tag{6}$$

This function has been proposed by Rihaczek (1968).

Johannesma et al. (1981) introduced it as a coherent spectro-temporal image of sound. Altes (1980) discussed it in relation to echolocation. Hermes et al. (1981) applied it to the spectro-temporal sensitivity of neurons. The relation of CoSTID to other second order functionals is discussed in Aertsen et al. (1983).

The CoSTID cannot be regarded as a physical entity but should be interpreted as a formal structure defined on a signal which by application of appropriate operators produces physical entities.

The phonochrome is a representational structure of a signal based on the CoSTID and intended for visual perception. Its realisation is formed through a representation of a complex function of two variables by a chromatic image. For an extensive discussion and presentation of phonochromes of different signals see Johannesma et al. (1981).

3 Identification and Localization

For identification and localization of an acoustic source of which only the air pressure variations can be observed, two receivers are needed. Not directly considering biological realisation or technical implementation it is possible to evaluate the CoSTID for this purpose. If the microphones receive signals x_1 (t) and x_2 (t) then their difference will be mainly in the phase of the signals. Now two signals are formed:

$$x_\pm = \frac{1}{2} (x_1 \pm x_2) \tag{7}$$

and for each the associated CoSTID $\Xi_\pm$:

$$\Xi_\pm = \frac{1}{2} \left\{ \Xi_{11} + \Xi_{22} \pm (\Xi_{12} + \Xi_{21}) \right\}. \tag{8}$$

Now take even and odd part

$$\Xi^e = \Xi_+ + \Xi_- = \frac{1}{2} (\Xi_{11} + \Xi_{22}), \tag{9a}$$

$$\Xi^o = \Xi_+ - \Xi_- = \frac{1}{2} (\Xi_{12} + \Xi_{21}). \tag{9b}$$

Then Ξ^e is based only on the autoCoSTID's
and Ξ^o is based only on the crossCoSTID's.

As a consequence Ξ^e is weakly and Ξ^o strongly dependent on the phase relations of x_1 and x_2 and as a consequence on the position of the source.

For active localization and identification an analogous way of reasoning may be applied. However now x_1 is the emitted signal and x_2 is the reflected signal. Comparison of Ξ_{22} with Ξ_{11} may lead to identification while evaluation of Ξ_{12} and Ξ_{21} may supply the clues for localization.

In this context the function of EE- and EI-neurons in the central auditory nervous system should be considered.

Acknowledgements. This investigation forms a part of the research program "Brain and Behaviour: Neural base of Localization and Orientation". The work has been supported by the Netherlands Organisation for Advancement of Pure Research (Z.W.O.).

References

Aertsen AMHJ, Johannesma PIM, Bruijns J, Koopman PWM (1983) Second order representation of signals. In: Varjú D (ed) Orientation and localization in engineering and biology. Proc 8th Cybernetics Congress, Tübingen, B.R.D.
Altes RA (1980) Models for echolocation. In: Busnel RG, Fish JF (eds) Animal sonar systems. Plenum Publ Corp, New York, pp 625–671
Bastiaans MJ (1979) Wigner distribution and its application to first order optics. J Opt Soc 69: 1710–1716

Claasen TACM, Mecklenbräuker WFG (1980) The Wigner distribution – A tool for time frequency signal analysis. I. Continuous time signals. II. Discrete time signals. III. Relations with other time-frequency signal transformations. Philips J Res 35:217–250; 276–300; 372–389

Flandrin P, Escudie B (1980) Time and frequency representation of finite energy signals: a physical property as a result of an Hilbertian condition. Signal Processing 2:93–100

Hermes DJ, Aertsen AMHJ, Johannesma PIM, Eggermont JJ (1981) Spectrotemporal characteristics of single units in the auditory midbrain of the lightly anaesthetised grassfrog (Rana temporaria L.) investigated with noise stimuli. Hear Res 5:145–179

Johannesma PIM, Aertsen AMHJ, Cranen B, Erning L v (1981) The phonochrome: a coherent spectro-temporal representation of sound. Hear Res 5:123–145

Rihaczek AW (1968) Signal energy distribution in time and frequency. IEEE Tr on Information Theory IT-14:369–374

Wigner E (1932) On the quantum correction for thermodynamic equilibrium. Phys Rev 40: 749–759

Wolf E (1982) New theory of partial coherence in the space-frequency domain I. JOSA 72–3: 343–351

Directional Hearing and the Acoustical Near-Field of a Human and an Artificial Head

R. WEBER and H. BELAU[1]

1 Introduction

Localization of sound sources and acoustical orientation with the aid of artificial heads are limited due to some restrictions in acoustic transmission properties.

If sound, which has been recorded by an artificial head, is reproduced with head-phones, it is very difficult to detect a sound source in *front* of the head. For frontal sound incidence very often *front-back* reversals occur, which means that the sound source seems to be in the *back* instead of in *front* where it actually is.

This is caused by linear distortions in the artificial-head-headphone system be-cause an ideal sound transmission system, which reproduces the original transfer functions from a sound source to a listener's ears, allows the correct determination of the source position. The distortions, which change the directional information of the ear signals, may be caused by either the artificial head or headphones used, or by both.

We investigated the differences between an artificial and a human head by com-paring the acoustical near-fields close to the head's surfaces.

These near-fields, however, possess such a complex structure, that we retreated to a simple geometric form – a sphere with head-like dimensions – in order to study the influence of small obstacles at the surface upon the near-field structures.

2 The Diffracted Acoustical Near-Field of a Human Head, an Artificial Head and a Sphere

2.1 The Near-Field Measuring Method

In an anechoic chamber, the test objects – human heads, an artificial head (Weber and Mellert 1978) and a sphere – are exposed to pseudo-random noise which comes either directly from the front ($0°$) or from the back ($180°$). In order to elude possible imperfect frequency responses in the measuring device, we measure the transfer func-tions (cross-power spectrum divided by auto-power spectrum) between two identical microphones.

---

[1] FB 8 – Physik, Universität Oldenburg, FRG

Localization and Orientation in Biology and Engineering
ed. by Varjú/Schnitzler
© Springer-Verlag Berlin Heidelberg 1984

The first microphone serves as a reference and is positioned between the test object and the loudspeaker at the height of the ear canal entrances (or the center of the sphere) about 60 cm away from the test object. The second microphone is moved around the test object in a horizontal plane at the same height in steps of $10°$ and its distance is held constant at about 2 cm from the diffracting surface. A whole set of 36 amplitude spectra of the transfer functions from a complete turn represents the acoustical near-field for a given direction of sound incidence. These amplitude data of the transfer functions are identical to those acoustical near-field spectra which would be produced by ideally white noise.

2.2 The Acoustical Near-Field of a Human Head and an Artificial Head

A typical acoustical near-field of a human head is shown in Fig. 1 for frontal sound incidence. The position of the near-field microphone is indicated by the angle φ (x-axis). The y-axis represents the frequency range f starting with the low frequencies in the rear and the amplitudes of the spectra in decibels (dB) are drawn on the z-axis. Figure 1, as well as the following figures, contains data which are already smoothed in the frequency and angle domain.

At frequencies below 5 kHz this and other near-fields of human heads show well noticeable characteristic structures which can be described in a first approximation as a row of curved ridges of hills which start from the sides and meet in the middle of the drawings. This means that e.g. level maxima in the near-field change their position on the surface of the head when the frequency changes. If e.g. the frequency is increased the maxima are shifted from the front to the back in the low-frequency range. This is typical for human heads in the case of frontal sound incidence.

The near-field of an artificial head is given in Fig. 2 and it exhibits quite a different structure. The clearly distinguishable curved ridges in the low-frequency range have vanished and one finds patterns which are nearly parallel to the frequency axis e.g. the ridge in the region of the ears at about $90°$ resp. $270°$. This indicates that there is a level increase at the fixed position of the ears over nearly the whole frequency range for frontal sound incidence.

Fig. 1. Acoustical near-field (2 cm distant from the surface) of a human head for frontal ($0°$) sound incidence. φ position of the near field microphone (nose position at $0°$; φ is clockwise increasing); f frequency and L level of the near field amplitude spectrum

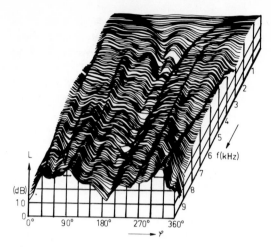

Fig. 2. Acoustical near-field for an artificial head

2.3 The Acoustical Near-Field of a Sphere with Obstacles

Figures 3 and 4 show the acoustical near-field of a wooden sphere with headlike dimensions (diameter: 17 cm) with different obstacles at the angle position 90° and 270°. In Fig. 3 this obstacle consists of a small plate (42 mm x 62 mm x 1 mm) and in Fig. 4 the sphere has a simple thin concha model only at the 270° position.

If one regards the position of the ears only one finds that the level increase in this region which was found in Fig. 2, for the artificial head bears a remarkable similarity to the case of the simple reflector (Fig. 3). On the other hand a simple concha model which is fixed at the sphere already begins to show a weakly curved amplitude maximum. This indicates the important role of a correct model of the human outer ear.

3 The Necessity of an Acoustically Correct Outer Ear Model

Former studies (Belau and Weber 1982) have shown that the near-field of a human head changes in a significant manner, if the sound comes from the rear. On the other hand, the near-field of the artificial head does not alter much, if the direction of the sound incidence is reversed. Furthermore, there is an astonishing similarity between the near-fields of the artificial head and that near-field of a human head for backward sound incidence. This fact corresponds to the observed front-back-reversals of the artificial head recordings.

The results presented here indicate that the acoustical properties of the ears of the artificial head are incorrect. They seem to work similar to sound reflectors. So an acoustically correct and exact model of the outer ear might mean a progress towards a correct directional mapping of the acoustical world.

Fig. 3. Acoustical near-field for a sphere with small reflectors at the 90° and 270° positions

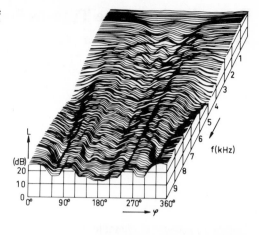

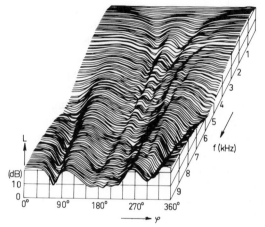

Fig. 4. Acoustical near-field for a sphere with a thin concha model only at the 270° position

References

Belau H, Weber R (1982) Schallbeugung im Nahfeld am menschlichen Kopf. In: Fortschritte der Akustik − FASE/DAGA '82, voll II, VDE-Verlag, Berlin, p 1199

Weber R, Mellert V (1978) Ein Kunstkopf mit "ebenem" Frequenzgang. In: Fortschritte der Akustik − DAGA '78, VDE-Verlag, Berlin, p 645

Determination of the Transfer Function of a Time-Varying Transfer Channel by a Correlation Method. Illustrated on Sound Propagation Through a Turbulent Medium

W. WILKEN and V. MELLERT[1]

1 Introduction

Time-varying channel characteristics (e.g. caused by changing meteorological conditions) lead to phase shifts in the transferred signal. Correlation or other long-term averaging methods applied to intensify the signal to noise ratio in determining the outdoor transfer characteristics are therefore difficult to apply. Under certain theoretical assumptions it is possible to account for the variations of the propagation channel, in such a way that transfer functions can be evaluated from a correlation measurement. It is of general interest to get to know the status of the turbulent medium. The turbulence can be detected by a passing sound wave. Especially noise emission outdoors is influenced by meteorological conditions. Further applications are air pollution, micrometeorology, acoustic detection, and remote sensing.

2 Theory

The variance of the phase of a wave, propagating through a medium of homogeneous fluctuating index of refraction, or velocity field, is in good approximation proportional to ω^2 (ω: frequency) (e.g. Monin and Yaglom 1981). Therefore, mainly the time lag of the wave travelling a given distance through the atmosphere will vary. If the wave propagates in different "channels" resulting from non-isotropic behaviour of the medium near the ground boundary, further phase variations will be seen due to interference of partial waves.

If there is some stationary condition during a time T, it is possible to measure the cross spectrum $\underline{G}_{AB}^T = \underline{A}^* \underline{B}$ between the transmitted signal a(t), and the received signal b(t). A procedure is given in Wilken and Mellert (1981). Assuming a sufficient ensemble average during T, the incoherent parts in $\underline{G}_{AB}^T$ can be neglected, at first:

$$G_{AB}^T = A^* A H^T \quad \text{with} \quad B^T = H^T \underline{A} . \tag{1}$$

(Disturbing noise in Eq. (1) will be eliminated by averaging many $\underline{G}_{AB}^T$ after processing as follows.)

---
[1] Fachbereich Physik, Universität Oldenburg, 2900 Oldenburg, FRG

Localization and Orientation in Biology and Engineering
ed. by Varjú/Schnitzler
© Springer-Verlag Berlin Heidelberg 1984

The transfer function H^T, which is stationary during T, is separated into:

$$\underline{H}^T = \underline{H}_{min} \exp[i\phi^T(\omega)] . \tag{2}$$

$\underline{H}_{min}$ is supposed to be time invariant at first and considers the whole channel with its average transfer characteristic being stationary (e.g. ground impedance). $\phi^T(\omega)$ accounts for the phase (or time) fluctuations, caused by varying index of refraction and the wind. The average time delay between a(t) and b(t) is eliminated by the measuring procedure (Wilken and Mellert 1981). Because of the impulse response of the system being real $-\phi^T(\omega) = \phi^T(-\omega)$. ϕ^T is developed into a series of linear independent functions of physical meaning, $f_j(\omega)$:

$$\phi^T = \Sigma_j a_j f_j .$$

We propose the following:

$$f_0 = a_0 \, \text{sgn}(\omega) , \tag{3a}$$

is a constant phase shift (e.g. due to an impedance transition),

$$f_1 = a_1 \omega, \text{ is a time delay} \tag{3b}$$

$$f_j = a_j \sin(2\pi\omega/\omega_j), \, j \geqslant 2 , \tag{3c}$$

is a weak interference of other sound rays. [Exact calculation see Mellert, Radek, Wilken (1983)].

a_j: amplitude ratio of two interfering sound rays ($a_j \leqslant 1$ by definition).

Every a_j is a slowly varying function of time and regarded stationary during T. Equation (1) can be written as

$$\underline{G}^T_{AB} = \underline{A}^*\underline{A} \, \underline{H}_{min} \exp[ia_0 \text{sgn}(\omega)]$$

$$\exp(ia_1\omega)\Pi_{j\geqslant 2} \exp[ia_j\sin(2\pi\omega/\omega_j)] . \tag{4}$$

The cross correlation function then becomes:

$$r_{ab} = r_{aa} \bullet h_{min} \bullet \Pi^\bullet_j f_j \quad (\Pi^\bullet: \text{product of convolutions}) \tag{5}$$

r_{aa}: autocorrelation function of signal a(t)
h_{min}: impulse response of the idealized system.

Calculating the inverse Fourier transform of Eq. (4) gives the impulse response of the filters, which provide the phase alterations developed by Eq. (4):

$$j = 0: h_0(t) = \delta(t) \cos a_0 - (\pi t)^{-1} \sin a_0 \tag{6}$$

that is a superposition of a directly travelling signal and a Hilbert-transformed signal (i.e. $\pi/2$ phase shifted).

$$j = 1: h_1(t) = \delta(t + a_1) \tag{7}$$

that is a time delay of the signal, by an amount $-a_1$.

$$j \geqslant 2: h_j(t) = \Sigma_n J_n(a_j) \delta(t + 2\pi n/\omega_j) \tag{8}$$

that is a series of time delays given by the acoustic path differences, $d_j = c/\omega$ [Eq. (3c)], resulting in a comb filter characteristic. Practically, it is sufficient to use only two terms of the series Eq. (8).

The observed phase fluctuations occurring during the outdoor propagation of the acoustic signal behave as if the signals is transmitted through a series of filters Eqs. (6), (7), and (8). The inverse filters are easily derived. Applying such a set of filters with proper characteristics a_j to the received correlation signal $b(t)$ will cancel the frequency dependent phase fluctuations (if they do not change too fast compared with T).

3 Experiment

Measurements with moderate wind show that modelling Eqs. (6) and (7) is sufficient to control the phase shift fluctuations. The filters are calculated on a computer (see Fig. 1). The original signals stem from a correlation measurement for determining an

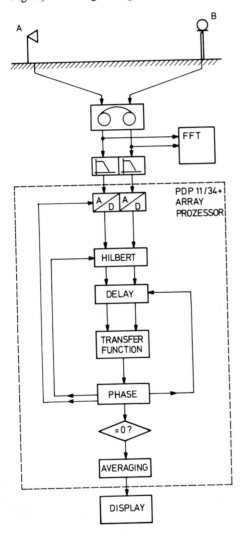

Fig. 1. Experimental setup

outdoor transfer function carried out with an experimental procedure described in Wilken (1983). The signals are processed on a fast minicomputer, practically in real time. Every 0.25 s the coefficients of the slowly varying filters [Eqs. (6) and (7)] are adjusted. The processing procedure is seen in detail in the flow diagram of Fig. 1.

4 Results

Figure 2 shows a series of succeeding transfer-functions, resulting from the filtering process discribed above, applied on a measurement with wind variations between 2 and 4 m/s. The variation of the positions of the interference minima shows that H_{min} is not time-invariant as supposed. These variations of positions of the inter-ference minima can be compensated for by controlling the sampling frequency of the A/D-converter. This adjustment of the sampling frequency is done adaptively by the coefficient of the varying slope of the phase function, and the varying position of the interference minima. The result is shown in Fig. 3.

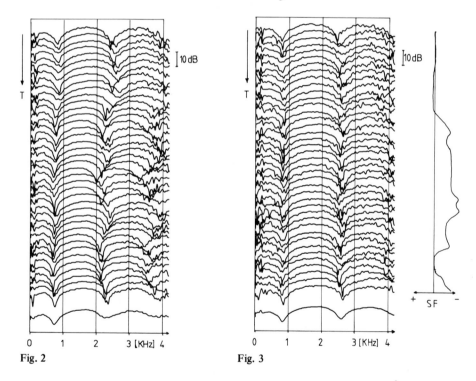

Fig. 2

Fig. 3

Fig. 2. A series of succeeding transfer functions resulting from the filtering process described above. Measurement over water. Wind speed: 2–4 m/s. Distance: 37.25 m. Height of transmitter: 1.73 m. Height of receiver: 2.27 m

Fig. 3. A series of succeeding transfer functions of the same measurement as in Fig. 2, but this time with control of the sampling frequency (SF). *Right curve* is proportional to the actual sampling frequency

References

Mellert V, Radek U, Wilken W (1983) Correlation measurement of an acoustic transfer function with time varying phase. 11th ICA, Paris, proc vol 1, pp 57–60

Monin AS, Yaglom AM (1981) Statistical fluid mechanics, vol 2, chap 9. Cambridge, Mass., London

Wilken W, Mellert V (1981) Measurement of fluctuating outdoor sound propagation using an analytical signal. INTER NOISE '81, proc pp 981–984

Wilken W (1983) Correlation of wind and turbulence with outdoor sound propagation. 11th ICA, Paris, proc vol 1, pp 37–40

Redundance of Acoustic Localization

J. MANTEL[1]

Acoustic localization of sound waves in nature is possible in that different incoming signals are received by two ears and then processed. The intensity, phases, and differences in tone allow the radial distance, angle in the horizontal plane, the radial velocity and the horizontal angular velocity to be determined. The vertical components likewise determined are based on less pronounced methods.

1 Binaural Localization

Most localization theories work on the basis that the two ears receive signals that are different

a) in sound pressure level,
b) in arrival time (phase difference)
c) in other properties of the incoming signal.

Several mechanisms have been put forward in the past for binaural localization, and equivalent-circuit diagrams (see Fig. 3 and Mantel 1982a) have also been produced. Circuit diagrams have also been developed in detailed areas such as intensity and phase localization (see Fig. 2 and Mantel 1982b). It has been demonstrated that a large number of possibilities for determining localization angles and distances are conceivable, which take different lengths of time to process in the auditory duct and the auditory organ. The method of first wave triggering is the fastest way, while localization based on comparison with data stored in the brain (e.g. differentiation in time) takes much longer.

The weakness of this localization theory is that it cannot explain the small localization errors at interaural time differences of about only 50 μs in animals with head diameters of just a few centimetres. A maximum neural spike of approx. 1,000 Hz and a relatively large jitter are in contradiction to the observations of localization error and interaural time difference in animals. The explanation of this apparent discrepancy in terms of certain neural phenomena, the role of monaural localization and localization by fine intensity discrimination are described here.

---

[1] Dr. Mantel & Partners GmbH., Ingenieurbüro für Akustik, Simeonistr. 11, 8000 München 19, FRG; Geoula Street 12, Haifa/Israel

Localization and Orientation in Biology and Engineering
ed. by Varjú/Schnitzler
© Springer-Verlag Berlin Heidelberg 1984

2 Localization by Fine Intensity Discrimination

Because of the short interaural time differences in animals with small heads it has been attempted to explain angular localization by intensity evaluation. It is known that this only functions at higher frequencies, only the envelope of the high-frequency signal being used for localization (Neu-Weiler pers. commun.).

Genuine phase localization is effective up to a frequency of 1,500 Hz, which is equivalent to the reciprocal of the neural pulse width. Above 1,500 Hz localization operates on differences in the envelope of the high-frequency signals. Even very small differences in intensity can be perceived since certain neurons have a higher spike rate with rising intensity, while others have a lower rate. A network of such neurons can react very sensitively to differences in intensity.

The discovery of neurons with two inputs which are very sensitive to differences in intensity in a similar manner reinforces the intensity evaluation theory, especially because such neurons often do not react to the interaural time differences.

3 Monaural Localization

Monaural localization has been completely neglected in previous localization theories, apart from various vestibular theories (Blauert 1974). Monaural localization mechanisms nevertheless deserve our attention for various reasons:

The sound wave reaches the head and skull a relatively long time before the inner ear is stimulated. The sound velocity in the skull is higher than the atmosphere, so that a precursor wave certainly arrives at the inner ear. As a rule, this wave is attenuated by 40 dB relative to the airborne sound wave.

a) At low frequencies, however, this attenuation can be below 40 dB, so that it can play a role in localization at high sound pressure levels.
b) Particularly with small animals the efficiency of this stimulations seems to be higher since it is inversely proportional to the diameter of the head.
c) When sound waves from a point source arriving at the ear are diffracted around the head the direction of propagation is not identical with the direction of vibration of the air particles (Rosenhouse pers. commun.).
d) The external auditory duct stimulated by such sources sets the eardrum in motion which is dependent (the dominant mode) on the angle of incidence of the sound wave. For the shapes of some of the normal modes of vibration of the circular membrane see Fig. 1. The first three circularly symmetrical modes of vibration in a circular membrane (a–c), and some other possible modes (d–f). The nodal lines are indicated by dotted lines.

In view of the different angles of incidence of the sound wave and different modes of stimulation of the eardrum it cannot be ruled out that the oval window is also differently stimulated. The result of this would be that the distribution of the travelling wave velocities along the basilar membrane would also depend to a small extent on the angle of incidence of the sound wave. Such secondary effects cause a localization

Fig. 1a–f. Vibration mode of a membrane

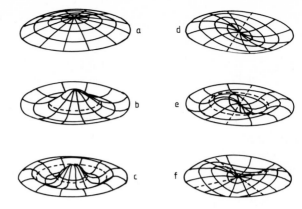

sensation (by learning) in each of the two ears. Comparison of the localization sensations in the two ears, after the time taken by the auditory organ to process the signal, leads to a better localization process.

The different localization perception when using earphones as opposed to a distant loudspeaker can be explained in terms of double monaural localization supplementary to the interaural time theory and the intensity theory.

4 Localization with Interaural Time Differences of Very Few µs

It is interesting to note that there are neurons which respond with a dead time if one of the inputs is stimulated. These neurons would fit in with models according to Mantel (1968) and (Blauert and Lindermann 1982), if it could be explained how with interaural time differences of very few microseconds the output of the neuron still reacts sensitively to the interaural time difference despite jitter and how the stimulation of the neuron arrives without jitter. The problem of jitter is important because in localizing pulsed sound the ear cannot perform averaging by the auto-correlation method. Already with short pulses it is possible to observe sharp localization in practice. Furthermore, averaging of signals will cause delay in localization sensation.

If one compares neural pulses with electric discharges or high-voltage breakdowns in capacitors, one sees that the frequency and the shape of the individual discharge pulses are dependent on the initial conditions and hence on the shape of the preceding pulse.

For the first pulse which can already be triggered, it is possible to give an exact starting time, especially when a bias voltage is applied. It is conceivable that the arrival times of the fronts of the first spike are precise and trigger control of the succeeding two-input neuron.

The discovery that various neurons with several inputs show different characteristics prompted new attempts to explain the short interaural time differences as the cause of localization. It was found that in the case of certain neurons (present in the

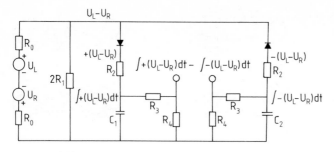

Fig. 2. Electrical model for localization (Mantel 1982[6])

$R_0 < R_1 \ll R_2 < R_4 \ll R_3$

$\tau_{CHARGE} \cong R_2 \cdot C \quad \sim \quad 0,1 \, msec$

$\tau_{DISCHARGE} \cong (R_3 + R_4) \, C \quad \sim \quad 10 \, msec$

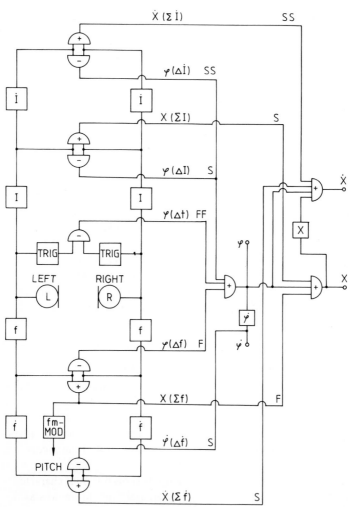

Fig. 3. Electrical block diagram of a localization model (Mantel 1982a) (without verification process)

MSO area of the bat's brain) the output of the neuron reacts with rising spike rate when the inputs are stimulated by the aural signal, whereas other neurons (in the LSO area of the bat's brain) block one another when the inputs are stimulated by both aural signals.

We assume here that a neuron (of the two-input species) with stimulating and blocking inputs is connected with aural signals. An output signal is then generated as a function of the interaural time difference, the voltage or charge of this signal being proportional to the interaural time difference.

5 Summary

It is shown that, in addition to likely localization models, there are some that are conceivable on the basis of new biological discoveries. The interaural time differences in the microsecond range have so far only been explained in a speculative manner and this speculation is further nourished by the discovery of new species of neurons, e.g. those that act discriminatorily to frequency modulations (sawtooth frequency modulation) in either a positive (frequency rise) or negative (frequency drop) manner.

References

Blauert (1974) Räumliches Hören. Hierzel-Verlag, Stuttgart

Blauert and Lindemann (1982) Zur Entwicklung von Modellen der binauralen Signalverarbeitung; Fortschritte der Akustik, S. 1161–1164, Göttingen

Mantel (1968) Correlation and cybernetic processes in speech perception, proceedings of the 5th Int. Congress on Cybernetic, Namur, pp 544–553

Mantel (1982a) Zeitliche Abläufe der Lokalisation; Fortschritte der Akustik, S. 1191–1194, Göttingen

Mantel (1982b) Ein Intensitäts-Phasen-Modell zur Lokalisation; Fortschritte der Akustik, S. 1177 bis 1180, Göttingen

Sound Localization in the Horizontal Plane by the House Mouse (*Mus musculus*)

G. EHRET and A. DREYER[1]

1 Introduction

Small mammals such as mice can be expected to have difficulty localizing a sound source, since (a) interaural arrival time differences (ΔT) are small; (b) interaural phase differences ($\Delta\Phi$) may be processed only up to about 5 kHz, where phase coding in the auditory nerve rapidly decreases (Rose et al. 1967), so that phase is not helpful for localizing in the high frequency range into which mouse hearing extends (Ehret 1974); (c) interaural intensity differences (ΔI) are nonexistent or small in the lower frequency range. Data on localization of tones and noise do not exist for small mammals (except opossum; Ravizza and Masterton 1972), and we therefore measured sound localization by the house mouse in behavioral tests in order to gain some information about localization acuity and mechanisms.

2 Methods

Eight female laboratory mice (*Mus musculus,* outbred strain NMRI, aged 2 months) with normal hearing were trained to run to a waterspout, situated in front of a speaker, in response to sound signals. During the localization tests the mice moved on a wire-mesh covered wheel (diameter 155 cm) in a sound-proof and anechoic room under dim red light ($\ll 1$ lx). The mice had to run from the center of the wheel toward 3 or 4 speakers mounted at head level beyond the outer rim of the wheel. Sound stimuli (tone bursts of 1, 15, 50, and 80 kHz and noise bursts of 15–80 kHz bandwidth) of 100 ms duration plus 10 ms rise and fall times (100 ms intervals) were presented until the mouse reached the margin of the wheel or the waterspout in front of the speaker, where it was then rewarded (closed loop condition).

Runs were videotaped for evaluation. The orientation during the runs was measured at five distances from the center using concentric circles. Measuring points (m) were at the intersection of the circles with the running track of the mouse. Orientation angles (a), the angle of deviation from the ideal line to the speakers, were measured between the line m_n and m_{n+1} and the line m_n and the center of the speaker. These angles were taken for all runs at each circle and for each sound signal, and were plotted

[1] Fakultät für Biologie, Universität Konstanz, Postfach 5560, 7750 Konstanz, FRG

Localization and Orientation in Biology and Engineering
ed. by Varjú/Schnitzler
© Springer-Verlag Berlin Heidelberg 1984

in frequency distribution with class widths of 5°. In addition, the total distributions (Σ of the distributions of the five circles) for each sound signal were plotted.

Using circular statistics, we calculated the directedness (vector r) of the distributions and the deviation angle ($\bar{a}$) of r from the hypothetical direction (0°). In addition, we·obtained another measurement for the peakedness of the distributions, the angle (β), which is the median of the angles (a) of the orientated runs (within ±90°) only. This angle (β) is taken as the average minimum audible angle (MAA). For further explanations and complete presentation of results see Ehret and Dreyer (1983).

3 Results and Conclusions

Figure 1 shows the distributions of orientation angles (a) for the 80 kHz tone bursts. All distributions are directed ($p < 0.01$) toward the hypothetical direction of 0° (speaker position). The directedness and peakedness increase with decreasing distance to the speaker (going from circle 1 to 5; distribution 2.1 to 2.5). In Fig. 2 the values of the vectors r of the total distributions (shown separately for orientated runs only and for all runs) and of the MAA's (β) are plotted against the different sound signals tested. In general, β decreases and r increases with increasing tone frequency while reaching lowest (highest) values with the noise band.

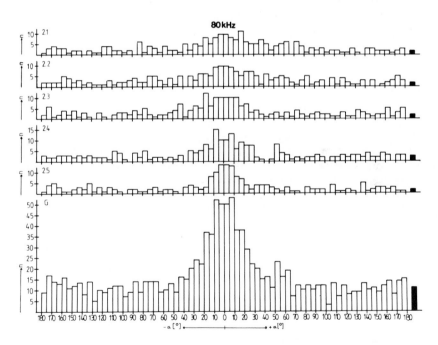

Fig. 1. Example of orientation acuity to 80 kHz tone bursts. Distributions of orientation angles (a) at the five distances (*circles* 2.1–2.5) from the center of the wheel, and total distribution (*G* sum of distributions 2.1–2.5). n number of angles (a) in a column. *Black column* average number of angles (a) in the columns outside ±90°

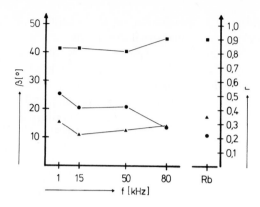

Fig. 2. Measures of directedness (vector r) and of peakedness (median localization acuity of oriented runs, angle β) in dependence on the sound signal. *Closed circles* β; *triangles* r, calculated from the total distributions of angle (α); *squares* r, calculated from the total distributions considering, however, only oriented runs (those within ±90° minus average level outside ±90°)

The following conclusions can be drawn: Sound localization ability in the mouse is well developed, especially for sounds in the high ultrasonic range and for a band of noise. Mechanisms involved: (a) Phase differences (ΔΦ) at the ears at 1 kHz (the ears are about 2 cm apart); (b) increasing intensity differences (ΔI) with increasing tone frequency, which explains the increasing acuity for localization of 15–50 and 80 kHz tone bursts; (c) spectral differences (ΔS), which lead to further increase of acuity when noise bursts are located. Arrival time differences (ΔT) between the two ears could also contribute, but since the temporal structure of the signals was always the same, they could not alone explain the measured differences in localization acuity for the different sound signals.

References

Ehret G (1974) Age-dependent hearing loss in normal hearing mice. Naturwissenschaften 61:506

Ehret G, Dreyer A (to be published 1983) Localization of pure tones and noise under closed and open loop conditions by the house mouse *(Mus musculus)*

Ravizza JR, Masterton B (1972) Contributions of neocortex to sound localization in opossum *(Didelphis virginiana)*. J Neurophysiol 35:344–356

Rose JE, Brugge JF, Anderson DJ, Hind JE (1967) Phase-locked response to low-frequency tones in single auditory nerve fibers of the squirrel monkey. J Neurophysiol 30:769–793

Localization of Water Surface Waves with the Lateral Line System in the Clawed Toad (*Xenopus laevis* Daudin)

A. ELEPFANDT[1]

1 Introduction

The clawed toad *Xenopus*, a frog living permanently in water, can detect the direction of surface waves running over its body and responds to such waves with a turn into the direction of the wave's origin. The underlying sensory capability has been assigned to the lateral line system (Kramer 1933, Dijkgraaf 1947, Görner 1973), which is retained by *Xenopus* after metamorphosis in accordance with its aquatic life. The lateral line system (LLS) consists of about 300 mechanoreceptive organs, the neuromast organs or stitches. They are distributed in groups and rows over the animal's body and are very sensitive to water movements. Each stitch has a sinusoidal directional sensitivity and projects into the medulla through two individual afferents (Görner 1963). However, *Xenopus* is capable of orienting to water waves even after destruction of all lateral line organs (Kramer 1933, Görner 1973). These findings raise some questions concerning the role of the LLS in the taxis response.

In the present study, behavioral tests on *Xenopus* with intact or lesioned LLS prove that the LLS is indeed involved in the determination of water wave directions and reveal important parameters used in this process.

2 Materials and Methods

Blinded adult *Xenopus laevis* Daudin with appropriate lesions of their stitches were tested in individual circular basins, 90 cm in diameter and filled with water, 7 cm deep. They were stimulated in open loop by slightly dipping a thin rod into the water at a distance of 10–12 cm. Angles of stimulation were multiples of $30°$ relative to the long axis of the frog in quasi-random sequence. Stimulation was continued until the frog had responded 10 times to each angle. Response angles were filmed with a video camera hanging above the basin and measured off line. For each lesion type the cumulative responses from 10–12 tests from at least three frogs were compared with the turn angles of frogs with intact system and after total lesion. For details see Elepfandt (1982).

---

[1]   Universität Konstanz, Fakultät Biologie, Postfach 5560, 7750 Konstanz, FRG

Localization and Orientation in Biology and Engineering
ed. by Varjú/Schnitzler
© Springer Verlag Berlin Heidelberg 1984

3 Results

3.1 Use of the LLS for Localization of Water Waves

With intact LLS, *Xenopus* can determine waves from all directions within ±5°. At higher stimulation angles, however, an increasing number of too small responses due to motivational and motor effects is observed. After destruction of all stitches only the responses to waves from anterior angles are oriented, but even they are significantly less accurate than with LLS. This strongly suggests localization of water waves with the LLS. In addition, after destruction of all stitches except around the left eye, significant bias to turn slightly left in response to frontal waves was found, and response amplitudes to the left side stimulation were larger than to the right side stimulation at all test angles. Also, in these animals a significantly higher percentage of turns to the contralateral side was found after stimulation from the right side than from the left side. This response asymmetry can only be explained by evaluation of wave directions in the LLS.

3.2 Mechanisms of Wave Localization in the LLS

Intensity gradient of the wave is not used. Stitches on the side of an incoming wave are stimulated more than those on the opposite side due to damping of the wave amplitude, and the frog could conceivably use this and turn to the side of maximal input. On the contrary to this hypothesis, frogs with all stitches on one side intact and all on the other side destroyed, i.e. with maximal asymmetry of the lesion, show less response asymmetry than those of the above example.

No orientation into the direction of maximal sensitivity of the stitches. Due to their sinusoidal directional sensitivity, stitches with maximal sensitivity in the direction of an incoming wave are stimulated greater than stitches with other alignment. If this effect would be used by *Xenopus* for orientation one should expect a tendency for 90°-turns in frogs in which only the dorsal occipital stitches are left intact. These stitches are aligned in two rows along the frog's long axis and their direction of maximal sensitivity is perpendicular to that axis. But frogs with only these stitches left intact respond to waves from any direction as accurately as unlesioned individuals. Even when only one row is left, the frogs perform normally at all angles except for 60°.

Temporal comparison of stitch activation is involved. The third possibility after exclusion of stitch position and orientation is temporal comparison of stitch activation. A wave running over the frog excites the stitches in a certain temporal sequence. If this were measured by the frog, one should expect reduced response accuracy if only one stitch is left intact. This was found. As shown above, frogs with one dorsal occipital row of stitches (i.e. 2–4 stitches) still respond with normal accuracy, but when these are destroyed except for one, response accuracy drops sharply and is similar to frogs with total lesion.

Accurate determination of wave direction is possible with small groups of stitches. Most lesions did not cause any deficit in the accuracy of wave determination. This was found even in lesions that destroyed up to 90% of the stitches (e.g. elimination of all stitches on the head and on one side of the body, or total elimination except around both eyes, or elimination except for the dorsal occipital stitches). This indicates a large redundancy in the system's capability of wave localization.

Local use of sine-cosine evaluation is possibly involved. The stitches on the sides of the animal are arranged in two adjacent longitudinal rows that show perpendicular stitch orientation to each other. Thus, corresponding stitches from these rows measure the sine- and cosine-components of local water movements respectively. To test whether this alignment is used in determining wave directions, frogs with intact stitches only along the upper lateral row were examined. These animals failed to respond at all for 2 months. Such a drastic reduction in responsiveness was not found after any other type of lesion – not even after total lesion – and seems to indicate severe "disorientation". This is what should be expected if sine-cosine analysis is made between these rows.

Acknowledgement. Supported by grant El 75/1 of the DFG.

References

Dijkgraaf S (1947) Über die Reizung des Ferntastsinns bei Fischen und Amphibien. Experientia 3:206–208

Elepfandt A (1982) Accuracy of taxis response to water waves in the clawed toad (*Xenopus laevis* Daudin) with intact or with lesioned lateral line system. J Comp Physiol 148:535–545

Görner P (1963) Untersuchungen zur Morphologie und Elektrophysiologie des Seitenlinienorgans vom Krallenfrosch (*Xenopus laevis* Daudin). Z vergl Physiol 47:316–338

Görner P (1973) The importance of the lateral line system for the perception of surface waves in the claw toad, *Xenopus laevis*. Experientia 29:295–296

Kramer G (1933) Untersuchungen über die Sinnesleistungen und das Orientierungsverhalten von *Xenopus laevis* DAUD. Zool Jb Physiol 52:629–676

Determination of Source-Distance by the Surface-Feeding Fishes *Aplocheilus lineatus* (Cyprinodontidae) and *Pantodon buchholzi* (Pantodontidae)

H. BLECKMANN[1], U. MÜLLER[2], and I. HOIN-RADKOVSKI[2]

Surface-feeding fish respond to wave signals with an orienting response, consisting of a directional and a translation component. If clicks (Fig. 1A) are presented, they determine the source distance up to 15–20 cm. The aim of the study was to look for the physical parameters within clicks used by surface-feeding fishes (*A. lineatus* and *P. buchholzi*) for distance determination.

1. The slightest disturbance of the water surface generates wave trains containing wave cycles of different amplitudes and frequencies. At a given source, distance and kind of wave production both the amplitude (or the amplitude modulation) and the frequency range of a signal rise with increasing source-intensity (Fig. 1C). Different kinds of wave production (e.g. struggling insect or fallen leaf) cause different kinds of signals with respect to amplitude and frequency range (Lang 1980).

Both, phase velocity (above 13 Hz) and damping of surface waves increase with increasing stimulus frequency [e.g. phase velocity 23 cm/s (10 Hz) vs. 40.4 cm/s (140 Hz); damping 1.67 dB/cm (10 Hz) vs. 8.57 dB/cm (140 Hz)]. This results in a regular decrease of amplitude, amplitude modulation, frequency range, and frequency modulation (always downward) of clicks during their propagation (Bleckmann and Schwartz 1982). While the amplitude, the amplitude modulation, and the bandwidth of clicks mainly depend on stimulus strength, their frequency modulation first of all reflects the source distance (Fig. 1B). In addition, independent of frequency range the curvature of concentric surface waves decreases with increasing source-distance.

2. If clicks are presented, the stimulus strength (variation of 14.6 dB tested) does not affect the precision of distance determination [slope M of regression swimming distance with source distance 0.79–1.18 *(A. lineatus)* and 0.73–1.1 *(P. buchholzi)*]. Independent of source distance and stimulus intensity, the relative error

$$r(X)\% = \frac{100}{n} \sum_{i=1}^{n} \frac{|X_i - Y_i|}{X_i},$$

where X_i = source distance, Y_i = swimming distance, n = number of trials) is 12–14% (the relative error was calculated only for *P. buchholzi*).

Therefore, if clicks are presented, the accurracy of distance determination does not depend on stimulus amplitude, amplitude modulation, and frequency range.

---

[1] Gruppe Sinnesphysiologie, Zoologisches Institut der Johann Wolfgang-Goethe Universität, Siesmayerstraße 70, 6000 Frankfurt, FRG
[2] Institut für Tierphysiologie, Justus Liebig-Universität, Wartweg 95, 6300 Gießen, FRG

Localization and Orientation in Biology and Engineering
ed. by Varjú/Schnitzler
© Springer Verlag Berlin Heidelberg 1984

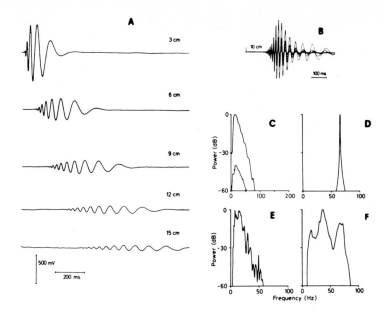

Fig. 1A–F. A Air-puff click at 3, 6, 9, 12, and 15 cm. **B** Clicks (source distance 10 cm) produced by touching the water surface with a rod (φ 2 mm). Their frequency modulation is independent of stimulus amplitude (actual pp-amplitudes 0.06–2.5 μm). **C** Power spectra of clicks of different intensities, **D** a 70 Hz cf-signal, **E** the simulated click, and **F** the ufm-signal. Source distance in **C** 10 cm; in **D, E,** and **F** 7 cm. In each case the highest power value was set at 0 dB

3. If single-frequency (cf) signals (30–100 Hz) (Fig. 1D) are presented, the estimated distances are too short. The error increases (a) with increasing source distance and (b) with increasing stimulus frequency. However, with cf-signals of 30 Hz (or 35 Hz for *P. buchholzi*) and 50 Hz (tested only for *A. lineatus*), the swimming distance still increase significantly with increasing source distance ($0.45 \geqslant M \geqslant 0.13$, $r \geqslant 0.46$, $p < 0.01$). With higher-frequencies (70 and 100 Hz) this was no longer observed (*A. lineatus*, $M \leqslant 0.08$) or at least much reduced (*P. buchholzi*, $M \leqslant 0.4$). The relative error is similar to that for clicks only up to 7–8 cm. Greater distances cause greater $r(X)$ values which increase with both increasing source distance ($n \geqslant 196$, $r \geqslant 0.391$, $p < 0.01$) and rising stimulus frequency (t-test, $p < 0.01$) (exception: 35 Hz stimulus. Here $r(X)$ is independent of source distance 19.7% ± 14.1 S.D.).

4. If we offer the fish a wave signal (source distance 6.5–7.5 cm) the frequency modulation of which at a source distance of 7 cm resembles that of an air-puff click at 15 cm (wave signal, see Bleckmann and Schwartz 1982; spectrum see Fig. 1E), they move significantly (t-test, $p < 0.01$) longer (*A. lineatus*: 10.8 cm ± 4.5 S.D., *P. buchholzi*: 13.3 cm ± 3.4 S.D.) than with clicks at the same distance (*A. lineatus*: 6.1 cm ± 1.9 S.D., *P. buchholzi*: 7.6 cm ± 1.4 S.D.). However, the swimming distance is not significantly different from those to clicks presented at 13.5–16.5 cm (*A. lineatus*: 11.8 cm ± 3.3 S.D., *P. buchholzi*: 13.3 cm ± 2.6 S.D.).

5. Clicks are always frequency modulated downward. Presentation of an artificial wave train, containing multiple frequencies (like clicks) but with upward frequency

modulation (ufm) (wave signal see Bleckmann and Schwartz 1982, spectrum see Fig. 1F) results again in a decrease in the estimated distance (*A. lineatus:* M = 54; *P. buchholzi:* M = 0.32). As with cf-signals, r(X) only in the near-field (up to 7 cm) is that obtained with clicks (12.5 to 14%). Above 7 cm r(X) rises to 50% (at 15 cm).

6. Even fish *(A. lineatus)* with only one head neuromast left intact determine roughly the source distance, if clicks are presented. Although the M of 0.44 is rather small, there is a strong correlation (n = 128, r = 0.72, p < 0.01) between swimming distance and source distance.

7. Conclusions: (1) Surface-feeding fish mainly use the frequency modulation of clicks for distance determination (paragraph 4). (2) The fact that such fish have some ability to estimate distance with cf-signals (which have no frequency modulation at all), and the ufm-signal indicates that they must also utilize parameters other than frequency modulation for distance determination. The suitable one is the curvature of concentric wave signals. The curvature (or the radius-source distance) can be determined by measuring the difference in arrival time between at least three neuromasts. It can be shown that the accurracy of such a mechanism should decrease with increasing source distance (because the curvature becomes progressively flattened) and increasing frequency (increasing phase velocity) of the stimulus (because of the decrease in time difference) (Hoin-Radkovski et al., in press). All these effects have indeed been observed (paragraph 3 and 5). (3) If the filter characteristics of the water surface are known, distance can be determined approximately by calculating the frequency-amplitude content of a stimulus. We interpret the fact that both *A. lineatus* and *P. buchholzi* swim for progressively shorter distances in response to cf-signals at increasing frequencies as an indication that the amplitude spectrum is also evaluated in distance determination.

Acknowledgements. Supported by DFG, grant Schw. 21/5.

References

Bleckmann H, Schwartz E (1982) The functional significance of frequency modulation within a wave train for prey localization in the surface-feeding fish *Aplocheilus lineatus (Cyprinodontidae)*. J Comp Physiol 145:331–339

Hoin-Radkovski I, Bleckmann H, Schwartz E (in press) Determination of source distance in the surface-feeding fish *Pantodon buchholzi* (Pantodontidae). Animal Behaviour

Lang HH (1980) Surface wave discrimination between prey and nonprey by the back swimmer *Notonecta glauca* L. *(Hemiptera, Heteroptera)*. Behav Ecol Sociobiol 6:233–246

Determination of Stimulus Direction
by the Topminnow *Aplocheilus lineatus*

G. TITTEL, U. MÜLLER, and E. SCHWARTZ[1]

1 Introduction

The topminnow *Aplocheilus lineatus* receives information on the direction as well as on the distance of centers of concentric surface waves, generated by objects touching the water surface signalling prey to these fish.

Perception of stimulus is achieved by the use of free-standing lateral lined neuromasts (Nm's) which are arranged in a species-specific pattern at the dorsal body surface of the fish, subdivided into a dorsal cephalic, temporal, and back line.

Reducing the number of Nm's results in a qualitative decrease in the fish's ability to correctly determine the stimulus direction. Here we are dealing with the problem of whether *A. lineatus*, with different numbers of Nm left intact, is able to differentiate stimulus directions.

2 Results

Typical directional responses of blinded fish with an intact lateral line system, demonstrate an almost perfect determination of stimulus direction, at least up to target angles of $\pm 150°$ ($0°$ in front of the fish).

When only the Nm's of the dorsal cephalic line were intact, we observed a further increase of uncertain and inaccurate reactions exclusively at target angles greater $150°$. We conclude that the dorsal cephalic line plays a major part in determining direction. This lateral line section consists of three Nm's and is termed, starting rostrocaudally, the nasal (I), supraorbital (II), and postorbital (III) group (see inset Fig. 1). If only the left part of the dorsal cephalic line remains intact, or additionally the Nm's of the temporal and back line of the left body side, *Aplocheilus* turns at all target angles to its left body side. In both cases, stimuli approaching the fish from the intact side still cause well-differentiated responses within a target angle range of $-150°$ and $-10°$.

For stimuli impinging first upon the contralateral body side, the obtained values scatter parallel to the x-axis showing, however, a dramatic change of the mean response angles from about $41°$ to $170°$ at target angles between $+110°$ and $+120°$.

[1] Institut für Tierphysiologie, Wartweg 95, 6300 Gießen, FRG

Localization and Orientation in Biology and Engineering
ed. by Varjú/Schnitzler
© Springer Verlag Berlin Heidelberg 1984

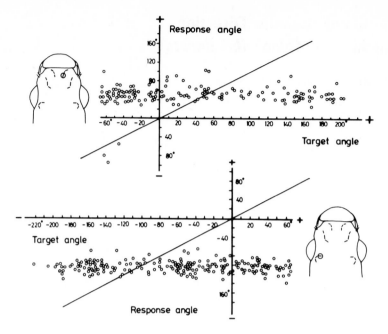

Fig. 1. Directional responses of fish with a single intact Nm (*encircled*)

This step is not quite that prominent in fish with only left dorsal cephalic line left intact.

Figure 1 clearly shows that fish with only one Nm left intact did not localize stimulus directions, however, they remain sensitive to surface waves from all directions. It seems that for each Nm, *Aplocheilus* oriented to an imaginary but stationary stimulus location. As our experiments show, this fictitious location varies for different individual Nm's. Beyond this, the position of the Nm on the fish's head determines the magnitude of the response angle: nasal neuromasts generally produce smaller response angles than more caudally situated ones (Müller and Schwartz 1982). How do those fixed Nm outputs, called here place values, participate in orienting reactions of intact topminnows?

Experiments with "two Nm systems" were proposed to approach this problem in its first step. Figure 2 shows the relationship between target and response angle for a fish, in which only the 3rd Nm of the supraorbital-group (II_3) and the 1st Nm of the postorbital group (III_1) of the left dorsal cephalic line was left intact. Within a target angle range of $-110°$ to $-25°$ an unequivocal correlation between target and response angle is obvious again. Such a range of persisting localization potency called "dynamic interval" is now delimited on both sides by the place values of the two intact Nm's, meaning left side turns greater than $120°$ (place value of Nm III_1), and smaller than $40°$ (place value of Nm II_3), are not obtained. Wave stimuli impinging within $-25°$ and $+180°$, again induce almost exclusively turns to the left (ipsilateral). The dynamic interval of response angles varies according to different combinations of neuromasts (Tittel 1982).

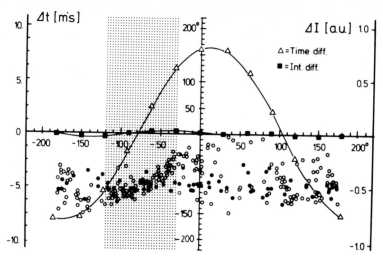

Fig. 2. Directional responses of a fish with two intact Nm's (II$_3$ L, III$_1$ L; = ○) and in a later experiment with only one Nm (III$_1$ L) *left* intact (●). The calculated differences of intensities (■ *right*) and arrival times (△ *left*) between the two Nm's are shown in relation to the target angles. The *dotted area* marks the "dynamic interval"

3 Discussion

Determination of stimulus direction by means of two Nm's may be based upon evaluation of time- rsp. intensity differences or, as a consequence of arrival time differences of the wavefront also phase differences. Calculations of the time and intensity differences (Fig. 2) between two neuromasts reveal changes in the same direction but of different scales. The calculated values are well correlated with behaviorally determined experimental results. The "dominance" of the "place value" of an individual Nm seems to depend on the lead of time and on the differences of stimulus intensity existing between two Nm's consecutively (see Fig. 2) activated. Points of simultaneous stimulation times (time diff. = 0) and equal intensities (int. diff. = 0) reveal for such Nm combinations few individual variabilities and lie in the midst of the "dynamic interval". Greater individual differences exist exclusively in the marginal zones of this dynamic range with respect to time and intensity differences. The latter ones especially depend on individually varying Nm alignment. At combinations where the two Nm's are standing almost parallel to one another and therefore showing only very small intensity differences, the precision of the localizing responses was only slightly reduced compared to individuals with the same Nm-combination but showing less parallel alignment of the Nm's. We interpret this result as follows: the time difference mechanism plays a major role in detecting target angles by fish carrying lateral line Nm's. Such a basic mechanism of directional responses to vibratory stimuli based on time differences has also been described in arthropods (Wiese 1974; Brownell and Farley 1979). Because of the small number of involved Nm's, it is possible for the first time in lateral line systems, to identify definitely, the mechanism of determination of stimulus direction.

References

Brownell P, Farley RD (1979) Orientation to vibrations in sand by the nocturnal scorpion Paruroctonus mesaensis: Mechanism of target localization. J Comp Physiol 131:31–38

Müller U, Schwartz E (1982) Influence of single neuromasts on prey localizing behavior of the surface feeding fish, Aplocheilus lineatus. J Comp Physiol 149:399–408

Tittel G (1982) Richtungslokalisation von Aplocheilus lineatus (Cyprinodontidae) bei stufenweiser Ausschaltung des Seitenliniensystems. Diplomarbeit, Gießen

Wiese K (1974) The mechanoreceptive system of prey localization in Notonecta. J Comp Physiol 92:317–325

Mathematical Description of the Stimulus for the Lateral Line Organ

El-S. HASSAN[1]

1 Introduction

When a fish glides through water, a field of currents is produced by the head displacing water and by suction in the tail area. This field of currents may be simulated by sources in the head area and sinks in that of the tail. Any obstacle along the fish's path will modify this field of currents and behavioral studies have shown that the fish is capable of detecting these alterations (v. Campenhausen et al. 1981). This paper describes a mathematical procedure developed to calculate changes in current velocity along the skin of the fish gliding past an obstacle. This is regarded as adequate stimulus for the lateral line organ.

2 Methods

The following simplifications are introduced: (1) the calculation is performed in two-dimensional space; (2) the obstacle is a cylindrical rod vertical to this plane; (3) the fish is modelled by a system of sources and sinks with different strengths (Fig. 1, right side). The calculation of the field of currents in the presence of the cylindrical obstacle is performed by the image method (Betz 1964, Curle and Davies 1968). The system of sources and sinks is mirrored in the obstacle, Fig. 1. If the original sources and sinks and its mirror image are operative at the same time, the currents in the field on the surface of the obstacle will be parallel to this surface. The velocity of the current and its alteration along the skin fish's due to the obstacle can be deduced from the field of currents and calculated as follows.

The calculations have to be performed in several steps. The first step of the image method preserves the surface of the obstacle but not that of the fish, i.e., the calculation may lead to theoretical currents crossing the surface of the fish. To avoid this systematic error, the current field when calculated has to be modified and this is achieved by introducing the mirror image from the obstacle into the fish. Before this step can be performed, the shape of the fish has to be transformed into a circle. After the new equation of velocity potentials has been established, the whole is retransformed. Now

[1] Arbeitsgruppe III (Biophysik), Institut für Zoologie, Universität, Saarstr. 21, 6500 Mainz, FRG

Localization and Orientation in Biology and Engineering
ed. by Varjú/Schnitzler
© Springer Verlag Berlin Heidelberg 1984

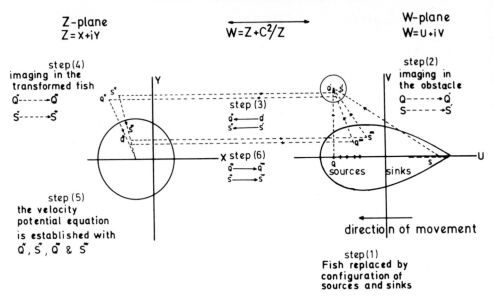

Fig. 1. Diagram of the computation process

the surface of the fish will not be crossed by the water currents, but the shape of the obstacle may be slightly deformed. By repeating this procedure this error can be reduced and virtually eliminated.

3 Results

Alterations in current velocity along the surface of the fish gliding past an obstacle are demonstrated in the drawing below Fig. 2: the fish glides continuously from position A through B to C. In Fig. 2a the ordinate shows changes in current velocity relative to the fish and the x-axis corresponds to the surface of the transformed shape of the fish in the z-plane (Fig. 1, left side). Current velocity of the tail end is not depicted in the diagrams because it can be calculated only by introducing additional methods which will be described elsewhere. The lowest line in Fig. 2a corresponds with the fish in position A, and the lines above show velocity distribution in sequential positions up to B. This demonstration is continued in Fig. 2b where the uppermost line corresponds to position B again, and finally the lowest line to position C.

The diagrams in Figs. 2a and b show that the obstacle causes a specific change in current velocity in space and time, which can be picked up by the lateral line organ. Current distribution of this type has been recorded in model fish and will be described elsewhere. The velocity distribution along the fish depends on the distance, Fig. 3, and the size of the obstacle, Fig. 4. These diagrams show only the velocity distribution

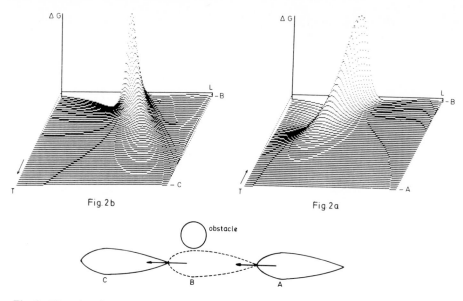

Fig. 2. Alterations in current velocity ΔG along the surface of the fish *L* gliding past an obstacle in dependence on time *T*

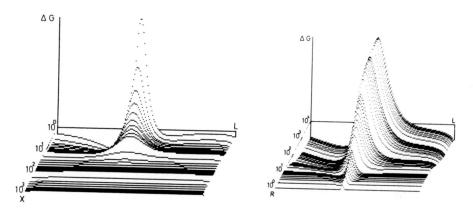

Fig. 3. Current velocity distribution ΔG along the fish in dependence on the distance of the obstacle X

Fig. 4. Current velocity distribution ΔG along the fish in dependence on the size of the obstacle R

for position B. To record this velocity distribution exactly, the fish has to record the course of the stimulus at many points on its skin. This may be one reason why there are so many neuromasts in the skin of the fish. It may be mentioned that the velocity of the gliding fish is introduced into the calculation by the strength of the sources and skins. This parameter governs the amplitude, but not the distribution, of the velocity along the fish. Velocity distribution, therefore, does not depend on swimming velocity. A closer look at the distribution in Figs. 3 and 4 reveals that theoretically the fish should be able to discriminate between size and distance of objects. This result has not yet been verified in behavioral experiments.

Acknowledgements. Supported by DFG Grant Ca 34/4.

References

Betz A (1964) Konforme Abbildung, 2nd edn. Springer, Berlin Heidelberg New York

Campenhausen C v, Riess I, Weissert R (1981) Detection of stationary objects by the blind cave fish *Anoptichthys jordani* (Characidae). J Comp Physiol Psychol 143:369–374

Curle I, Davies HJ (1968) Modern fluid dynamics, vol I. Incompressible flow. D van Nostrand Company, London

Image Sequence Analysis for Target Tracking

K-H. BERS, M. BOHNER, and P. FRITSCHE[1]

1 Introduction

In the past, systems have been built where object detection for target tracking was based on a simple evaluation of the intensity values of the object and its environment. After preprocessing of the sensor signals the maximum intensity level or other similar features like centre of gravity are used to detect the position of a selected object. Those systems only work successfully if a high contrast between object and background is given. Objects on the ground are often lost especially in the European landscape.

For those situations a tracking system based on correlation has been designed, simulated on a digital computer and tested with an equipment for real time application (Bohner 1976). The basic shortcomings of a correlation tracker system (like image modification by foreground and background distortions) have been eliminated in combining correlation and target detection methods, where parts of the moving object and mainly foreground and background objects are detected and identified within a period of time.

2 Tracker System

Figure 1 shows the principle mode of operation of a system based on the correlation function. An operator detects an interesting target (the car), which is stored in a memory as a reference and in the next image the correlation function between the intensity values of the actual scene $t(n+1)$ and the reference is computed. The reference image has to be shifted within a search area which results in a field of correlation values $K(x,y)$. These values are used to detect the target in the actual scene by a decision criterion (e.g. maximum of the correlation values) and to update the reference memory according to changes in the representation. For such a system a number of different decision and update criterions have been tested and the optimal system parameters have been evaluated. The software simulation at FIM as well as an

[1] Forschungsinstitut für Informationsverarbeitung und Mustererkennung (FIM/FGAN) (Research Institute for Information Processing and Pattern Recognition), Eisenstockstr. 12, 7505 Ettlingen 6, FRG

Localization and Orientation in Biology and Engineering
ed. by Varjú/Schnitzler
© Springer Verlag Berlin Heidelberg 1984

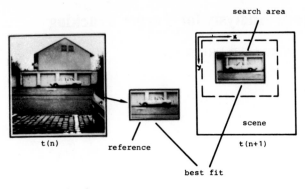

a) image sequence and reference

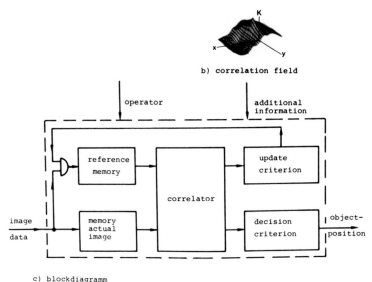

b) correlation field

c) blockdiagramm

Fig. 1 a–c. Correlation-tracker (principle)

online hardware system of a German company revealed the basic problem of the method, demonstrated in Fig. 2. In the moment $t(n)$, the reference memory contains the target without any foreground objects, whereas in the actual scene the car is partially hidden by the trunk of a tree. The maximum of the correlation function decreases but the object position can still be found with a high probability. If the update criterion is active in this moment, the trunk is learned into the memory as a part of the object. Therefore in the moment $t(n + 1)$ two peaks in the field of the correlation values are obtained (matches of car and trunk). Dependent on the size and contrast of the trunk a wrong object position may be found. A similar problem arises by big contrast changes in the background of the target.

To solve this problem at FIM a higher sophisticated tracking system has been developed, in which additional features (pointed out in Fig. 3) have been added to the original correlation system (Gerlach 1979).

Fig. 2. Influence of foreground objects

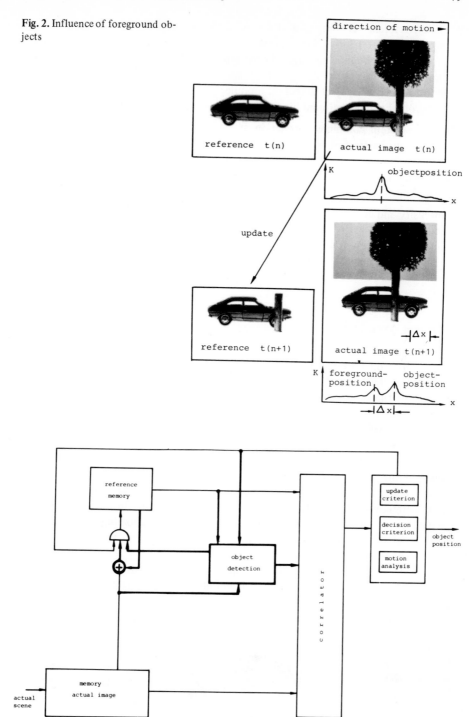

Fig. 3. FIM-Tracker-System (picture processing unit)

1. A new reference is composed of a part of the old reference and the actual image. Because of the relative speed between a moving target and the stationary background, only the moving target is well represented, whereas background information is blurred.
2. Monitored by an object detection module, parts of the moving object and mainly foreground and background distortions are detected and identified.

Thus this system combines correlation and object detection methods (Bers 1980). The object to be tracked is mainly located by the correlation of the actual scene and a reference. However the evaluation of the correlation values and especially the decision for the final object position is highly influenced by object detection methods, which are based on features like

— intensity and contrast levels
— image differences of "consecutive" images
— distribution of contour lines
— shape of special areas near the moving object
— relative speed between objects.

3 Object Approach

When approaching a target (sensor on board), the scale of the object representation is constantly increased. The homing systems applied until now especially analyse the intensity of the sensor signals and the approach of the missile to the target normally is compensated by a variable amplification of the sensor signal. As for the systems with simple analysation of intensity this kind of tracking system is very susceptible to false targets.

The method developed at FIM is based on the correlation tracker system as described before. A new element has been added which continuously adapts the size of the actual incoming object to a stored reference (Bers 1981). Because of the sensitivity of the correlation function to changes in scale the object size in the image has to be reduced to an original value. This adaption can be done by zooming or electronically by a digital size transformation. The principle of determining the scaling factor is shown in Fig. 4.

To evaluate the degree of size-reduction it is necessary to detect and classify parts of the object or marked details in the object although the target is changing in size. Therefore a reference including the object is generated at time t(n), which is systematically divided into several subreferences, the so-called drift-references. Depending

t(n)   t(n+1)   t(n+1)

Fig. 4. Determination of the scaling factor

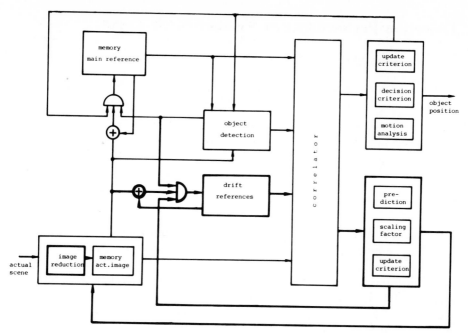

Fig. 5. FIM-Homing-System (picture processing unit)

on the speed of the missile the size of the target increases at $t(n+1)$ and the position of the sub-references drifts apart. According to the general system for target tracking the object position in the actual scene is found by evaluating the correlation function between the actual scene and the main reference. The difference between the position of the drift-references and the main reference is a measure for the change in size. The scaling factor is computed by analyzing these geometrical proportions and the results of the evaluation of the correlation values of main reference and drift-references (Fig. 5).

4 Results

The simulation with scenes in the visual and IR-range has been very successful and has demonstrated the advantage of the FIM-Tracker-System to other tracking systems when used in a complex scenario.

References

Bers K-H, Bohner M, Gerlach H (1980) Object detection in image sequences. Proc ICPR, Miami
Bers K-H, Doll H, Fritsche P, Stroh H (1981) Automatische Kompensation der Bildvergrößerung für Zielanflugverfahren. FIM-Bericht 97
Bohner M, Gerlach H (1976) Simulation eines Korrelationstrackers. FIM-Bericht 44
Gerlach H (1979) Digitale Bildfolgenauswertung zum Wiederfinden von Objekten in natürlicher Umgebung. FIM-Bericht 68

Detection Performance in Visual Search Tasks

A. KORN and M. VOSS[1]

1 Introduction

In a visual search task one or several targets are to be found among confusing objects. Visual search proceeds in a sequence of fixations and saccades where information is acquired during the intersaccadic intervals (glimpses). Any procedure to predict search performance must be based on measurements of the detection probability per glimpse and the searching strategy. The amount of the physically measurable properties determining the detection probability at very short presentation times (approx. 0.2 s) is called "visual conspicuity". As an experimental quantity for the "visual conspicuity", the size of the visual field around the fixation point is used in which the relevant object can be detected without any a priori knowledge of the position. Three methods of measuring the visual conspicuity or lobe area are known.

a) A tachistoscopical approach (Engel 1976, Voss 1982). Here the subject has to fix the center of a screen. The target appears, for a short time, in different peripheral positions. The detection probability can be calculated as a function of eccentricity.
b) Measurement of the distance of the last fixation before target detection.
c) Artificial limitation of the visual field by electronic means (Korn 1981). From the dependence of the performance parameters, i.e. the recognition rate and/or the search time within the size of the preset visual field, conclusions can be drawn about the useful visual field.

Besides a theoretical basis for the detection probability per glimpse some experimental results are presented in this paper for the methods a and c.

2 Theoretical Basis for Detection Probability per Glimpse

In signal detection theory, the relation between signal to noise ratio (SNR) and detection probability P_g of a signal can be expressed as a cumulative Gaussian probability distribution. For practicability a good approximation is given by (Voss 1982)

$$P_g(SNR) = 1 - \exp\left(-\frac{SNR}{a_1}\right)^{a_2} . \tag{1}$$

---

[1] Fraunhofer-Institut für Informations- und Datenverarbeitung, Karlsruhe, FRG

Localization and Orientation in Biology and Engineering
ed. by Varjú/Schnitzler
© Springer Verlag Berlin Heidelberg 1984

Using the visibility level VL_o (defined as ratio of actual and threshold contrast at eccentricity $0°$) and the relative contrast threshold $R(\beta)$: $= VL_o/VL_\beta$ as function of the eccentricity angle β, Eq. (1) leads to (Inditsky et al. 1982)

$$P_g(VL_o, \beta) = 1 - 2^{-\left(\frac{VL_o}{R(\beta)}\right)^a} \quad , \quad a = \text{steepness} . \tag{2}$$

Furthermore the influence of additional target background interference and the influence of additional informational load in the central field of vision (which is in effect additional noise for the detection task) can be characterised by noise factors $b, c \geqslant 1$, which are defined in terms of brains internal noise (Voss 1982); they transform $R(\beta)$ in Eq. (2) into $b \cdot c \cdot R(\beta)$.

Based upon this equation and assuming some scanning behaviour (e.g. random scanning) the cumulative distribution of detection time can be calculated in dependence of search task parameters (Inditsky et al. 1982).

Apparatus

Used in method a: For measuring the influence of informational load and of external noise on the detection performance of peripheral light stimuli the experimental set-up shown in Fig. 1 is used (Voss 1982). In addition to the primary task subjects have to detect light stimuli (short flashes with 0.3 s duration) peripherally by an adjustable, light weight spectacle frame (Fig. 2). In this method, eye movements related to the head and head movements have no influence on the detection rate averaged over a measurement period.

Used in method c: Experiments with different window sizes have been performed with the experimental set-up shown in Fig. 3. The picture on the TV monitor appeared to the subject only partly within a small square the position of which coincided exactly with his visual axis when eye movements are used to change the position of

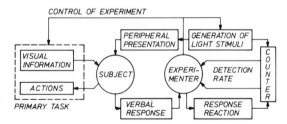

Fig. 1. Schematic diagram of the experimental set-up for measuring the influence of informational load

Fig. 2. Adjustable, lightweight spectacle frame presenting peripheral light stimuli

Fig. 3. Experimental set-up for measuring scan-paths for restricted visual field sizes

the window. The two dimensional movements of the subjects eye were detected by the corneal reflection method using a silicon TV camera for registration of the IR-light which is reflected from the subjects eye. The simultaneous measurement of eye and head movements in order to obtain the eye position relative to the background automatically is possibly by a combination of the NAC-Eye-Mark-Recorder with the S.A.M. image analysis system developed in our Institute.

Results

Referred to method a: As an example for a field experiment the detection performance of a subject during road crossing (without light signals, having no priority) is shown in Fig. 4. In relation to the areas of approach and drive away, the detection rates decreased strongly in the stop and cross areas (Voss 1982). Based on Eq. (2) the noise factor related to informational load during crossing can be calculated to 3,5 in the mean. This noise factor characterizes the amount of subjective informational load for this situation; furthermore for practical application it can be interpreted as contrast multiplier needed for constant detection performance in this situation. As can be seen by this example, the temporal resolution of the measurement method is in the order of a few seconds.

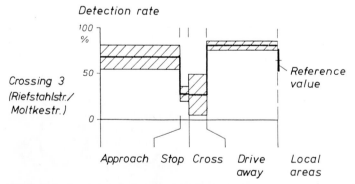

Fig. 4. Detection rate as a function of informational load during road crossing in a field experiment

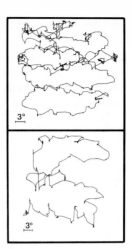

Fig. 5   **Fig. 6**

Fig. 5. Ocular scanning in realistic terrain (scanpath)

Fig. 6. Scanpath of eye movements for a 3 deg (*top*) and a 6 deg visual field size (*bottom*)

Referred to method c: To analyze the eye movement behaviour the scanpath has been recorded for each visual field size. In Fig. 5 the record of eye movements with a x-, y-plotter is superposed on the picture of an open field with three targets (vehicles). Here the visual field size of 18 x 24 deg was not restricted. Starting point was the center. Two examples of eye movement records are shown in Fig. 6 for two different visual field sizes. Here a tank must be detected in an open field. For a 2 x 3 deg visual field size (top) the detection time was 100 s and for a 4 x 6 deg visual field size (bottom) only 16 s. Figure 7 shows the detection time as a function of the visual field

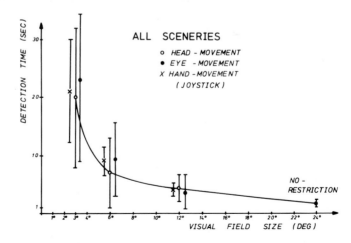

Fig. 7. Detection time as a function of the visual field size for different modes of control (max. target size 1.5 deg)

size for different modes of control. The position of the window was changed not only by eye movements but in other experiments also by head – or hand – movements. The average detection time for the 3 deg visual field size and nonrestricted search differ by a factor of about 10. Taking into account the slope of the curve the size of the conspicuity area is obviously at least 12 deg. The mechanism of motor control does not play a significant role for the detection performance.

Acknowledgements. This report was supported in part by the German Federal Ministry of Defense.

References

Engel FL (1976) Visual conspicuity as an external determinant of eye movements and selective attention. Thesis, Eindhoven TH

Inditsky B, Bodmann HW, Fleck HJ (1982) Elements of visual performance; lighting research & technology, vol 14, no 4, pp 218–231

Korn A (1981) Visual search, relation between detection performance and visual field size. Proc of the 1st European Annual Conference on Human Decision Making and Manual Control. Delft, University, pp 27–34

Voss M (1982) Aufnahme und Verarbeitung peripherer visueller Information unter dem Aspekt der Beanspruchungsmessung. Dissertation, Universität Karlsruhe

Prediction of Human Pattern Discrimination

K.R. KIMMEL[1]

1 Introduction

The problem of prediction of human pattern discrimination performance has a great relevance for many practical questions, as the specification of pattern sets for indicators in view of a low confusability. For the description of the aspect of human form discrimination, a number of different approaches applying different metrics has been used. Haller (1979) and Suen and Shiau (1980) used the Hamming distance, implementing it in the object space. This paper relies on the approach of Gagnon (1977) who used the psychophysically measured human modulation transfer function (MTF) enlarging it to arbitrary contrast conditions.

2 Method

The response of the visual system indicates its ability to place attention selectively on different regions of the range of available frequencies. This capability is the concept used to measure the visual response function for the tasks of foveal view, peripheral view, and foveal form perception. The prediction algorithm uses the MTF as given by Gagnon (1977) who refers to Cowger (1973) and Hilz and Cavonius (1974). Their measurements can be summarized as follows using a spatial frequency f and a point of regard θ

$$MTF = (f/f_0) \exp(1-f/f_0) \exp(-\theta/7) ; \qquad (1)$$

with $f_0 = 8 \exp(-2\theta/30)$, the frequency of maximal sensitivity and θ being the angle between the visual axis and a point of interest in the periphery.

This relation, representing essentially a bandpass, is used to filter the Fourier-transformed image data. The Euclidean distance d_{ij} between the filtered vectors X_i ($i \neq j$) of a given set of stimuli is computed and converted to a "proximity"-matrix by

$$d_{ij}^p = \begin{cases} 1 - d_{ij} & d_{ij} \leq 1 \\ 0 & d_{ij} > 1 , \end{cases} \qquad (2)$$

[1] Forschungsinstitut für Antropotechnik, Königsstraße 2, 5307 Wachtberg Werthhoven, FRG

Localization and Orientation in Biology and Engineering
ed. by Varjú/Schnitzler
© Springer Verlag Berlin Heidelberg 1984

giving the confusability prediction according to the basic assumption that two stimuli
are the easier to be confused the closer they are to each other.

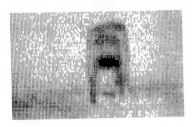 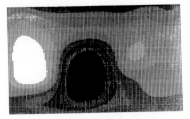

Fig. 1. Line printer plot of one digital thermal image

The experimental data were taken from two reports of van Meeteren and Schipper
(1980, 1981) who investigated the effects of different parameters of a night vision
system. The used material consisted of thermal images of four vehicles in frontal view,
presented to the subjects as slides in a laboratory experiment. The stimuli were seen
in a natural way with both eyes under an angle of 3 x 4.5 deg. As an example Fig. 1
shows a line printer plot of one of the images digitized in a format of 128 x 128 dots
in original (left) and MTF-filtered state (right).

3 Results

Table 1 contains the experimental human response data and the proximity-measures
according to Eq. (2) both ordered such that to each stimulus (in column 1) the closest
stimulus is placed in rank 1 (column 2), the next closest in rank 2 and so on. The
rank correlation between both matrices gives the following results:

rank number N	1	2	3
$\rho(N)$	0.92	0.89	0.89

(The rank correlation coefficient ρ is obtained by correlating the values of Table 1
after reordering the left matrix in the same way as the right one.) If we compute
for the rows of both matrices of Table 1 the rank correlation coefficient of Spearman,

Table 1. Experimental results (left) and model predictions (right)

	M1	M2	M3	M4		M1	M2	M4	M3
M1	0.47	0.30	0.13	0.07	M1	1.000	0.814	0.743	0.736
	M2	M1	M3	M4		M2	M1	M3	M4
M2	0.62	0.12	0.12	0.12	M2	1.000	0.814	0.766	0.689
	M3	M2	M1	M4		M3	M2	M1	M4
M3	0.40	0.29	0.20	0.11	M3	1.000	0.766	0.736	0.736
	M4	M2	M3	M1		M4	M1	M3	M2
M4	0.50	0.17	0.16	0.05	M4	1.000	0.743	0.736	0.689

r_s, in assigning the ranks 1 to 4 to the columns, we receive for rows 1 and 4 a value r_s of 0.8, thus confirming us that there exists a significant dependence between human response and model prediction on a 5%-level. (The lines 2 and 3 have an r_s of 1.0 because there are no rank differences.)

4 Discussion

The conclusion drawn from the above results is that the chosen distance measure between patterns of a given set is a reasonable predictor of the visual form recognition task. There are several reasons for the differences between experimental results and model predictions. The first one is that the effects of the random noise were not modeled. A second reason is that the prediction algorithm is not response biased, while the human response data are biased. The expected responses of the subjects are not equiprobable across all stimuli. The consequence is that the human response matrix contains some errors which are not related to what was seen but to what was expected or was thought to be a good response because it was not used for a while.

References

Cowger RJ (1973) A measurement of the anisotropic modulation transfer function of the extra foveal human visual system. AD 777, 853, WPAFB, OH: Air Force Inst of Techn
Gagnon RA (1977) A predictor of visual performance at selected visual tasks. Proc of the National Aerospace Electronic Conference (NAECON): 666–671
Haller R (1979) Gestaltung von Bildzeichen. PDV-Bericht KfK-PDV 174
Hilz R, Cavonius CR (1974) Functional organization of the peripheral retina: Sensitivity to periodic of the peripheral retina: Sensitivity to periodic stimuli. Vision Res 14:1333–1337
Meeteren A van, Schipper J (1980, 1981) Herkennungsproeven met warmtebeelden. Deel II, III, Instituut voor Zintuigfysiologie TNO Soesterberg, NL, Rapport-IZF 1980, 1980-14, 1981-5
Suen CY, Shiau C (1980) An iterative technique of selecting an optimal 5 x 7 matrix character set for display in computer output systems. Proc of the SID, vol 21, no 1:9–16

On the Application of Associative Neural Network Models to Technical Control Problems

E. ERSÜ[1]

1 Introduction

The application of the principles of biological information processing to real technical tasks has not been attempted very often except in a few cases of without promising any generality. However, the results of psychological and neurological research on the human brain are not only interesting from the point of view of providing physical explanations to mental effects, but also from a cybernetical viewpoint of providing design principles for man-made systems.

Optimized by evolution, the human brain – which can be regarded as a highly sophisticated, self-organizing system with tremendous abilities to learn, adapt, associate and influence or change its environment – displays properties needed for the control of complex systems.

2 General System Representations

Many attempts to model the human memory result in associative neural networks to be classified from the technical point of view as associative memory systems. In these systems, information processing is represented as a process of associations. This kind of mappings is also the basis of the psychological association theory for the explanation of mental processes in the human brain. It is assumed that for each afferent information (stimulus) an efferent one is associated as a response. The process of thinking is then the goal-directed cognitive organisation of these associations.

This stimulus-response type (S–R–) of associative mappings can easily be transferred to system theory for system representations, i.e. a system description by associative mapping of a n-dimensional input vector I onto a m-dimensional output vector O:

$$S: I \to O .\tag{1}$$

Assuming $I \in R^n$ and $O \in R^m$, it is clear that miscellaneous mappings such as linear, nonlinear, logical or binary can be represented by Eq. (1). Thus an associative neural

---

[1] Fachgebiet Regelsystemtheorie, Technische Hochschule, Darmstadt, FRG

Localization and Orientation in Biology and Engineering
ed by Varjú/Schnitzler
© Springer Verlag Berlin Heidelberg 1984

network model, i.e. an associative memory system, can be utilized to store the relationship between input and output of the system.

3 Human Problem-Solving Procedure as General Control Strategy

In the investigated control concept, the S-R-type system representations using trainable neuron-like memory systems are structurally combined with principles of human cognitive problem solving. The human ability of self-organizing control, i.e. developing a control (problem solving) strategy to achieve a given goal, is based on the cognitive interaction of the following basic elements:

a) a given goal which presents the problem to be solved
b) a predictive model of the environment which is learned by interactive behaviour with the environment itself
c) a planning decision scheme using the predictive model to learn the control strategy, a sequence of meaningful control actions, which achieve the goal.

The scheme illustrated in Fig. 1 as a control system is the most fundamental in human intelligent behaviour, as found in Dörner (1974) and Newell et al. (1972).

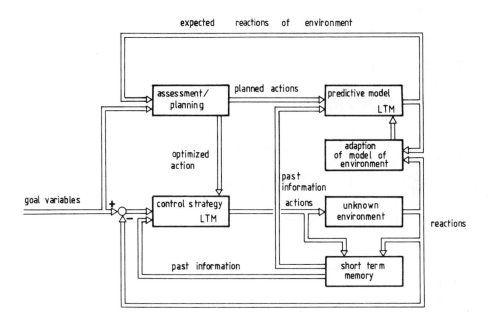

Fig. 1. Human problem-solving procedure as control system. *LTM* long-term memory

4 The Control Concept with Associative Memory Systems

The concept incorporates two associative memory systems, one for the mapping of the predictive model and one for the mapping of the control strategy. The choice of the type of associative memory system to be used depends on the abstraction level of the control problem at hand. For different control problems, different types of memories or neural network models are necessary. For the task of controlling the temperature of a chemical reactor, for example, a memory system is required which can process reveal variables, and for the task of object handling for manipulation problems in robotics, the used memory system must have decisive character, i.e. it must be able to process binary information.

The control concept in each case uses a so-called output predictive algorithmic scheme by which in each time cycle

a) the predictive model is updated by measured system and control data,
b) an optimization scheme is activated when necessary for searching for an appropriate control action which matches the goal, by utilizing the predictive model as state predictor for the different control actions,
c) the so found control actions are then memorized in the control memory to be used in similar situations as the best answer, making (b) superfluous in the long range.

$$\dot{y}_1 = \frac{5-y_1}{12,5}\, u_1 - 18,828 \cdot 10^{33} \cdot y_1 \cdot e^{-75,2315/y_2}$$

$$\dot{y}_2 = \frac{1}{400}\left[(24-32y_2)\cdot u_1 + 242,88 \cdot 10^{33}\cdot y_1 \cdot e^{-75,2315/y_2} + 28,8\,\frac{3,73-4y_2}{5+0,92u_2}\,u_2\right]$$

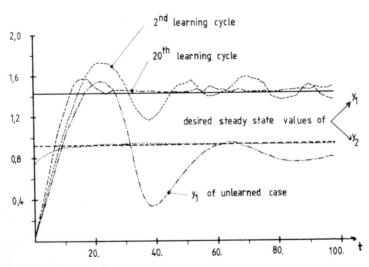

Fig. 2. Application example: chemical reactor

5 Applications

The applicability of the above control concept for technical processes has already been successfully tested on simulations of a chemical reactor, a neutralisation process (pH control) (Ersü et al. 1983), a robot arm (dynamic), and other problems. The associative memory system incorporated for this type of control applications is a perceptron – like model of the human cerebellar cortex in *Albus* (1975), being further modified and implemented for realtime applications (Ersü et al. 1982). Another, more abstract case is under study. It concerns the planning of object goal configurations in a multi-object scene for robotics applications. The associative memory used here is based on the system theoretical neural network modelling approach of Kohonen, which is already used for image recognition tasks in Kohonen (1977).

Figure 2 demonstrates the simulation example of the chemical reactor.

References

Albus JS (1975) The cerebellar model articulation controller. Trans ASME, Series G, vol 97, no 3
Dörner D (1974) Die kognitive Organisation beim Problemlösen. Hans Huber Verlag, Bern
Ersü E, Mao X Control of pH using a self-organizing control concept with associative memories.
    Int IASTED Conf on "Applied Control and Identification", June 28, 1983, Copenhagen
Ersü E, Militzer J (1982) Software implementation of a neuron-like associative memory system
    for control applications. Proc ISMM of 8th Int Symp on MIMI, March 2–5, 1982, Davos
Newell A, Simon HA (1972) Human problem solving. Englewood Cliffs, NJ Prentice Hall
Kohonen T (1977) Associative memory. Springer, Berlin Heidelberg New York

Control of Hand-Movements Towards a Target

A.G. FLEISCHER[1]

1 Introduction

Skilled performance of hand movements often requires a high level of accuracy. Discrete hand movements between a clear home position and a fixed target have been a preferred subject of numerous investigations (Howarth and Beggs 1981). Most studies of discrete movements have concentrated on the relationship between speed, distance, and accuracy. By contrast, very little work has been done on the manner in which skilled movements are executed in three-dimensional space. In particular, there are only a few data available which reveal the strategy which underlies free hand movements towards a target during the performance of a complex task (Soechting and Lacquiniti, 1981).

It is evident that movement towards a target starts with an acceleration phase and finishes with a deceleration phase. Information about the termination of a movement is the important contribution made by vision. In order to make visual feedback possible, the velocity of the hand during the deceleration phase must not exceed a certain limit (Keele and Posner 1968), which decreases as the distance from the target diminishes. However, there is still a controversy whether, in comparison with the acceleration phase, visual feedback leads to a longer decerleration phase (Beggs and Howarth 1972, Morasso 1981). The velocity of the hand has been compared therefore with the transverse dispersion of movement traces in this study, to determine the onset of visual feedback. It was also necessary to investigate the rotation of the hand about the wrist with respect to the forearm, since this rotation could influence the recorded kinetics of the hand movement. Furthermore, a new analytical method was designed to analyze free hand movements during the performance of a complex task.

2 Methods

An electroacoustic method was developed to analyze hand movements in space (Fleischer and Lange 1983). The recording of hand movements was based on the

---

[1] Bundesanstalt für Arbeitsschutz und Unfallforschung, Vogelpothsweg 50–52, 4600 Dortmund, FRG

Localization and Orientation in Biology and Engineering
ed by Varjú/Schnitzler
© Springer Verlag Berlin Heidelberg 1984

principle, that the X-, Y-, Z-co-ordinates of a moving point can be computed with respect to the reference system of the laboratory from the distances between a moving point and three fixed reference points. This principle has been realized by measuring the transmission time of an ultrasonic pulse from a small moving piezocrystal source to three microphones. In order to record the angles of the hand and of the forearm, three ultrasonic sources were strapped onto the right hand, the wrist, and the forearm, respectively (see insets of Fig. 3a).

The experiment is represented in Fig. 1. The subject was asked the beginning of the experiment to take a plastic chip with the right hand from pile P1 and put it down on pile P2, to take a second chip from pile P6 and to be put down on pile P3 and so forth. The full sequence is noted in Fig. 1. A recording of the hand positions in the X-, Y-plane during the performance of this task is shown in Fig. 2a. The small dots represent single recordings of the hand position and the large dots represent the position of the piles P1–P6.

The main methodical problem in analyzing the complex movement pattern described was to select, for further computations one-directional movements between two piles from the continous process. This could be achieved by forming a class of movement traces which leave, for instance, a defined area around maximum M6 of the distribution of hand positions and enter a defined area around maximum M2 (Fig. 2b). In this paper movements from M6 to M2 are selected and analyzed; forty experiments have been performed.

3 Results and Discussion

Analysis of movement characteristics during the approach of a target started with the computation of a linear regression line r (Fig. 3c) from all X-, Y-co-ordinates sampled for movements from maximum M6 to M2. The regression line r was divided into intervals 10 mm wide, and the angles of the hand and the forearm, the Z-co-ordinate of the hand, the hand position, and the velocity was averaged within every interval for all movement traces from M6 to M2. This resulted in the distributions shown in Fig. 3.

Comparison of the hand angle with the forearm angle reveals that the wrist is nearly stiff during the performance of the described task (Fig. 3a). This leads to the conclusion that even close to the target the control of the forearm plays the major part and that the position of the ultrasonic sound source has little effect on the kinetics recorded.

The selected hand movements between two piles can be characterized by two different phases. During the first phase (Fig. 3b, right side), which covers nearly one third of the distance between maximum M6 and M2, the hand is raised about 2 cm and reaches its maximum speed (Fig. 3d). The increase of the transverse dispersion of the movement traces during the fast acceleration phase (Fig. 3c) corroborates the hypothesis that the acceleration of the hand is ballistic with no or little visual feedback. During the second, phase the hand is lowered slowly (Fig. 3b, left side) and during the slow deceleration (Fig. 3d) the mismatch between the current and the

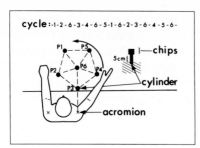

Fig. 1. The analyzed task. The subjects were asked to move plastic chips between the piles *P1* to *P6* in the sequence noted

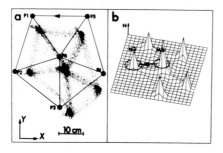

Fig. 2. a Recording of the hand positions during the performance of the described task. Recording period 15 min, sample rate 25 Hz. **b** Histogram of the hand positions

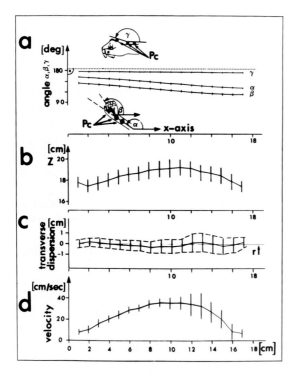

Fig. 3. The hand movements are represented from *right* to *left*. Average from 55 movement traces. **a** Angles of the hand and the forearm as defined in the insets. **b** Elevation of the hand above the test table. **c** Average of the hand positions and the corresponding transverse dispersion (standard deviation, *dashed curves*) of the movement traces. *r* regression line. **d** Velocity distribution for hand movements from pile P6 to pile P2

target position is successively reduced (Fig. 3c). It seems very likely that the decrease in the transverse dispersion is caused by visual feedback, since pure motor memory cannot reduce the mismatch developed during the first phase (Laabs and Simmons 1981). For all experiments performed the asymmetry of the acceleration and the deceleration phase remained in a proportion which allowed for visual error information processing.

References

Beggs WDA, Howarth CI (1972) The movement of the hand towards a target. Q J Exp Psychol 24:448–453

Fleischer AG, Lange W (1983) Analysis of hand movements during the performance of positioning tasks. Ergonomics 26:555–564

Howarth CI, Beggs WDA (1981) Discrete movements. In: Holding DH (ed) Human skills. Wiley & Sons, New York Chichester, pp 91–117

Keele SW, Posner MI (1968) Processing of visual feedback in rapid movements. J Exp Psychol 77:155–158

Laabs GJ, Simmons RW (1981) Motor memory. In: Holding DH (ed) Human skills. Wiley & Sons, New York Chichester, pp 119–151

Morasso P (1981) Spatial control of arm movements. Exp Brain Res 42:223–227

Soechting JF, Lacquaniti (1981) Invariant characteristics of a pointing movement in man. J Neurosci 1:710–720

How the Primate Brain Perfoms Detection and Pursuit of Moving Visual Objects

R. ECKMILLER[1]

1 Introduction

The design and implementation of technical devices to pursue moving visual targets, or to direct the optical axis quickly towards a briefly presented stationary target, forms a formidable challenge to engineers. Yet the primate oculomotor system performs both tasks superbly, using different strategies and different sets of neural elements to acquire the appropriate information from the retina and to generate the required neural control signals for the oculomotor motoneurons (for review see: Robinson 1981). A framework for these two oculomotor subsystems, the pursuit system and the saccadic system, is proposed here on the basis of neurophysiological findings in alert monkeys and is arranged into a sequence of Spatio-Temporal Translation, Motor Program Generation, and Neural Integration.

2 Components of the Eye Movement System in Primates

For the sake of simplicity we will disregard many important features of the primate eye movement system, such as vestibular input, binocularity, vertical and torsional movements, and accommodation. We are left then with the simplified system of a single eye ball capable of being rotated in the horizontal plane by two horizontal muscles in response to a small visual target. Figure 1 shows the "saccade path" for saccades to the left and the "pursuit path" for movements to the right between the retina of the right eye and the corresponding horizontal eye muscles. The neural elements for control of movements in the opposite directions are indicated by broken lines.

Extensive neurophysiological studies have shown that the eye movement control system in the primate brain stem receives only signals encoding eye movement rather than eye position. For this reason a step of Neural Integration has to be postulated, which transforms eye movement signals ($\dot{\Theta}_e$) into eye position signals (Θ_e) and holds the most recent eye position signal in the absence of eye movement signals, i.e. during fixation.

---

[1] Physikalische Biologie, Abteilung Biokybernetik, Universität Düsseldorf, FRG

Localization and Orientation in Biology and Engineering
ed. by Varjú/Schnitzler
© Springer Verlag Berlin Heidelberg 1984

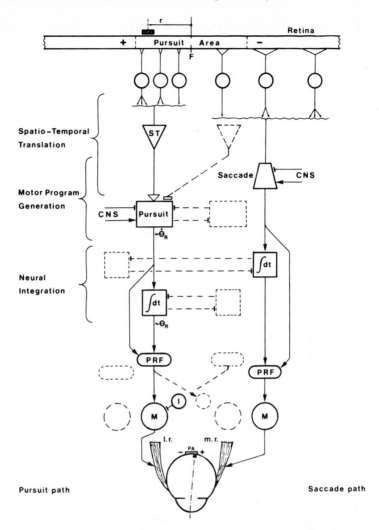

Fig. 1. Model of two primate oculomotor subsystems: the saccade path and the pursuit path. The right eyeball with the two horizontal eye muscles is indicated as output

The neural control signals for saccadic or pursuit eye movements usually depend on visual stimuli. However, single neurons in the afferent visual system cannot themselves encode the required eye movement signals. A neural structure for Motor Program Generation is therefore required to produce specific eye movement signals by means of appropriate neural impulse rate time courses.

The retina must be able to "tell" the eye movement system how to move in order to relocate a stationary target from an extrafoveal location onto the fovea, or to relocate a slowly moving target from an extrafoveal location onto the fovea and minimize the position error relative to the foveal center (F) during subsequent foveal pursuit eye movements (PEM). The necessary task of the postulated step of Spatio-

Temporal Translation is to turn the spatial value of target eccentricity relative to the fovea into a neural signal in the temporal domain so as to assure the appropriate eye movement.

Possible neural representations of these stages are discussed in the following paragraphs.

3 Spatio-Temporal Translation

The concept of a Spatio-Temporal Translator (STT) for the control of movements was first put forward by Braitenberg and Atwood in 1958 in an attempt to assign a functional role to the unique architecture of the cerebellar cortex. A more recent theoretical paper on saccadic eye movements (Robinson 1973) placed the STT in the brain stem. The simplest model for a neural realization of the STT consists of two sets of neurons which monitor the retinal eccentricity r of a visual target location in one horizontal direction (left or right of the fovea). Figure 1 shows such a Spatio-Temporal (ST) neuron in the pursuit path on the left (connected with the nasal part of the retina of the indicated right eye) and a "Saccade" neuron in the saccade path on the right.

As indicated in this figure it is assumed that both classes of neurons receive a stronger excitation from visual neurons (large open circles) with receptive fields in more eccentric locations than from those with receptive fields closer to the foveal center (Eckmiller 1981), due to the synaptic connectivity pattern (the number of synapses increases with eccentricity).

The STT of visually induced saccades is likewise assumed to consist of two sets of neurons, each of which monitors retinal eccentricity in one horizontal direction. In this case, however, the range of eccentricities is not restricted to the central retina. For this reason the receptive fields of the corresponding afferent visual neurons must also lie outside the fovea and are likely to increase in diameter with increasing distance from the fovea (Fig. 1). The 'Saccade' neurons (equivalent to medium lead burst neurons), however, are not assumed simply to increase the impulse rate (IR) if a visual stimulus moves away from the fovea, like ST neurons in the pursuit path, but to encode eccentricity in the duration and slope of IR decay in a burst of impulses.

4 Motor Program Generation

Many movements in higher vertebrates are controlled by internally generated motor programs which do not necessarily require (but can be guided and modified by) sensory information (Arbib 1981, Miles and Evarts 1979).

Foveal pursuit eye movements (PEM) require a pair of Motor Program Generators (MPG) which produce the neural signal for $\dot{\Theta}_e$ to the left and right, respectively. The aim of PEM is to maintain vision of a small moving target by keeping the target projection continuously on the center of the fovea. During PEM, the retinal slip velocity is minimal and of instantaneously varying direction because both the target and eye must move with approximately the same velocity, or else the target projection would slip outside the fovea. In macaques the fovea centralis and parafovea (similar to the Pursuit Area in Fig. 1) are 3 deg and 6 deg in diameter, respectively, whereas in humans they are slightly larger (5 deg and 8 deg, respectively). After sudden disappearance of stimulus during PEM these monkeys (similar to humans in this regard) continued to perform smooth post-pursuit eye movements (PPEM) for more than 1 s. It was therefore concluded that the time course of $\dot{\Theta}_e$ must be internally generated. This task is assigned to the MPG. The input of ST neurons in Fig. 1 serves to "update" the internally generated motor program. This updating process is likely to function as a leaky integrator which changes the signal for $\dot{\Theta}_e$. Once the movement time course is memorized (by an unknown mechanism), the MPG needs only occasional updating signals. Note in Fig. 1 that both MPG's receive inhibitory inputs from the contralateral ST neurons which can happen during PEM to the right or to the left whenever the target projection is in the contralateral half of the Pursuit Area (PA). Pursuit neurons, which are the only possible candidates to represent the MPG in the brain stem, have been discovered recently in trained macaques (Eckmiller and Mackeben 1980) close to the abducens nuclei.

The generation of saccadic eye movements requires quite different features and involves different sets of neurons. One class of pre-motor neurons, the medium-lead burst neurons, is generally accepted to encode both amplitude and velocity of saccades. The fine structure of these neural bursts and their relationship to saccadic parameters in monkeys was recently analyzed (Eckmiller et al. 1980). Figure 2 displays the relationship between various parameters of individual bursts and associated saccades. The upper left diagram exemplifies the time course of the instantaneous impulse rate IR(t) of a typical burst neuron controlling three saccades to the left. During large, high velocity saccades the slope m is low whereas it is high during brief, low velocity saccades. A typical characteristic of the average saccade velocity $\dot{\Theta}$ versus slope m (the burst characteristic) is shown in the upper right diagram. It is assumed that a burst neuron can be characterized by its IR_{max}, IR_{min}, as well as its burst characteristic with the constants P and V. The resulting characteristic for $\dot{\Theta}$ versus $\Delta\Theta$ fits the physiological data well (lower right diagram). The lower left diagram gives the schematic time courses of three saccades which could have been elicited with participation of the three bursts in the diagram above using the same time scale.

The Renshaw Model. In Fig. 3 some data on Renshaw cell bursts (Eccles et al. 1961) have been replotted (left diagram) to illustrate one finding which might be applicable to the synaptic input and function of our medium-lead burst neurons. Single volley stimulation was applied either to the anterior biceps muscle (AB; filled circles at maximal stimulus amplitude) or to the posterior biceps muscle (PB; open circles at maximal and stars at submaximal stimulus amplitude). The three bursts from a single Renshaw cell in the diagram have many features in common with those of medium-lead burst neurons, namely: IR_{max} and IR_{min} are about constant and IR decays relatively linearly after an initial drop from IR_{max}. The striking feature for our purposes is that a single uniform impulse (stimulus) can trigger bursts with different slopes m by either selecting another input line (PB versus AB) or by changing the

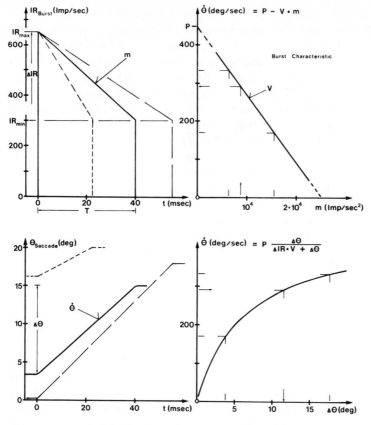

Fig. 2. Quantitative relationship between various burst and saccade parameters. The scheme describes the hypothesis of MPG which encodes velocity, amplitude, and duration of a saccade only by the slope m of the corresponding burst of a burst neuron

stimulus amplitude (open circles versus stars using input line PB). The schematic drawing of a Renshaw cell was added in order to interpret the plotted data. This Renshaw cell demonstrates a stronger "effectiveness" for the AB versus PB input in terms of the number and location of the synapses with respect to the cell soma. This Renshaw Model, if applied to medium-lead burst neurons, could provide a physiological explanation of why different bursts are generated when the same target excites different loci on the retina, i.e. by sending a brief, uniform signal via different, retinotopically ordered, input lines.

5 Neural Integration

When a target appears at an eccentric location on the retina, the afferent visual system is able to measure its distance from the fovea. To generate the required change in the

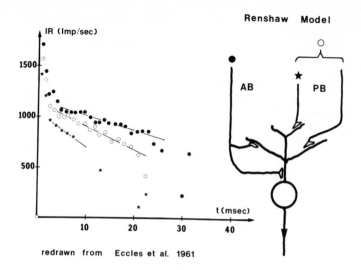

Fig. 3. *Left side* replotted data of Renshaw cell bursts by Eccles and co-workers. The bursts were elicited by single stimulation volleys of two muscles (*AB* and *PB*). *Right side* Renshaw Model = scheme to interpret Eccles' findings that single impulses on various neural input leads to bursts with different slopes, similar to those of medium-lead burst neurons

initial eye position Θ_{in} on the basis of the available $\dot{\Theta}$ signal, the hypothesis of a Neural Integrator (NI) is necessary. Models of neural networks which are able to hold the activity (and consequently the muscle tone) at a given level, which is an essential feature of any integrator circuit, were proposed about 50 years ago by Forbes (delay path theory), Ranson and Hinsey (theory of reverberating circuits), Barany, and by Lorente de Nó (1933). A pontine reticular formation (PRF) neuron (Fig. 1) encodes eye position signal pulse eye velocity signal in one direction only. Each motoneuron M receives the control signals for eye position and eye velocity in one direction from the ipsilateral PRF neuron and in addition for eye velocity in the opposite direction from the contralateral PRF neuron via an inhibitory interneuron I.

Acknowledgements. This work was supported by the Deutsche Forschungsgemeinschaft, Sonder-forschungsbereich SFB 200.

References

Arbib MA (1981) Perceptual structures and distributed motor control. In: Brookhart JM, Mount-castle VB, Brooks VB (eds) Handbook of physiology, section 1: The nervous system, vol II. Motor control, part 2, chap 33. Williams & Wilkins, Baltimore, pp 1449–1480

Eccles JC, Eccles RM, Iggo A, Lundberg A (1961) Electrophysiological investigations on Renshaw cells. J Physiol 159:461–478

Eckmiller R (1981) A model of the neural network controlling foveal pursuit eye movements. In: Fuchs AF, Becker W (eds) Progress in oculomotor research. Elsevier, North Holland, Amsterdam New York, pp 541–550

Eckmiller R, Blair SM, Westheimer G (1980) Fine structure of saccade bursts in macaque pontine neurons. Brain Res 181:460–464

Eckmiller R, Mackeben M (1980) Pre-motor single unit activity in the monkey brain stem correlated with eye velocity during pursuit. Brain Res 184:210–214

Lorente de Nó R (1933) Vestibulo-ocular reflex arc. Arch Neurol Psychiat 30:245–291

Miles FA, Evarts EV (1979) Concepts of motor organization. Ann Rev Psychol 30:327–362

Robinson DA (1973) Models of the saccadic eye movement control system. Kybernetik 14:71–83

Robinson DA (1981) Control of eye movement. In: Brookhart JM, Mountcastle VB, Brooks VB (eds) Handbook of physiology, sect 1: The nervous system, vol II. Motor control, part 2, chap 28. Williams & Wilkins, Baltimore, pp 1275–1320

Non-Linear Interaction of Vestibulo-Ocular Reflex and Pursuit Eye Movements

O. BOCK[1]

During head motion, the vestibulo-ocular reflex (VOR) and the pursuit system interact in the control of eye movements. Head rotation induces opposite-directed eye movements via VOR which stabilize the gaze. Any shortcoming of VOR is compensated in the presence of visual fixation targets by the pursuit system, which keeps targets stable on the retina.

While it is known that interaction of VOR and the optokinetic system is linear (Robinson 1977, Schmid et al. 1980, Collewijn et al. 1981), no comparable data are available on the interaction of VOR and the pursuit system. We therefore investigated this interaction in two experiments by inducing a conflict between the two systems as to the appropriate direction of eye movements.

In the first experiment, ten subjects were rotated manually at different frequencies in a rotating chair (Amplitude 8 to 12 deg). Through a head-fixed mirror the subjects watched a stationary target light (LED) in otherwise complete darkness. The mirror reversed relative LED motion: the pursuit system had to move the eyes in phase with the body in order to stabilize the retinal image. VOR, however, is expected to move the eyes opposite to the body, as normally.

Figure 1 shows original recordings. At low stimulus frequencies (below 0.8 Hz), the eyes moved in phase with the body, suggesting a dominance of the pursuit system in oculomotor control. At high stimulus frequencies (above 1.5 Hz), however, eye movement was reminiscent of pure vestibular nystagmus, with the smooth component directed opposite to the body. Of particular interest are the findings at intermediate frequencies. Here, no gradual transition from in-phase to opposite-directed eye movements was found. Rather, the eyes moved at times unequivocally in phase with the body, and at other times unequivocally opposite to the body (arrows in Fig. 1b).

The gain of pursuit-dominated eye movements (● in Fig. 2) dropped steadily with increasing frequency. Additional tests with the same ten subjects investigating the pursuit system in the absence of vestibular stimulation revealed a similar behavior of the gain.

The gain of VOR-dominated eye movements (▲ in Fig. 2) is frequency-independent, but lower than in control tests of the same ten subjects which stimulated VOR in the absence of visual inputs (0.6 to 0.8).

Taken together, these findings are in sharp contradiction to a theory of linear interaction of VOR and pursuit systems. They suggest that interaction is governed by

1 Abt. Biokybernetik, Universität Düsseldorf, 4000 Düsseldorf, FRG

Localization and Orientation in Biology and Engineering
ed by Varjú/Schnitzler
© Springer Verlag Berlin Heidelberg 1984

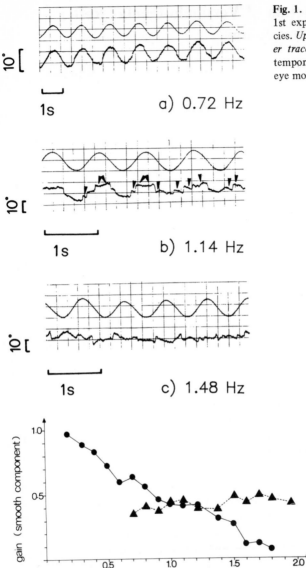

Fig. 1. Original recordings from the 1st experiment at 3 sample frequencies. *Upper traces* body position. *Lower traces* horizontal eye position (bitemporal EOG). *Arrows* in **b** indicate eye movements opposite to the body

a) 0.72 Hz

b) 1.14 Hz

c) 1.48 Hz

Fig. 2. Gain of eye movements in phase with (●) and opposite to (▲) the body. At intermediate frequencies, two gain values corresponding to the two-phase relationship were calculated. Data are the means of ten subjects

a switch, connecting oculomotor control to the pursuit system at low, to VOR at high, and oscillating between these two modes at intermediate frequencies. If so, however, one would expect to encounter this type of non-linear behavior in other experimental paradigms as well. This was the rationale for our second experiment.

In the second experiment, the same ten subjects were rotated while watching an LED which rotated with them (head-fixed). In this case, the pursuit system should keep the eyes stationary in the head in order to stabilize the retinal image, while VOR should move the eyes opposite to the body, as normal.

Fig. 3. Original recordings from the 2nd experiment at 3 sample frequencies. Traces as in Fig. 1. *Arrows* in **b** indicate time intervals when eyes remained stationary

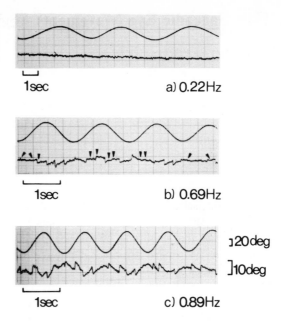

1sec   a) 0.22Hz

1sec   b) 0.69Hz

⌐20deg

⌐10deg

1sec   c) 0.89Hz

Figure 3 shows original recordings. Up to about 0.4 Hz the eyes remained stationary, suggesting a dominance of the pursuit system. Above 0.85 Hz vestibular nystagmus-like eye movements were found. At intermediate frequencies, the eyes remained at times stationary, and at other times they moved opposite to the body. The gain of these eye movements, which are presumably vestibularly dominated, was similar to that in the first experiment.

Such behaviour of oculomotor control is what would be expected given the results of the first experiment. They support the conclusion that the interaction of pursuit and VOR is governed by a switch.

As mentioned earlier, the interaction of VOR and the optokinetic system is presumably linear. Therefore, our present findings are in line with the theory that the optokinetic and the pursuit system use different mechanisms to interact with VOR, probably corresponding to different anatomical loci.

Acknowledgements. Thanks are due to Prof. O-J. Grüsser for valuable discussions and to Dipl.-Phys. F. Behrens, cand. med. G. Curio, Dr. med. M. Pause, and Dr. med. U. Schreiter for encouragement and discussions. Figures 1 and 2 in part reproduced from: O. Bock (1982) Non-linear interaction of the vestibular and the eye tracking system in man. Exp Brain Res 47:461–464.

References

Collewijn H, Curio G, Grüsser O-J (1981) Interaction of the vestibulo-ocular reflex and sigma-optokinetic nystagmus in man. In: Fuchs AF, Becker W (eds) Progress in oculomotor research. Elsevier, Amsterdam

Robinson DA (1977) Linear addition of optokinetic and vestibular signals in the vestibular nucleus. Exp Brain Res 30:447–450

Schmid R, Buizza A, Zambarbieri D (1980) A non-linear model for visual-vestibular interaction during body rotation in man. Biol Cybern 36:143–151

How Saccadic Eye Movements Find their Target

H. DEUBEL and G. HAUSKE[1]

Normally, we perform saccadic eye movements continuously to explore our visual environment, selecting objects in the visual field in order to redirect our gaze. The target once selected, the oculomotor system has to compute its retinal eccentricity to determine the amplitude of the goal-directed saccade. After the saccade, the oculomotor system has to check the result of the movement and, eventually, a correction must be performed. We investigated how visual information is processed to compute saccadic amplitudes and to correct motor errors.

A first issue of our experiments concerned how the amplitude of the goal-directed saccade is determined from the spatial properties of the target. For this purpose our subjects had to perform saccades to targets which consisted of two LED's (T_1, T_2) with different luminances (L_1, L_2). Experimental procedure and results are given in Fig. 1.

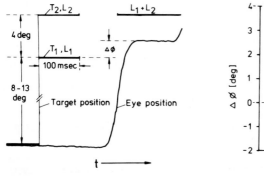 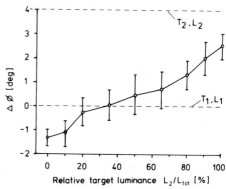

Fig. 1. *Left* the experimental sequence. Initially the subject fixates on a primary target position. After a random delay, two peripheral targets T_1 and T_2, with luminances L_1 and L_2, appear for 100 ms. The angular distance $\Delta\phi$ of the eye from target T_1 is measured after the primary saccade. In these experiments the target luminances had a constant sum ($L_1 + L_2 = L_{tot}$, $L_{tot} = 60$ cd/m²). Background luminance was 18 cd/m².

The data points on the *right* represent the mean angular distance of the eye position from T_1 after the primary saccade plotted as a function of the relative target luminance L_2 in percent of L_{tot}. The data show that in addition to a constant undershoot of 1.5 deg the virtual saccade goal is determined by the physical center of gravity of the input luminance distribution

[1]   Lehrstuhl für Nachrichtentechnik, Technische Universität München, FRG

Localization and Orientation in Biology and Engineering
ed by Varjú/Schnitzler
© Springer Verlag Berlin Heidelberg 1984

Consistent with recent results from Findlay (1982), the data show that having two targets as saccade goal the eye lands at an intermediate position. More specifically the data demonstrate that the physical centre of gravity of the luminance distribution acts as virtual goal for the saccade (Deubel and Wolf 1982).

In our view, this kind of spatial integration in the sensory-motor system is a direct consequence of its function. Convergence and integration of information over a large spatial window (the retinal hemifield) are principal properties of a mechanism which has the task of reducing complex, spatially coded excitation at its input in order to form direction and amplitude of a goal-directed motor response. A corresponding concept in the sense of a Spatio-Temporal Translator was recently developed by neurophysiologists (e.g. Eckmiller 1981) who postulate a neural mechanism performing an integration of retinal input weighted by its retinal eccentricity. The invariance against the absolute target luminance demands an additional, normalizing operation.

Taken together, weighted spatial integration and a normalization on the mean input intensity lead to a system which calculates the physical centre of gravity of the input signal.

The second issue of our experiments concerned the correction of motor errors which frequently occur because large primary saccades to peripheral targets usually fall short of the target. Experiments with isolated targets on homogeneous background demonstrate that the refixation error is then eliminated by short-latency corrective saccades which are programmed on basis of visual feedback (Deubel et al. 1982). Problems, however, arise with the ecological validity of this simple mechanistic concept. In order to correct a motor error remaining after the primary saccade, the oculomotor system must be able to relocalize the initially selected target within a complex visual environment. The question arises if perceptual reidentification of the target is a necessary condition for the programming of the corrective saccade.

In our experimental approach to this question, our subjects had to perform horizontal goal-directed eye movements on extended pseudo-noise patterns of vertical bars. The saccadic target here was defined by a dark/bright inversion of a limited area in the periphery. During the primary saccade an artificial refixation error was induced by displacing this target area in order to examine the effect of the visual reafference on the execution of the corrective saccade. Procedure and results from this experiment are given in Fig. 2.

Since target area and background have similar structure, it is not possible after the primary saccade to reidentify the target perceptually, nor do the subjects perceive its intrasaccadic displacement (Bridgeman et al. 1975). Surprisingly, however, involuntary correction saccades occur which accurately eliminate the remaining refixation error. Consequently we have to postulate a low-level, automatic corrective mechanism based on visual input and independent from perception: We assume that visual information about the object selected as target for the saccade is stored. After the execution of the primary saccade, this stored information is compared with the actual visual reafference in order to relocalize the target for the correction saccade.

Evidence for the postulated interaction between pre- and postsaccadic visual information was already given in former experiments (Wolf et al. 1980). The results are an indication that the oculomotor system has access to information which is not available to perception.

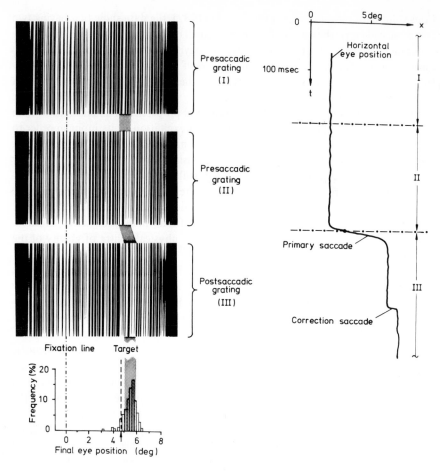

Fig. 2. The pseudo-noise patterns of *vertical bars* were generated on a display and changed for each trial. The figure shows different phases of the experimental sequence:

(*I*) The subject fixates on a given fixation line;

(*II*) In order to elicit the goal-directed saccade, a target is defined by a dark/bright inversion of a small area of the grating 5 deg in the periphery. The target area is indicated by the *hatched area* in the figure.

(*III*) Triggered by the primary saccade this target area is displaced 0.5° to 1° either into the same or into the opposite direction as the primary saccade. The final eye position is determined after the corrective saccades.

In control experiments without target displacement, the mean final eye position was found to be at about 4.8 deg as indicated by the *vertical dashed line*. The data with target area displacement show that the final eye position corresponds to the shifted target area. This means that the artificially induced refixation error is corrected by the corrective saccade

Acknowledgement. This investigation was supported by the Deutsche Forschungsgemeinschaft as a project of the Sonderforschungsbereich 50 Kybernetik.

References

Bridgeman B, Hendry D, Stark L (1975) Failure to detect displacement of the visual world during saccadic eye movements. Vision Res 15:719–722

Deubel H, Wolf W, Hauske G (1982) Corrective saccades: Effect of shifting the saccade goal. Vision Res 22:353–364

Deubel H, Wolf W (1982) Secondary saccades induced by altering visual feedback. Invest Ophthalmol Vis Sci 20, Suppl 26

Eckmiller R (1981) A model of the neural network controlling foveal pursuit eye movements. In: Fuchs AF, Becker W (eds) Progress in oculomotor research. Elsevier North Holland, Amsterdam, New York, pp 541–550

Findlay JM (1982) Global processing for saccadic eye movements. Vision Res 22:1033–1045

Wolf W, Hauske G, Lupp U (1980) Interaction of pre- and postsaccadic patterns having the same coordinates in space. Vision Res 20:117–125

Visual Orientation of Flies in Flight

C. WEHRHAHN[1]

1 Introduction

Flies are able to track small, fast moving objects, a task which requires high spatial and temporal resolution in information processing of the visual system and aerobatic capabilities of the flight motor system. This review briefly outlines which basic perceptual principles are used by flies during tracking.

2.1 Free Flight Analysis of Female Flies

Houseflies kept in a cage will fly around as they do in houses. Both female and male flies will track other flies differing, however, greatly in their success. Their flights can be filmed in the cage simultaneously from two sides and subsequently the flight trajectories reconstructed in three dimensions from the films through frame by frame analysis (Wehrhahn et al. 1982). As is known from everyday experience a fly (female or male) may cruise through a contrasted environment, keeping its course without being visibly influenced by the flow of visual contrast, generated by the flight motion. In some cases, however, a fly will locate a resting object and land on it or follow a moving object and perhaps even try to catch up with it. Nothing can be said within this analysis about the processes causing a transition from the cruising state into other states. Apparently some properties of the moving objects are necessary but not sufficient to elicit specific responses as described below.

Figure 1 shows a flight episode from two female houseflies, the first fly apparently flying without a specific goal, the second, however, clearly trying to keep track with the first. More specifically the second fly tries to align itself with respect to the first fly so that its body axis points towards it. A quantitative evaluation of the turning response $\dot{a}_v$ as a function of the position ψ_A of the target fly in the horizontal plane shows that a "best fit" is achieved when the turning response is shifted by 20 to 30 ($\pm$10) msec with respect to the target position. Thus the time required by the fly to compute a turning response from the target position and to generate a response is very short. If we assume a delay of about 8 msec for the photoreceptors (Hardie unpubl), and assume that the signal has to cross at least four synaptic junctions (of about 1 msec each) before leaving the head, we are left with a lower limit of about

---

[1] Max-Planck-Institut für biologische Kybernetik, Spemannstraße 38, 7400 Tübingen, FRG

Localization and Orientation in Biology and Engineering
ed by Varjú/Schnitzler
© Springer Verlag Berlin Heidelberg 1984

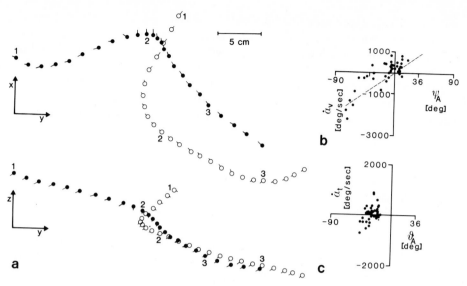

Fig. 1. a Flight episode between two female flies in the horizontal (xy)-plane (*upper*) and the vertical (xz)-plane (*lower graph*). The leading fly is represented by the *empty dots* the tracking one by *full dots. Lines* represent body axis. *Numbers* indicate equal times. Interval between two successive points is 20 ms. (Wagner unpubl). Definition of coordinates used see inset of Fig. 4a. **b** Turning response around the vertical axis $\dot{a}_v$ of the tracking fly plotted as a function of the error angle ψ_A in the horizontal plane under which the leading fly is seen by the tracking fly shifted by 20 msec in time. The turning response depends on the position of the target. If we approximate the relation between $\dot{a}_v$ and ψ_A by $\dot{a}_v = - k\psi_A$, then $k \approx 20$ s^{-1}.
c Turning response around the horizontal axis perpendicular to the direction of flight (pitch) $\dot{a}_t$ plotted as a function of the vertical error angle ϑ_A. It can be seen that the target is fixated in the lower frontal part of the field of view ($\vartheta_A < 0$)

12 msec for synapses in the thoracic ganglia and the generation of the motor response. On the other hand, the velocity of a fly is about 1 msec thus in 20 msec the fly moves 2 cm. For a reasonable tracking performance in free flight such a fast response is very important. The turning response apparently depends on the position ψ_A of the target as can be concluded from Fig. 1b. Finally Fig. 1c shows that during the whole pursuit the target is held by the tracking fly in the lower frontal part of its field of view.

2.2 Tethered Flight. Open-Loop Experiments, with Female Flies

The turning response of a fly can also be determined in tethered flight with a highly sensitive torque meter (Fermi and Reichardt 1963). Flies which are attached to such a device (and are thus fixed in space) will fly up to several hours and their intended flight torque to visual stimuli, whose parameters under these circumstances are well defined, can be measured in real time. The fly is completely decoupled from its environment under these circumstances, hence its responses do not change its visual input: the "motor loop" of the animal is "opened". Such a situation is often called an "open loop" experiment (Hassenstein 1951; Mittelstaedt 1950). Figure 2 shows averaged open loop torque responses of 20 houseflies to a narrow vertical black stripe on a bright cylinder which was rotated clockwise (upper trace) and counterclockwise (lower trace) around the test animals. The responses in Fig. 2 clearly depend on the position of the stripe. Many experiments have shown that this is in part due to a direction insensitive response component (Pick 1974, Reichardt 1979, Strebel 1982) and a direction sensitive response component (Wehrhahn and Hausen 1980, Wehrhahn 1981), that is, both response components depend on the position of the respective stimulus in the receptive field. The contribution of the respective response components to the fixation and tracking process is at present not completely clear (see Bülthoff and Wehrhahn Chap. II.6 this Volume). Theoretical considerations have shown that either component would be suitable for a fixation and tracking mechanism of the kind used by free-flying houseflies (Reichardt and Poggio 1976, Wehrhahn and Hausen 1980, Poggio and Reichardt 1981).

The orientation behaviour of tethered flying male and female houseflies has been compared recently by Strebel (1982). Both sexes behave very similarly in tethered flight, in contrast to free flight (see Sect. 4).

3 Conclusions from Experiments with Female Flies

a) Female flies fixate and track objects in free flight.
b) The object is held preferably in the lower frontal part of their field of view.
c) The turning response around the vertical axis in free flight depends on the position of the tracked object.
d) Open loop experiments with tethered flying houseflies show that their torque signal depends on the position of the stimulus.

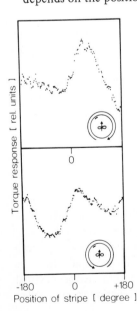

Torque response [rel. units]

-180   0   +180
Position of stripe [degree]

Fig. 2. Average flight torque of female flies recorded as a function of the position of a dark vertical stripe on a bright background rotated clockwise (*above*) and counterclockwise (*below*) around the fly. The response strongly depends on the position of the target. (From Wehrhahn and Hausen 1980). The average luminance of the panorama is 60 cd m^{-2}. The stripe was 3° wide and 45° high. The response amplitude at $\psi = 45°$ is about 10^{-7} Nm. Average of 20 flies

4 Male Chasing

Much higher performance is achieved by the male chasing system than by the female tracking system. As can be seen from Fig. 3, the flight path of the leading fly is very tortuous including three more or less sharp bends in the beginning and a spectacular looping. All manoeuvers are followed easily by the chasing fly. A performance like that would rarely be seen observing female flies.

A closer look at some properties of the chasing system is possible by examining a reconstruction of a shorter chase which now includes the body axis of the flies. It is easy to recognize that the body axis of the chasing fly and flight trajectory coincide only in the minority of the observed instances (Fig. 4a). Quantitative evaluation of this chase shows that the turning response around the vertical axis $\dot{a}_v$ of the chasing fly is strongly dependent upon the horizontal position ψ_A of the target fly (Fig. 4b). In this case the best fit is achieved when the turning response is shifted by 12.5 ms with respect to the target position. Thus for the male system the total delay is even shorter as that for the female system.

In addition the slope of the regression line connecting the points is much steeper than that of Fig. 1b. Thus the gain in the relation error angle/turning response is much higher in males than in females, greatly improving tracking performance. The delay contained in this feedback system represents a potential problem of instability. If k is the value relating $\dot{a}_v$ and ψ_A and ϵ the delay involved in the system, the "aperiodic limit", that is, when no oscillations occur, is given by $k = e^{-1}/\epsilon$ (Land and Collett 1974). For ϵ = 12.5 ms this gives $k \approx 30$ s^{-1}. The values of k in the chase evaluated until now are between 40 and 80 s^{-1} which means that slight oscillations or, overshoots are to be expected in the flight response of the chasing fly and are indeed found (Land and Collett 1974, Wehrhahn et al. 1982). However, for small amplitudes, this error becomes very small and, in addition, the turning response also depends on the error rate (Fig. 4c). The small amount of instability is diminished probably by this effect. Thus the operation of the chasing system seems to be adapted to turning as fast as possible without becoming unstable. In contrast, the female system seems to operate at longer delays and, correspondingly, lower values of k (Fig. 2b).

Figure 4d shows the turning response around the horizontal axis (pitch) plotted as a function of the vertical error angle ϑ_A. It can be seen from these data that the target is held by the chasing male in the upper frontal part of the field of view. In

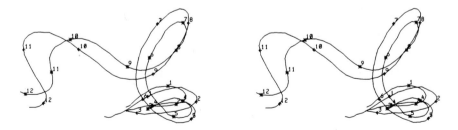

Fig. 3. Three-dimensional reconstruction of a chase between two male flies seen from above. The numbers correspond to 100-ms intervals. The plot should be observed with standard stereo glasses. (From Wehrhahn et al. 1982)

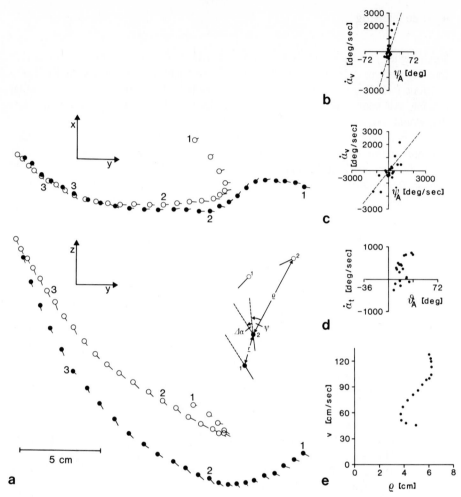

Fig. 4. a Reconstruction of a shorter chase between two male flies in the *horizontal xy-plane* (*upper*) and in the *vertical zy-plane* (*lower graph*). *Empty dots* leading fly; *full dots* chasing fly. Interval between two successive points is 12.5 ms. *Inset* Defintion of coordinates used.
b Turning response around the vertical axis $\dot{a}_v$ plotted as a function of the error angle ψ_A. $\dot{a}_v$ is delayed by 12.5 ms with respect to ψ_A. The turning response strongly depends on the target position. If we approximate the relation between $\dot{a}_v$ and ψ_A with $\dot{a}_v = -k\psi_A$ then $k \approx 80$ s^{-1}.
c Turning response around the vertical axis $\dot{a}_v$ plotted as a function of the error rate $\dot{\psi}_A$. $\dot{a}_v$ is delayed by 25 ms with respect to $\dot{\psi}_A$. The turning response also depends on the error rate.
d Pitch response (see legend to Fig. 2b) plotted as a function of the vertical error angle ϑ_A. The target is fixated in the upper part of the field of view ($\vartheta_A > 0$).
e Relation between forward velocity v and distance ρ for a shift of 72.5 ms

addition the visual field of sex-specific photoreceptors found in the retina, and the sex-specific neurons found in the lobula, are located in the upper frontal part of their field of view. Thus it is tempting to speculate that the sex-specific neuronal structures found in male flies may be the physiological counterpart for the behaviour of male flies. During chasing, the target is seen by the males against the sky, optimizing the contrast and dispensing the male fly from solving the figure-ground problem (Reichardt et al. 1983).

Finally, another property of the male chasing system should be mentioned. Plotting the three-dimensional forward velocity v against the distance ρ to the target, reveals a linear relation $v = \chi \cdot \rho$ (Fig. 4e). v was delayed with respect to ρ by 72.5 ms in this plot. This is long compared to the 12.5 ms used by the turning system. Thus male flies, at least in some cases, may regulate their forward velocity as a function of the distance to their target and this represents a further optimization of the system (Wehrhahn et al. 1982). In female flies a type of regulation of the forward velocity is also observed (Wagner unpubl.)

The fact that distance is used by the chasing fly to regulate its forward velocity raises the question as to how this might be achieved. The distance on the compound eyes between corresponding ommatidia is about 1 mm. The angular resolution of about 1.5° would produce very large errors if the target distance would be computed from triangular measurements. Collett and Land (1975) have proposed a mechanism for surphid flies which comprises objects of specific size as targets for the chasing system, an assumption implying that only objects of the "right" size (= species) are tracked properly.

A problem connected to the obvious specificity of the male system for small objects is the sensitivity to large field stimuli and possible interactions with the "female" system described above which is also present in males. This can be shown by comparing the closed loop behaviour in fixed flight of male and female flies, which is entirely the same (Strebel 1982). In particular, stimulation with a periodic grating of the upper frontal part of the field of view only, elicits equally weak yaw torque responses of tethered flying female and male flies, irrespective of the direction of motion (Wehrhahn unpubl.).

Obviously the male chasing system is triggered by small dark objects (Collett and Land 1975). It is known from surphid flies, that the optomotor system is in full operation during chasing (Collett 1980).

5 Conclusions from Observations with Male Flies

a) Male flies chase other flies.
b) The target is held during chases in the upper frontal part of their field of view.
c) The turning response around the vertical axis depends on the position of the tracked target.
d) The forward velocity of the chasing fly depends on the distance to the target.
e) The male chasing system is physiologically different from the female tracking system.

Acknowledgements. I am grateful to H. Wagner for the permission to use the unpublished data of Fig. 2 and together with K. Hausen and W. Reichardt for comments on the manuscript, also to H. Hadam for drawing the figures, and I. Geiss for typing the manuscript.

References

Bülthoff H, Wehrhahn C (1983) Computation of movement and position in the visual system of the fly (Musca): Experiments with uniform stimulation. This volume

Collett TS, Land MF (1975) Visual control of flight in the hoverfly, Syritta pipiens L. J Comp Physiol Psychol 99:1–66

Collett TS (1980) Angular tracking and the optomotor response. An analysis of visual reflex interaction in a hoverfly. J Comp Physiol Psychol 140:145–158

DeVoe RD (1980) Movement sensitivities of cells in the flies medulla. J Comp Physiol Psychol 138:93–119

Fermi G, Reichardt W (1963) Optomotorische Reaktionen der Flige Musca domestica. Kybernetik 2:15–28

Hausen K, Strausfeld NJ (1980) Sexually dimorphic interneuron arrangements in the fly visual system. Proc R Soc Lond [Biol] 208:57–71

Hassenstein B (1951) Ommatidienraster and afferente Bewegungsintegration. Z vergl Physiol 33:301–326

Land MF, Collett TS (1974) Chasing behaviour of houseflies (Fannia canicularis). A description and analysis. J Comp Physiol Psychol 89:331–357

Mittelstaedt H (1950) Physiologie des Gleichgewichtssinnes bei fliegenden Libellen. Z vergl Physiol 32:422–463

Pick B (1974) Visual flicker induces orientation behaviour in the fly Musca. Z Naturforsch 29c: 310–312

Poggio T, Reichardt W, Hausen K (1981) An neuronal circuitry for relative movement discrimination by the visual system of the fly. Naturwissenschaften 68:443–446

Reichardt W (1969) Movement perception in insects. In: Reichardt W (ed) Processing of optical data by organisms and by machines. Rendiconti SIF, XLIII. Academic Press, London New York, pp 465–493

Reichardt W (1979) Functional characterization of neural interactions through an analysis of behaviour. In: Schmitt FO, Worden FG (eds) The neurosciences, fourth study program. MIT Press, Cambridge, England, pp 81–103

Reichardt W, Poggio T (1976) Visual control of orientation behaviour in the fly. Part I. A quantitative analysis. Q Rev Biophys 9:311–375

Reichardt W, Poggio T (1981) Visual fixation and tracking in flies. Biol Cybern 40:101–112

Reichardt W, Poggio T, Hausen K (1983) Figure-ground discrimination by relative movement in the visual system of the fly. Part II. Towards the neural circuitry. Biol Cybern 46 (Suppl):1–30

Strebel J (1982) Eigenschaften der visuell induzierten Drehmomenten-Reaktion bei fixiert fliegenden Stubenfliegen Musca domestica L. und Fannia canicularis L. Dissertation, Eberhard-Karls-Universität, Tübingen

Wehrhahn C (1979) Sex-specific differences in the chasing behaviour of houseflies (Musca). Biol Cybern 32:239–241

Wehrhahn C (1981) Fast and slow flight torque responses in flies and their possible role in visual orientation behaviour. Biol Cybern 40:213–221

Wehrhahn C, Hausen K (1980) How is fixation and tracking accomplished in the nervous system of the fly. Biol Cybern 38:179–187

Wehrhahn C, Poggio T, Bülthoff H (1982) Tracking and chasing in houseflies. An analysis of 3-D flight trajectories. Biol Cybern 45:123–130

Chapter II Orientation and Path Control

Roll-Stabilization During Flight of the Blowfly's Head and Body by Mechanical and Visual Cues

R. HENGSTENBERG[1]

1 Movement in Space

Flying animals have six degrees of freedom of movement in space: three of translation along their body axes (lift, slip, thrust), and three of rotation about these axes (yaw, pitch, roll). For aerodynamical reasons, however, they maintain on average a characteristic flight attitude, i.e. a particular orientation of their body with respect to the gravity field. Translatory and yaw movements have little influence upon flight stability, but pitch and roll movements involve the risk of crashing. Consequently such movements must be rigidly controlled, especially in highly manoeuvrable animals like flies.

Flying, like any other form of locomotion, causes the image of the environment to drift across the retina, particularly when objects are nearby, and especially during rotatory movements. Since the visual field of a fly comprises virtually the whole solid angle around the animal, any flight manoeuvre generates a complex optical flow-pattern in the compound eyes. Consequently different kinds of visual disturbances occur simultaneously in different parts of the fly's visual field. Rolling, for example, causes maximal retinal slip speed along the transverse meridian of the eyes, but much less straight ahead of the fly. At the same time, rolling disturbs the angular alignment of the visual field with the environment, but predominantly so in frontal parts of the visual field. Rolling, however, does not directly disturb the fly's orientation e.g. towards an object of interest.

Obviously, all such disturbances of vision can be simultaneously counteracted by compensatory head/eye movements and/or corrective flight manoeuvres. This article deals mainly with the question how the blowfly *Calliphora erythrocephala* L. stabilizes its head during flight with respect to the environment. From the arguments above it is evident that this task is closely interrelated with the more general task of maintaining the whole fly's orientation with respect to the vertical.

2 Methods

Video recordings were taken head on, from flies flying stationarily in a wind tunnel. Either the fly or its visual surround could be rolled to simulate self-motion of the fly

---

[1]   Max-Planck-Institut für biologische Kybernetik, Spemannstraße 38, 7400 Tübingen, FRG

Localization and Orientation in Biology and Engineering
ed. by Varjú/Schnitzler
© Springer-Verlag Berlin Heidelberg 1984

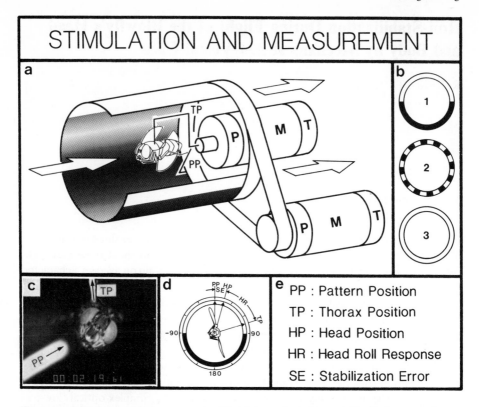

Fig. 1a–e. *Stimulation and measurement.* **a** Stimulator and wind-tunnel. The fly is mounted on to the shaft of motor (*M*) and surrounded by one of the pattern cylinders (see **b**) driven by a second motor. Fly and pattern can be turned independently in position- or velocity-servo-mode, using potentiometer (*P*) or tacho generator (*T*) respectively. This assembly is diffusely illuminated from outside, and the housing acts as a suction wind tunnel (*arrows*). **c** Video frame, taken through the entrance nozzle of the wind tunnel, including pattern pointer (*PP*), thorax pointer (*TP*), and time display of the frame. **d** Characteristic angles and measuring conventions, as defined in **e**

or relative motion between fly and environment (Fig. 1; for details see Hengstenberg et al. in prep.). The angle HR = HP – TP denotes the roll response of the head with respect to the thorax.

3 Head Movements

Unlike many other insects (e.g. beetles), flies have a thin neck, and their head is highly movable. During tethered flight in an airstream, and in the absence of any time-variable stimuli, flies perform spontaneous head movements about all three body axes: yaw and pitch movements are comparatively small (±10°), but rolling may cover ±90°. The velocity of such head movements is usually less than 100°/s, but occasionally there are large (up to 120°) and rapid (up to 1,000°/s) movements which resemble saccadic eye movements of vertebrates. Their origin and significance is so far little understood.

When a fly is passively rolled during tethered flight in a stationary visual surround, it generates, after a small delay, head movements which counteract the imposed motion (Fig. 2). For Fig. 2a, the fly is sinusoidally rolled by its thorax (TP) at 1 Hz through 180° within pattern 1. In response, the head (HR) rolls in antiphase relative to the body, thus reducing its net movement relative to the environment. Such head roll could be due to inertia if the head was lightly suspended, but this does not apply (Hengstenberg et al., in prep.). For Fig. 2b, the fly is intermittently rolled at constant angular velocity of TV = 180°/s, within a stationary striped pattern of λ = 30° spatial wavelength. When movement starts, the fly tries to counteract the stimulus by rolling its head against the imposed motion, but after 2–3 s, the response levels off because the head reaches its maximal excursion. For Fig. 2c and d, flies were alternately held either upright (Fig. 2c), or rolled by 60° (Fig. 2d) within pattern 1. When the transient responses to the change of orientation had declined, measurements of the head orientation were taken. The negative mean value of the distribution of Fig. 2d in response to an imposed tilt of +60° indicates that the flies try to compensate for their misalignment with the environment. The control histogram of Fig. 2c proves that this is neither accidental nor the remainder of a response to the change of orientation.

Such experiments demonstrate that flies produce compensatory head movements when their normal flight attitude is disturbed by an imposed motion. In tethered flight, the amplitude of compensatory head movements is usually not sufficient to compensate for the imposed movements. In free flight, however, compensatory flight manoeuvres of the body add to the overall movement of the head, thus improving the degree of compensation.

4 Analysis of the Sensory Channels

The experiments above were made to simulate as closely as possible the natural flight situation: flies were rolled in a structured visual surround. The results do not reveal therefore whether flies perceive their self-motion independent of the environment, or the relative motion between them and the environment. From the stimulus arrangement (Fig. 1a) one can distinguish between the two possibilities by either rolling flies in a homogeneous visual surround, or rolling a structured surround around a stationary fly. The analysis is presented elsewhere in detail (Hengstenberg et al. in prep.): its main results are shortly presented below, and are summarized in Fig. 3.

4.1 Perception of Self-Motion

When flies are rolled in a visually homogeneous surround, they still respond to sinusoidal motion, and to rotation at constant angular velocity, proving that they perceive self-motion. They do not respond to static deviations from the vertical, i.e. to the direction of gravity during flight, although they do so during walking (Horn and Lang

EVOKED HEAD-ROLL-MOVEMENTS

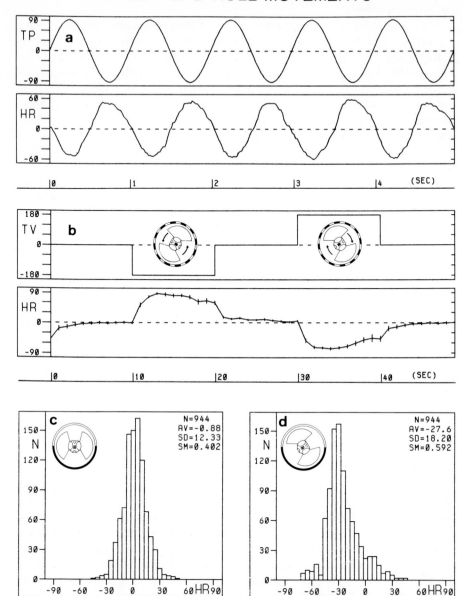

Fig. 2. *Evoked head roll movements.* Flies are passively rolled during flight within a stationary visual surround. **a** Sinusoidal motion (*TP*) of 1 Hz, 180° pp within pattern 1 (see inset of **c**) elicits roughly sinusoidal, compensatory head movements (*HR*). **b** Rolling at constant angular velocity (*TV*) of 180°/s within pattern 3 elicits a steady head roll in compensatory direction until the head roll saturates. **c, d** Steady angular displacement of the fly by +60° elicits a steady realignment of the head (*HR*)

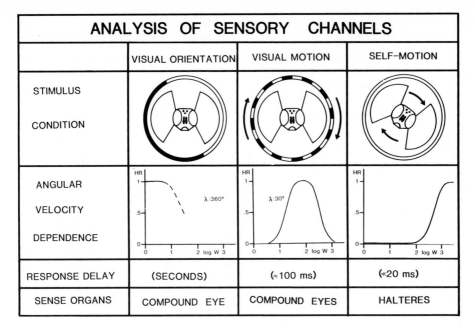

Fig. 3. *Analysis of sensory channels.* Different sensory mechanisms are isolated by rolling either the fly in a homogeneous surround (self-motion), or different patterns at different speeds around a stationary fly (visual orientation and motion). Note the different angular velocity characteristics [w($^{\circ}$/s)], and their correlations with different response delays. Selective elimination of halteres, ocelli, and compound eyes identifies sensory inputs

1978). The response to self-motion depends dramatically on the angular velocity: it is small up to w = 100°/s, and saturates beyond w = 1,000°/s (Tracey 1975, Hengstenberg unpublished). The latency of these responses is apparently small (< 20 ms), and may even be smaller than can be resolved in video recordings (Fig. 3). With selective elimination of the halteres, all responses to self-motion are completely abolished.

The halteres of diptera correspond to the hind wings of other insects. They are small organs, buried in the cleft between thorax and abdomen (Fig. 4a). A halter looks like a small pendulum: its base is attached to the thorax by a horizontal hinge, and a thin stiff stalk carries a distal bulbus (Fig. 4b, c). When a fly rests the halteres stick out laterally, but in flight, they oscillate vertically at wing beat frequency (ca. 150 s^{-1}) through an arc of about 170°. The planes of oscillation of the two halteres enclose an angle of about 130°.

Halteres were recognized by Pringle (1948) as "gyroscopic sense organs": When oscillating, the mass of the bulbus generates a periodic inertial force acting in the plane of oscillation, i.e. vertically. If the fly rotates in any plane not coinciding with that of the haltere, a periodic Coriolis force is additionally generated which acts normally to the plane of haltere oscillation, i.e. horizontally. Frequency, amplitude, and phase of the Coriolis force, referred to the haltere oscillation, depend upon the direction and angular velocity of the fly's rotation. This orthogonal pair of forces causes a periodic, and specifically directed strain in the haltere base. This contains

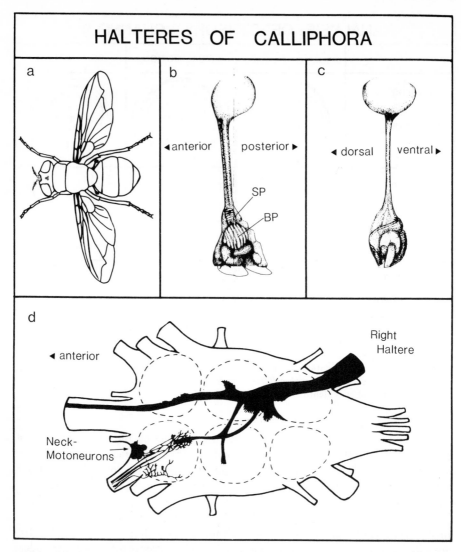

Fig. 4a–d. *Halteres of Calliphora.* **a** Fly seen from above; halteres are buried between thorax, abdomen and hind legs; **b** dorsal; and **c** posterior view of the right haltere (1 mm long); *SP* scapal plate; *BP* basal plate denote two of seven clusters of mechanoreceptors (Schneider 1953); **d** thoracic compound ganglion of *Calliphora* showing projections of haltere sensilla after mass impregnation of the right haltere nerve (≈ 400 fibres), and the motoneurons of neck muscles after staining of the frontal prothoracic nerve. (After Sandeman and Markl 1980.) Note the abundance of haltere projections, and the direct connections of haltere sensilla with motoneurons

about 400 mechanoreceptors, mainly organized into seven clusters (Pflugstaedt 1912). At present, however, neither the time-dependent strain distribution in the haltere base nor the directional sensitivity of the different mechanoreceptor fields are sufficiently well known to ascertain their role in the perception of yaw, pitch, and roll movements of the fly.

Figure 4d shows the thoracic compound ganglion of *Calliphora* seen from above, and the primary projections of haltere sensilla after mass staining of the right haltere nerve with cobalt (from Sandeman and Markl 1980). The haltere receptors have profuse projections to pro-, meso-, and metathoracic neuropils on the ipsi- and contralateral side, as well as a prominent projection ascending to the brain. Such mass impregnations are not suited to establishing the connection between specific haltere receptors and a particular projection site; more refined techniques are required to resolve these relationships. Nevertheless, Fig. 4d shows that there are direct projections of haltere receptors into those areas of the ipsi- and contralateral prothoracic neuropil, where the motoneurons of the neck muscles have their dendritic arborizations. The fast action of halteres upon head movements during self-motion is likely to involve such direct pathways.

4.2 Perception of Relative Motion

Visually mediated responses are revealed when a structured visual surround is rolled around a stationary fly (Fig. 1a). With sinusoidal oscillation of pattern 1, the fly produces roughly sinusoidal following movements of its head, which reduce the retinal slip speed, and thus act in a compensatory manner. The nature of these responses remains however obscure, because pattern 1 is ambiguous: the fly could either consider pattern 1 as the lower limiting case of a periodic grating with a spatial wavelength of $\lambda = 360°$, and generate a direction-specific optomotor response; or it could fixate and track prominent features of pattern 1, namely the horizon or the dorsoventral brightness gradient; or it could use all of these cues, possibly via different neural pathways. The respective influence of these different mechanisms can be explicitly tested by use of further patterns combined with the appropriate stimulus movement. The striped pattern 2 does not contain any asymmetry, but a wealth of equally distributed contours. When rolled around a fly at constant angular velocity, the fly's head follows the movement until it reaches its maximum excursion, and stays there until the pattern motion stops. Preliminary experiments with a pattern of $\lambda = 30°$ and a limited range of pattern motion ($< \pm 90°$) show that maximum responses are obtained at an angular velocity of about $w = 100°/s$. The response declines towards higher and lower pattern velocities. This bandpass characteristic, and the location of its maximum near the contrast frequency $w/\lambda = 3/s$ (Fig. 3) correspond very well with various other optomotor characteristics (cf. Hengstenberg 1982), and in particular with the roll-torque generated by the wings under similar stimulation in other flies (Srinivasan 1977, Blondeau and Heisenberg 1982).

The visual system of flies consists, as in other insects, of the two large compound eyes and three small ocelli on the vertex of the head (see Fig. 4a). Selective blinding of the ocelli does not noticeably influence visually evoked head movements, but blinding of the compound eyes abolishes them completely. Thus the neural circuitry which extracts visual motion information to control compensatory head rolling must be located in the optic lobe. The neurons performing elementary movement detection have still not been identified. It is however known that this process requires a nonlinear interaction between two adjacent small-field elements, whose arrangement with respect to the visual field coordinates determines their preferred direction (Buchner

1976). Such local movement signals can however be caused by different flight manoeuvres of a fly. Correct interpretation of the optical flow pattern across the retina therefore requires that local movement signals are selected from different parts of the visual field and are specifically summed up by wide-field tangential neurons (cf. Hausen 1983).

The lobula plate of flies (cf. Wehrhahn, this Vol.) contains a class of 11 giant tangential neurons, the "Vertical System" (VS). Each VS-neuron has a characteristic shape, occupies a distinct area in the retinotopic lattice of input elements, and has a specific stratification in depth. These features are largely invariant in different individuals (Fig. 5a,b; Hengstenberg et al. 1982). Intracellular records from VS-neurons in brightness show an abundance of postsynaptic potentials, caused by a multitude of independent input elements (Fig. 5c,d, middle traces). VS-neurons respond to the coherent motion of extended patterns in the visual field, but each in a characteristic manner (Hengstenberg 1981): some respond best to pitch stimuli, and others (VS2 to VS5; Fig. 5b, asterisks) to simulated roll motion (Fig. 5e). Pattern movements in directions that are inappropriate for a given VS-neuron elicit no response (Fig. 5c). This particular case arises from the convergence of a distinct fraction of local movement signals onto any particular VS-neuron, and from their wide-field integration by this cell. The motion responses of VS-neurons depend upon contrast frequency in a similar manner as the head movements (Fig. 3). It is therefore probable that moto-

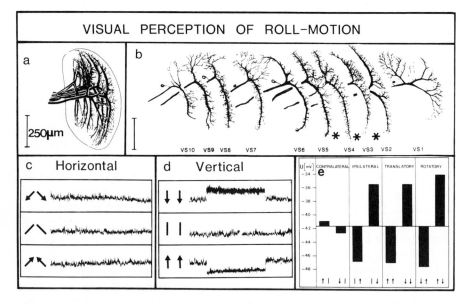

Fig. 5a–e. *Visual perception of roll motion.* Wide-field movement-sensitive interneurons (VS-cells) of the lobula plate (third visual neuropil): a In situ, as reconstructed from one fly; b drawn apart, after intracellular staining in different individuals. Axon terminals are omitted; scale 250 µm. c–e Intracellularly recorded responses from neurons of the group VS2–VS5 to movements of a periodic grating. They respond to vertical, but not to horizontal movements; they are excited by ipsilateral downward and much less by contralateral upward motion. Reverse motions act inhibitory. Maximal responses are elicited by apparent roll-motion

neurons of the neck muscles (Fig. 4d) receive input from VS-neurons via so far un-identified descending fibres.

4.3 The Fly's Subjective Vertical

It was mentioned above that flies may not only respond to self-motion or to visual motion, but also to the orientation of prominent landmarks in the visual field. Figure 2c,d proves that flies respond to a static misalignment with their environment. This result does not, however, reveal whether the fly responds to its inclination relative to the gravity vector or relative to a supposedly horizontal environment, or both. This question can be answered by either tilting the fly in a visually homogeneous surround, or conversely, by tilting pattern 1 around the fly. Figure 6a shows that flies do not respond at all to the direction of the gravity field while flying (data from 14 flies). This applies not only to head movements but also to the flight torque (Fig. 6e: Srinivasan 1977). Figure 6b shows that flies do respond to a static misalignment of pattern 1 (data from 8 flies). Apparently flies take this pattern to indicate the normal orientation of the world and try to realign themselves with the environment. Figure 6b does not, however, reveal whether flies take the contrast borders of the pattern to indicate the "horizon", or whether they take the brightness gradient, normal to the horizontal plane, to indicate the "vertical". Two additional patterns make it possible to distinguish between these alternatives: in one pattern, the horizon is marked by heavy black lines ($15°$ width), but there is no brightness differences between the upper and lower half. If this pattern is tilted $60°$, flies do not respond at all, i.e. they disregard the apparent orientation of a "horizon" in the lateral visual field (Fig. 6c; 8 flies). The complementary pattern consists of a sinusoidal optical density distribution of $360°$ spatial wavelength and 95% contrast along the circumference. When normally oriented, this pattern provides a dorsoventral brightness gradient but no sharp contours at the horizon. Tilting this pattern by $60°$ evokes a significant response (Fig. 6d, 8 flies). The response is about half of that in Fig. 6b, probably because pattern 1 is on average twice as luminant. This response to the brightness distribution in the environment is well known as "doral light reflex" in insects, crustacea, and fishes (Buddenbrock 1915b). When compared with the movement-dependent responses described above, it is very unsteady and sluggish. When a fly starts flying with pattern 1 already tilted, it may take seconds until the head starts to be turned. The dorsal light reflex is exclusively mediated by the compound eyes, but neither the type of photoreceptors nor any of the neurons involved are presently known.

Srinivasan (1977) describes another visual orientation response in the housefly *Musca*: when faced with a circular pattern of parallel stripes in the frontal visual field, the fly generates a roll-torque which can be used to control the angular orientation of the pattern. Under such closed loop conditions, *Musca* aligns the stripes so that they appear vertical. Mounting the fly obliquely yields the same results (Fig. 6e), proving that gravity has no influence upon the flight torque either. Preliminary experiments with this stimulus, and observation of head movements in *Calliphora*, suggest that in this case also parallel stripes are taken to be vertical.

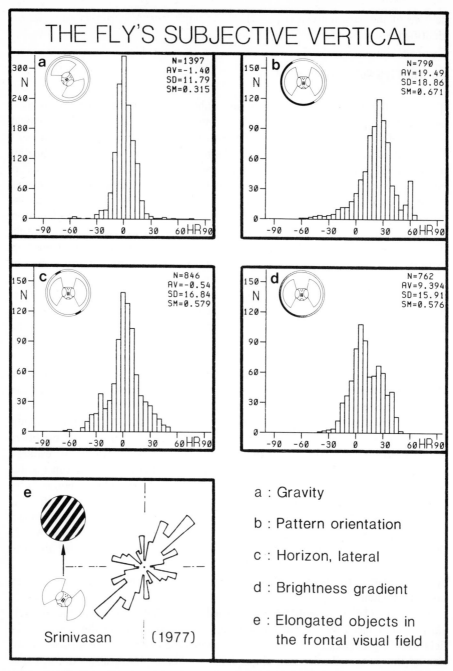

Fig. 6 a–e. *The fly's subjective vertical.* Flies are confronted with a static misalignment of either themselves relative to the direction of gravity (**a**), (**e**), or of different patterns with the fly oriented normally (see insets). They do not respond to gravity (**a**), but to the orientation of an asymmetric pattern in the *lateral* visual field (**b**). There, they do not respond to the displacement of the horizon (**c**), but to the orientation of the brightness gradient (**d**). *Musca* takes long things in the *frontal* visual field for vertical (**e**), and orients itself in closed loop accordingly, irrespective of gravity (see text)

Taken together, these results show that during flight flies do not recognize the vertical by the direction of gravity but only by visual cues. The horizon is apparently not considered as a reliable indicator, but elongated things, possibly such as trees, in the frontal visual field are taken to be vertical. This cue is of course ambiguous with respect to up and down, but the overall brightness gradient usually defines this direction unambiguously.

4.4 Dynamic Complementation

The analysis of the sensory channels (Fig. 3) shows that the mechanosensory channel via the halteres, and the visual channels, differ in their dynamic properties. Although the latency measurements so far lack the required accuracy, and the contrast frequency dependence of the visual motion response is based on preliminary data, the following features nevertheless emerge: The mechanosensory input via the halteres acts apparently fast ($\ll 20$ ms) and at high angular velocities ($> 200°$/s); visual motion acts moderately fast (≈ 100 ms) and at intermediate velocities, depending upon the spatial wavelength spectrum of the environment; the dorsal light reflex seems very sluggish (seconds) and operates at very low angular velocities including zero. Apparently the different sensory channels which control the potentially hazardous roll motions of the fly are complementary and supply the fly with sufficient information over the whole range of possible angular velocities, provided that the different sensory signals converge upon the neck muscles motoneurons without conditional interactions (gating). By selective elimination of the different sense organs, it can be demonstrated that this requirement is satisfied.

Blinding of all ocelli and compound eyes does not prevent compensatory head movements in response to haltere stimulation. Conversely, removal of both halteres does not noticeably affect any of the visually mediated responses. It is therefore to be expected that at different angular velocities the different sensory mechanisms contribute variable proportions to the overall control signal, depending upon their particular velocity characteristic (cf. Fig. 3).

5 Relationship of Head and Body Movements

Corrective flight manoeuvres and compensatory head movements are in principle entirely independent tasks: flight manoeuvres are required to ensure stability, and compensatory head/eye movements improve visual perception during self-motion. These tasks are however interrelated (a) by the congruence of their control requirements, and (b) by the fact that the head is carried by the body. It was noted earlier that compensatory head movements, as observed during tethered flight, usually undercompensate for the imposed stimulus. Since head movements are normally accompanied by simultaneous changes in wing beat amplitude and wing pitch, one must assume that in free flight head movements and flight manoeuvres occur at the same time. In this case they act functionally in series, and the effective stabilization of the head with respect to the environment must be better than that observed in tethered flight.

The primary task not to crash is apparently difficult for a fly: (a) most probably it cannot stabilize its flight attitude passively; (b) it has apparently no information about the direction of gravity, and therefore the body has no independent information about its orientation in space; (c) the true vertical must therefore be inferred from visual cues although flies cannot "know" a priori what the environment looks like (cf plain open country and forest); (d) even if the head be perfectly aligned with the surround, the body may still be misaligned by 90° because of the large range of head rolling. The following points emerge from these considerations: (1) only the fly's head can be directly stabilized with respect to the vertical; (2) the visual system must extract from the surround all the information that may be suitable for estimating the true vertical; (3) the body must be indirectly stabilized by reading its orientation relative to the head. Neck mechanoreceptors like the prosternal organ (Horn and Lang 1978) and probably other mechanoreceptive hairs of the neck are suited for this purpose; (4) such indirect control of the body via head/thorax coordination would fail if the head were artificially fixed to the thorax. However, Srinivasan (1977) has shown that *Musca* generates a roll-torque with visual stimulation when the head is fixed. Similarly, Wienrich (1979) has shown in *Calliphora* that halteres directly influence steering muscles when the head is fixed. Consequently both mechanosensory signals of the halteres and visual signals from the compound eyes control body rolling along two parallel pathways: one direct and the other via the neck mechanoreceptors.

Figure 7 schematically illustrates what is presently known about the roll-control of the head and body of *Calliphora* during flight. Boxes denote conceptually distinguishable entities, and arrows the flow of information. Squares denote perceptual processes which extract specific information from sensory signals (VM = visual motion, BG = brightness gradient, PO = pattern orientation, SM = self motion, HR = head/thorax angle, SR = spontaneous rolling). Rectangles (head/body control) show summation of the different signals and co-ordination of the output to the respective groups of muscles. Execution of compensatory head and flight movements physically reduces the different error signals: visual signals are affected by head and body movements, but halteres (SM) are only affected by body motion. Direct control of head and body is achieved via the respective lines; vertical alignment of the body is effected indirectly via head control, neck muscles, head movements, and neck mechanoreceptors which measure the HR.

This figure is mainly based upon the results of behavioural studies but so far only a few of the biological structures behind the boxes and lines are known. It will be necessary to identify the missing elements, and to determine their mode of operation.

6 Concluding Remarks

This study shows that a supposedly simple animal, the blowfly *Calliphora*, has a complex and sophisticated system for maintaining its normal flight attitude, and for

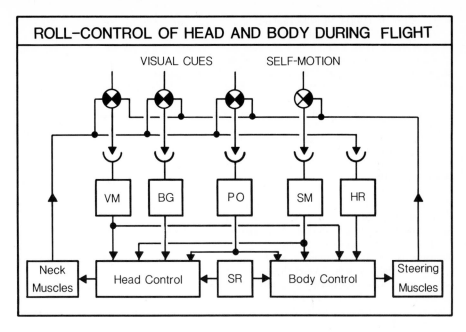

Fig. 7. *Roll-control of head and body during flight.* Summary diagram of the reflex organization, as known at present. *Squares* yield specific information about roll movements, *control boxes* summate such signals and coordinate the action of muscle groups. Compensatory movements physically reduce the error signals

stabilizing its visual field during flight. The convergence of mechanosensory and visual signals and their dynamic properties bear a striking similarity to the respective control systems for equilibrium and eye movements in crustacea (Sandeman 1977) and vertebrates (Carpenter 1977). The basic biological problems of equilibrium control and of visual perception during self-motion have previously prompted similar solutions during evolution, despite their largely different biological realization.

References

Blondeau J, Heisenberg M (1982) The three-dimensional optomotor torque system of *Drosophila melanogaster*. Studies on wildtype and the mutant *optomotor blind*[H31]. J Comp Physiol 145:321–329

Buchner E (1976) Elementary movement detectors in an insect visual system. Biol Cybern 24: 85–101

Buddenbrock W von (1915b) Über das Vorhandensein des Lichtrückenreflexes bei Insekten sowie bei dem Krebse *Branchipus grubei*. Sitz Ber Heidelberger Akad Math Nat Kl Abt B5, 1–10

Carpenter RHS (1977) Movements of the eyes. Pion, London

Götz KG, Hengstenberg B, Biesinger R (1979) Optomotor control of wing beat and body posture in *Drosophila*. Biol Cybern 35:101–112

Hausen K (1983) The lobula complex of the fly: Structure, function, and significance in visual behaviour. In: Ali MA (ed) Vision in invertebrates. Plenum, London New York

Hengstenberg R (1981) Visuelle Drehreaktionen von Vertikalzellen in der Lobula Platte von *Calliphora*. Verh Dtsch Zool Ges 1981, S. 180. G Fischer, Stuttgart

Hengstenberg R (1982) Common visual response properties of giant vertical cells in the lobula plate of the blowfly *Calliphora erythrocephala*. J Comp Physiol 149:179–193

Hengstenberg R, Hausen K, Hengstenberg B (1982) The number and structure of giant vertical cells (VS) in the lobula plate of the blowfly *Calliphora erythrocaphala*. J Comp Physiol 149: 163–177

Hengstenberg R, Sandeman DC, Hengstenberg B (in preparation) Compensatory head roll of the blowfly *Calliphora*, elicited during flight by mechanical and visual stimuli

Horn E, Lang HG (1983) Positional head reflexes and the role of the prosternal organ in the walking fly *Calliphora erythrocephala*. J Comp Physiol 126:137–146

Pflugstaedt H (1912) Die Halteren der Dipteren. Z Wiss Zool 100:1–59

Pringle JWS (1948) The gyroscopic mechanism of the halteres of Diptera. Philos Trans R Soc Lond [Biol] B 233:347–384

Sandeman DC (1977) Compensatory eye movements in crabs. In: Hoyle G (ed) Identified neurons and behaviour of arthropods. Plenum Press, New York

Sandeman DC, Markl H (1980) Head movements in flies *(Calliphora)* produced by deflexion of the halteres. J Exp Biol 85:43–60

Schneider G (1953) Die Halteren der Schmeißfliege *Calliphora* als Sinnesorgane und als mechanische Flugstabilisatoren. Z Vergl Physiol 35:416–456

Srinivasan MV (1977) A visually-evoked roll response in the housefly. Open-loop and closed-loop studies. J Comp Physiol 119:1–14

Tracey D (1975) Head movements mediated by halteres in the fly *Musca domestica*. Experientia 31:44–45

Wienrich M (1979) Untersuchung des neuromotorischen Erregungsmusters in direkten Flugmuskeln von Fliegen, die während des Fluges um ihre Hochachse gedreht werden. Diplomarbeit, Math Nat Fak Univ Düsseldorf

Activation of Flight Control Muscles by Neck Reflexes in the Domestic Pigeon (*Columba livia* var. *domestica*)

D. BILO, A. BILO, B. THEIS, and F. WEDEKIND[1]

Groebbels (1929) propounded the hypothesis that neck reflexes on wing and tail muscles played an important role in flight control of birds: the bird automatically follows its beak by the action of neck reflexes. However, this hypothesis could not be tested or verified until now as pure neck reflexes on wing and tail muscles cannot be initiated by neck flexion in birds, e.g. the domestic pigeon sitting or standing in calm air. This is drastically changed if the pigeon is blown upon from frontally. Now the activity of wing and tail muscles is strongly influenced by galvanic stimulation of the labyrinth, by optokinetic stimuli (Bilo and Bilo 1978) and by static laterad neck flexion as will be shown in the following.

The left-right response pattern of wing and tail muscles is regularly correlated with the direction of static laterad neck flexion: Fig. 1 shows the responses (impulse rates) of the left and right M. abductor indicis and of the left and right M. lateralis caudae to trapezoidal deflections of the head (B) or the body (A) [$\varphi_H(t)$ head-to-body angle]. The yawing moment of the head, $M_H(t)$ (L towards the left, R towards the right), opposes the forced neck flexion and is superimposed with a nystagmic component. As a rule, laterad neck flexion is correlated with increase of activity in the ipsilateral M. lateralis caudae and M. extensor digitorum communis (not shown) and in the contralateral M. abductor indicis and M. extensor metacarpi radialis (not shown). The activity of the corresponding opposing muscles simultaneously decreases. There is no difference in this response pattern whether the head or the body is deflected. The response pattern is frequently disturbed by the lack of responses of one or more muscles whose activity is zero or is not correlated with the stimulus (e.g. right M. lat. caud. in Fig. 1B).

In stepped neck flexions with increments of $12°$ (duration of static deflection 15 s; maximal deflection $±60°$) the correlation between muscle activity and stimulus going in on-direction is often greater than in off-direction. Different muscles of the same pigeon often show quite different stimulus-response relationships which are frequently more or less linear within certain limits.

The left-right activity pattern of muscles is the same in forced (passive) and in spontaneous (active) head deflections; correlation between head-to-body angle and finger muscle activity is maximal at zero time displacement between head and muscle signals.

[1]   Fachbereich Biologie der Universität des Saarlandes, 6600 Saarbrücken, FRG

Localization and Orientation in Biology and Engineering
ed. by Varjú/Schnitzler
© Springer-Verlag Berlin Heidelberg 1984

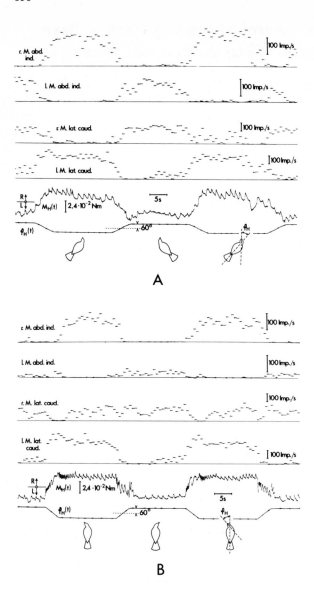

Fig. 1a,b. Responses of left and right M. abductor indicis and M. lateralis caudae to trapezoidal deflections of the head

Based on these results we assume that neck reflexes on wing and tail muscles exist which are "switched on" when the pigeon is blown upon, and that these reflexes might act as a servo-mechanism which, as proposed by Groebbels (1929), causes the body to follow its beak during flight.

Does this apply to the flying pigeon? – Does it always follow its beak? According to films taken at 55 frames/s of free flying pigeons, this is only true in slow flight, particularly during landing flight. In this case there is a significant correlation between the head-to-body angle $\varphi_H(t)$ and the angular velocity $\dot{\varphi}_B(t)$ of the body in the horizontal plain. The cross-correlation coefficient $\hat{\rho}_{\varphi_H, \dot{\varphi}_B}(\tau)$ of these two signals is maximal at the time displacement $\tau = 1/55$ s, i.e. the main component of $\dot{\varphi}_B(t)$ lags

Fig. 2A,B. Head oscillations in the horizontal plane

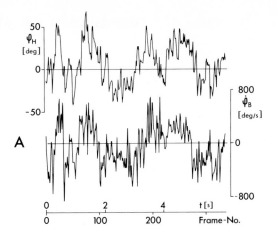

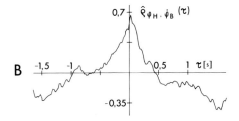

the main component of $\varphi_H(t)$ by 1/55 s. This applies to all four pigeons tested. When the nasal region of the pigeon's visual field is masked, the pigeon's head oscillates in the horizontal plain during flight and its body follows these head oscillations. In this case the correlation between $\varphi_H(t)$ and $\dot{\varphi}_B(t)$ is particularly strong with its maximum at $\tau = 1/55$ s (Fig. 2) as it is in untreated pigeons ($+\varphi_H$, $+\dot{\varphi}_B$ clockwise rotation).

There also is a significant correlation between φ_H and $\dot{\varphi}_B$ in the frequency domain at low frequencies: The coherence function $\hat{\gamma}(f)$ amounts to 0.7 at zero frequency and to 0.5 at $f \approx 2$ Hz, reducing to an insignificant level at higher frequencies.

The courses of the head and body change simultaneously during fast curving flight if the pigeon doesn't intend to land within the curve and is obviously not correlated at all when the bird is "far afield" where difficult flight manoeuvres are not performed. We therefore conclude that neck reflexes on wing and tail muscles are in action only if the bird has precisely to control its actual angular position and course, e.g. in landing reliably on a defined target, or in matching other difficult flight manoeuvres. The neck reflexes otherwise may be "switched off".

References

Bilo D, Bilo A (1978) Wind stimuli control vestibular and optokinetic reflexes in the pigeon. Naturwissenschaften 65:161–162

Groebbels F (1929) Der Vogel als automatisch sich steuerndes Flugzeug. Naturwissenschaften 17:890–893

Dynamics of Groundspeed Control in *Drosophila*

C.T. DAVID [1]

Free-flying *Drosophila* keep their groundspeed constant by adjusting their airspeed so that there is a constant "preferred" angular velocity of image movement past their eyes (David 1979, 1982). That is, if their groundspeed is too high, the angular velocity of image movement past their eyes will be higher than the "preferred" value, and the flies will reduce their airspeed. If their groundspeed is too low, the angular velocity of image movement will be too low, and the flies will increase their airspeed. The value of the "preferred" angular velocity of image movement at which the flies neither accelerate nor decelerate depends partly on the physiological state of the flies (David 1978), partly on their height (David 1982), and partly on other factors such as the presence of attractant odours or the light intensity.

The response time of this speed control reaction was determined by moving the pattern in a barber's pole wind tunnel (David 1982) sinusoidally up- and down-wind while video-recording the position of flies flying upwind inside the tunnel. The wind speed was always much greater than the speed of movement of the pattern, so the flies remained at all times facing upwind. This is because the flies fly upwind through turning responses to image drift at oblique angles to their body axis (Kennedy 1940, David 1982b), and this drift is always in the upwind direction if the pattern moves slower than the wind.

The best correlation between any parameter of the stripe motion and any parameter of the fly motion occurred, as expected, when the speed of the stripe relative to the fly was compared with the subsequent acceleration of the fly. The best correlation between these was at a time delay of between 0.05 and 0.15 s. The correlation between the speed of the stripe relative to the fly and the subsequent speed of the fly was less good, and the best time delay was ca. 0.3 s at 0.66 Hz.

The delay of the speed control reaction is thus much longer than that of the turning reactions of free-flying flies (Collett and Land 1975, Collett 1980) (about 0.02 s). This might be accounted for by a longer nervous reaction time for a translationary rather than a rotational visual stimulus, but, more likely, by the fact that the flies have to change their body angle as well as their thrust to change speed without changing height (David 1978), and this may take more time than the production of the differential in thrust between the two sides of the body that produces a turn.

---

[1]   A.R.C. Insect Physiology Group, Dept. of Pure & Applied Biology, Imperial College of Science and Technology, London, SW7 2 AZ, UK

Localization and Orientation in Biology and Engineering
ed. by Varjú/Schnitzler
© Springer Verlag Berlin Heidelberg 1984

References

Collett TS (1980) Angular tracking and the optomotor response. An analysis of visual reflex interaction in a hoverfly. J comp Physiol 140:145–158

Collett TS, Land MF (1975) Visual control of flight behaviour in the hoverfly, *Syritta pipiens* L.
J comp Physiol 99:1–66

David CT (1978) The relationship between body angle and flight speed in free-flying *Drosophila*.
Phys Ent 3:191–195

David CT (1979) Height control by free-flying *Drosophila*. Phys Ent 4:209–216

David CT (1982) Compensation for height in the control of groundspeed by *Drosophila* in a new
"barber's pole" wind tunnel. J comp Physiol 147:485–493

David CT (1982b) Competition between fixed and moving stripes in the control of orientation by
flying *Drosophila*. Phys Ent 7:151–156

Kennedy JS (1940) The visual responses of flying mosquitoes. Proc zoo Soc Lond A 109:221–242

Control of Flight Speed by Minimization
of the Apparent Ground Pattern Movement

R. PREISS and E. KRAMER [1]

When a male moth flies towards a chemically alluring female, he determines both the upwind direction and his forward progress from the ground pattern movement beneath him (Kennedy 1940, Heran 1955, Schneider 1965, Kennedy and Marsh 1974, Miller and Roelofs 1978). Thus, the insect completely compensates for its wind-drift, and maintains a constant speed over ground (Marsh et al. 1978). In order to reveal the underlying mechanism by which this information is processed, we measured the flight speed of a moth over a moving ground pattern.

Male gypsy moths (*Lymantria dispar* L.) were tethered in a flight mill and forced to fly in a circular path (Fig. 1). The mill was constructed to allow them to choose their own flight speed in a manner closely resembling the free-flight condition. They flew over a ground pattern, which itself was moved, simulating either forwards or backwards wind-induced drift.

When the ground pattern speed was modulated in a periodic "triangular" manner (0 m/s–4 m/s) the moth responded with a sequence of at least three different types of flight behaviour, which followed one another in a systematic manner (Fig. 2).

Landing response (L): at medium pattern speeds the moth kept the same flight speed as the pattern, resulting in a speed over ground of 0 m/s. In addition, the moth showed intentions to land.

Constant progression (P)/regression (R): at low and high pattern speeds the moth maintained a constant positive and negative speed over ground of 2–3 m/s respectively. The posture and behaviour of the moth were typical of normal flight.

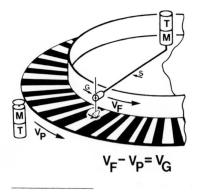

$$V_F - V_P = V_G$$

Fig. 1. Flight mill in which a tethered moth is forced to fly in a circular path (dia. = 175 cm) at 17 cm altitude over a striped movable ground pattern (random sequence of pairs of black/white stripes (stripe width: 10°, 16.5°, and 22.5°). The moth is not loaded with artificial friction and inertia. A servo-system rotates the suspension (S) according to the output of a thrust meter (G). The moth can achieve any normal flight speed since thrust and drag are balanced as in free-flight. Flight speed (V_F) and pattern speed (V_p) are monitored by tachometers (T), the difference between the two speeds gives the speed over ground (V_G). M motors drive suspension and ground pattern

[1]   Max-Planck-Institut für Verhaltensphysiologie, 8131 Seewiesen, FRG

Localization and Orientation in Biology and Engineering
ed. by Varjú/Schnitzler
© Springer Verlag Berlin Heidelberg 1984

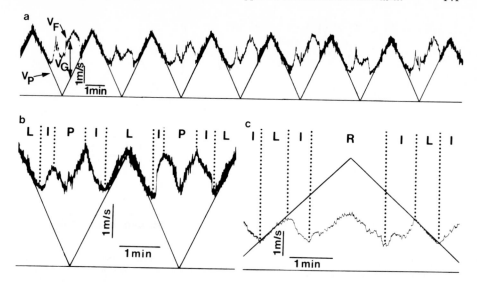

Fig. 2. Original recordings of flight speed (V_F) in response to a periodically increasing/decreasing ground pattern speed (V_p). The moth alters its mode of flight-control in a fixed sequence (**a**). Three types can be distinguished: no progression with respect to the ground (*L* in **b** and **c**); constant progression (*P* in **b**) or constant regression (*R* in **c**) and inverse flight (*I*) during the transitions between *R*, *L*, and *P*

Inverse flight (I): the moth responded to increasing pattern speed by proportionally decreasing its flight speed and vice versa. This behaviour characterizes the transitions between the landing response and constant progression/regression. It always occurred when the pattern-induced flight speed reached an upper or lower limit.

How could these flight behaviours be explained? Two visually induced responses, found by studying the optical orientation with respect to movements around the vertical axis have been previously described. Both stabilize the optical surroundings relative to the animal (for ref Reichardt and Poggio 1976): (1) the optomotor response, cancelling out the relative motion between object and the insect and (2) the position-sensitive response, causing turns towards an object.

The landing response can be explained by the optomotor response. With constant progression/regression, however, the angular velocity W of the ground pattern is kept constant at values of $700°/s$–$1,000°/s$. At such high values of W, the optomotor torque response, as measured in other insects, is far beyond its maximum and almost approaches zero again (Hassenstein 1959, Kunze 1961, Fermi and Reichardt 1963).

We therefore conclude that during constant progression *L. dispar* minimizes its optomotor signal not by deceleration but rather by speeding up until this signal reaches a critical low value. If it falls below this value, the moth decelerates. The opposite happens at constant regression.

So far, we cannot explain how optomotor and position-sensitive responses must be modified and superimposed to cover the whole set of responses. We can, however, reconstruct the thrust characteristics versus speed over ground from the experimental results (Fig. 3, inset). We assume that the whole set of responses is due to one and

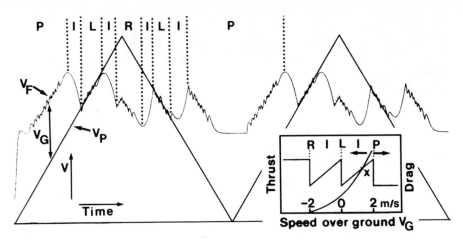

Fig. 3. Computer simulation of the observed relationship between flight speed (V_F) and pattern speed (V_p) as in Fig. 2. This simulation requires only one feedback system if a thrust versus speed over ground (V_G) characteristic (inset) interacts with the drag function. In order to obtain inverse flight (I), parameters of the drag function and the thrust characteristic must be chosen such, that in the point of intersection (x) the former is always steeper than the latter when shifted against each other

the same control circuit. This assumption is favoured by the continuous visual control of speed over ground. A computer simulation of the interaction of the thrust characteristic with a drag function – parabolically dependent on flight speed – produced all features found in the real recordings (Fig. 3).

References

Fermi G, Reichardt W (1963) Optomotorische Reaktion der Fliege *Musca domestica*. Kybernetik 2:15–28

Hassenstein B (1959) Optokinetische Wirksamkeit bewegter periodischer Muster (nach Messung am Rüsselkäfer *Chlorophanus viridis*). Z Naturforsch 14:659–674

Heran H (1955) Versuche über die Windkompensation der Bienen. Naturwiss 42:132–133

Kennedy JS (1940) The visual responses of flying mosquitoes. Proc zool Soc A 109:221–242

Kennedy JS, Marsh D (1974) Pheromone-regulated anemotaxis in flying moths. Science 184: 999–1001

Kunze P (1961) Untersuchung des Bewegungssehens fixiert fliegender Bienen. Z vergl Physiol 44:656–684

Marsh D, Kennedy JS, Ludlow AR (1978) An analysis of anemotactic zigzagging flight in male moths stimulated by pheromone. Physiol Entomol 3:221–240

Miller JR, Roelofs WL (1978) Sustained-flight tunnel for measuring insect responses to wind-borne sex pheromones. J Chem Ecol 4:187–198

Reichardt W, Poggio T (1976) Visual control of orientation behaviour in the fly, Part I. A quantitative analysis. Q Rev Biophys 9:311–346

Schneider P (1965) Vergleichende Untersuchungen zur Steuerung der Fluggeschwindigkeit bei *Calliphora vicina* Rob.-Desvoidy (Diptera). Z wiss Zool 173:114–173

Short-Term Integration: A Possible Mechanism in the Optomotor Reaction of Walking Crickets?

H. SCHARSTEIN[1]

The progression of crickets of the species *Gryllus campestris* as well as *Gryllus bimaculatus* is characterized by short walking periods of a few seconds' duration which are interrupted by short stops (duration: about 0.5 to a few seconds, see e.g. Schmitz et al. 1982).

The animals were tested on a locomotion compensator (Kramer sphere, which continuously compensates the translatory movements, but not the walking direction of the cricket) for their behavioural response to a rotating vertically striped cylinder. While walking, the angular velocity of the cricket is identical to the angular velocity of the striped pattern. During a stop within the run, the animal remains stationary without any head movements, while the pattern moves on. Upon starting again, the cricket performs a fast turn in the direction of the drum movement. It seems that the animal very quickly reaches its previous angular relationship to the cylinder, which is maintained up to the next stop. This would mean that the crickets are able to determine the total angle of drum movement by integrating the angular velocity during the stops. In this case, the accuracy of integrating mechanisms in biological systems can be estimated by a detailed analysis of such a behavioural performance.

Therefore, in addition to the data obtained from the Kramer sphere, (translatory movement of the animal in X- and Y-direction and accordingly the calculated velocity and walking direction) the cricket was filmed with a 16 mm camera at 25 frames per second, providing the direction of the longitudinal body axis via single frame analysis.

Figure 1 shows a 35 s section of such an experiment. Stopping and starting of the cricket can be detected with the accuracy of a single frame (40 ms) and is marked by crosses in the upper trace. During these stopping intervals, which coincide with stops in the velocity trace, the direction of the cricket's body axis remains constant. At the start of a walking period there is a fast change in direction, thus roughly restoring the previous angular relationship to the stripped pattern.

The accuracy of these corrective turns and their possible mechanisms were analyzed. In Fig. 2 all stops of two runs [angular velocity of the drum: +50 deg/s (counterclockwise) and −50 deg/s (clockwise) respectively] are shown from two points of view. In the left diagrams the angles between body axis and drum at the beginning of stops are set to zero. Thus, these diagrams show the time course of the deviation from this direction before and after stops, as well as the rotation of the drum and the spatial

[1] Zoologisches Institut der Universität zu Köln, Lehrstuhl Tierphysiologie, Weyertal 119, 5000 Köln 41, FRG

Localization and Orientation in Biology and Engineering
ed. by Varjú/Schnitzler
© Springer Verlag Berlin Heidelberg 1984

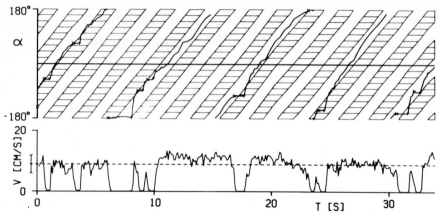

Fig. 1. A 35-s example of a cricket's run in a rotating vertically striped drum. *Upper trace* time course of the angle of the longitudinal body axis, measured by single frame analysis. *Lower trace* time course of velocity, as measured by the movement of the Kramer sphere. Spatial pattern period: 120 deg, angular velocity of the drum: 50 deg/s

period of the striped pattern (120°). Obviously the angles following the fast turns do not coincide exactly with the angles at the beginning of the stop.

This mismatch may be due to the fact that at the beginning of the stop the cricket did not walk in the "aspired" direction with respect to the pattern, and that this deviation is also corrected during the turn. Correspondingly, the right diagrams in Fig. 2 show the actual relationship between the cricket's direction and the pattern of the drum. For clarity, the black stripes are hatched, but still the result remains unclear. Indeed, before a stop the cricket does not maintain a specific angle with respect to the pattern (with rotation to the left a preference of the middle of a black or a white stripe might be anticipated, with rotation to the right a preference to the edges). Following a stop, however, the matching is not improved. This picture remains when applying different pattern periods (40° or 60°) and different angular velocities (10 deg/s or 20 deg/s).

It is possible that the walking directions within a rotating striped drum do not show a clear preference, because in crickets there is an unexpected fixation response towards single stripes. Lambin et al. (after Jeanrot et al. 1981) report that crickets of the species *Nemobius sylvestris* fixate the edge of a black stripe either 20° left or right to the sagittal plane. Within a striped pattern, this behaviour would result in walking directions difficult to follow up. Thus in *Gryllus bimaculatus* the frequency distribution of walking directions at presentation of stationary single or double stripes as well as a stationary striped drum were evaluated.

In single stripe experiments black stripes of 10, 20, 30, 90, 120, 150, and 180 degrees width were used. The crickets either fixate frontally (in these cases the distributions of walking angles show a steep maximum, superimposed by a back-ground of rather large deviations), or they walk away from the stripe (here the frequency distributions are rather broad).

What happens when presenting two black stripes at different angular distances? A stripe width of 30 deg and distances between the centers of the stripes of 60, 90, and 120 deg were tested. Corresponding to the steep maximum of the distribution of

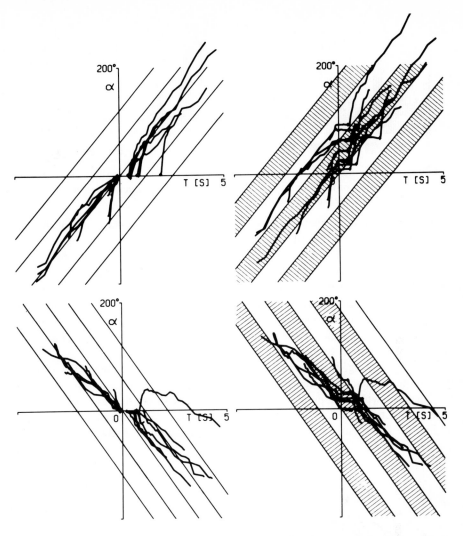

Fig. 2. Time course of the body angle before, during, and after the stops during 2 runs. *Upper diagrams* angular velocity +50 deg/s; *lower diagrams* –50 deg/s. *Left column* angular change relative to the direction at the beginning of the stops. *Right column* angular relationship to the rotating pattern. Spatial pattern period: 120 deg

walking angles, which was obtained when offering a single stripe, the distribution may already show two peaks at 60 deg stripe distance (see also Reichardt and Poggio 1976). In one of 5 cases tested still a unimodal distribution was found. In the remaining cases, the peaks diverge corresponding to the center distance of the two stripes; a mutual influence between the stripes was not found. In this situation too, the animals may walk away from the center of gravity of both stripes.

Examples of frequency distributions of walking directions with presentation of a stationary striped drum are shown in Fig. 3 (spatial period of the pattern: 60 deg). The first example (Fig. 3 left) shows fixation of the black stripes, but the peaks are shifted towards one edge of the stripes. In this optical surrounding, there also may be

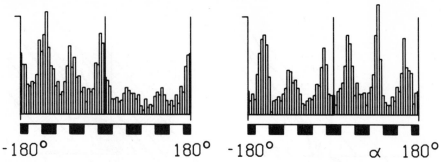

-180°   180° -180°   α   180°

Fig. 3. Two examples of the frequency distribution of walking angles in the presence of a stationary striped drum of 60 deg spatial pattern period

a preference of white stripes (as shown in Fig. 3 right) – this behaviour probably corresponds to the unimodal distributions sometimes occurring in two stripe experiments.

The distributions found in a stationary striped drum show that even though the crickets exactly fixate single black stripes, there are directional preferences when presenting periodic patterns, which do not necessarily coincide with the specific centers of the black stripes. Regarding these results obtained from stationary patterns, we can understand why there is no regular match in the time course of walking direction in Fig. 2: neither before after stops is an unequivocal preference of a specific direction with respect to the striped pattern to be expected.

In the following, the turns after stops are analyzed more closely. How does the magnitude of these turns depend on the duration of the preceding stop and thus on the angle by which the pattern has moved? Figure 4 shows the time courses of the direction of the animal's longitudinal axis following a stop (same examples as shown in Fig. 2). The angular position of the drum is presented by a straight line. The fast

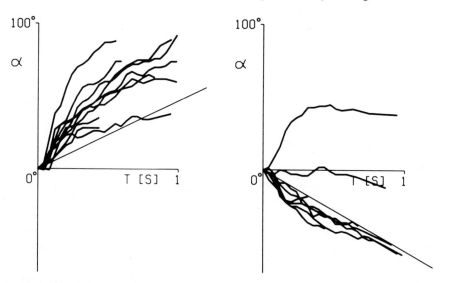

Fig. 4. Time course of the corrective turns following a stop. Pattern period: 120 deg, angular velocity 50 deg/s (left) or –50 deg/s (right)

turn is terminated in less than 0.5 s, at which point the body angle takes up with the drum movement. The magnitude of these fast turns was determined in the following way: The mean angular deviation between the animal's body axis and the increasing (or decreasing) angle of the drum after the cricket has started again (the straight lines in Fig. 4) is calculated from 0.5 s to 1.0 s after this start. This mean represents the magnitude of the turns following a stop with respect to the striped pattern.

The dependence of these turning angles on the duration of the stops in a rotating drum of 60° pattern period is shown in Fig. 5. If the crickets completely compensated for the rotation of the cylinder during the stops, the values should be placed on the corresponding straight lines. Figure 5, however, shows that there is no dependence on stop duration: short stops are overcompensated and the turns following long stops are too small. With increased angular velocity of the drum, the corrective turns also increase, in addition the turns at 60° period (shown in Fig. 5) seem to exceed those at 120°.

Finally, I would like to point out a striking feature of the dynamics of these turns: When looking at the time course of the angle between cricket and the continuously rotating drum, it turns out that these curves merely differ in amplitude, (in the specific corrective angle) but not in their shape. For clarity, the angles of the turns investigated were divided by their respective final value. These normalized angles are plotted according to the applied test conditions (Fig. 6). Independent of the pattern and the angular velocity of the drum, after a phase of acceleration of about 40 ms (1 frame) the final angle is reached exponentially with a time constant of about 0.4 s. This corresponds to a correction velocity of 2.5 deg/s per degree of the final angle. This finding suggests that these fast turns are merely controlled by the inherent properties of the locomotory system and not by the optic input.

The behaviour of the crickets, which previously suggested that the continuously increasing angle of the striped pattern is measured during the stops by integrating the

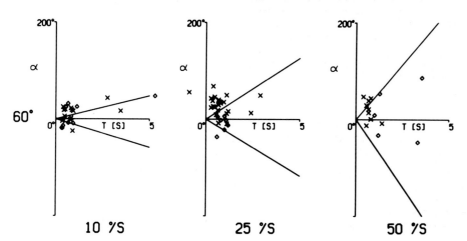

Fig. 5. Dependence of the size of corrective turns on the duration of stops at different angular velocities of the drum. The values obtained at counterclockwise drum rotation are indicated by *crosses*, those at clockwise rotation by *rhombs*. Pattern period 60 deg

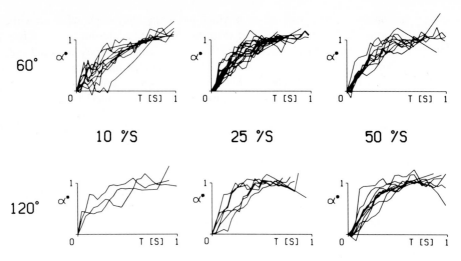

Fig. 6. Time course of the normalized angles between body axis and the rotating drum for the different angular velocities and two pattern periods

angular velocity, turns out to be caused by a trivial interplay of optomotor response and fixation: The angular velocity of the pattern during a stop determines the magnitude of the corrective turns; the angular relationship between animal and cylinder thus obtained is maintained via dynamic fixation (Varjú 1975).

One result of the analysis presented here, however, is remarkable: Although the crickets fixate single stripes sagitally within a small angular range, at presentation of a periodic pattern they are not only able to fixate each of the black stripes in their sagittal plane but also the center between two stripes and probably also each transition between these two extremes. Whether this behaviour can merely be deduced from the results of single stripe experiments is unclear at present and will be clarified by further experiments.

Acknowledgement. Supported by Deutsche Forschungsgemeinschaft (Scha 264, We 719)

References

Jeanrot N, Campan R, Lambin L (1981) Functional exploration of the visual field of the wood-cricket, *Nemobius sylvestris*. Physiol Entomol 6:27–34

Schmitz B, Scharstein H, Wendler G (1982) Phonotaxis in *Gryllus campestris* L. J Comp Physiol 148:431–444

Varjú D (1975) Stationary and dynamic responses during visual edge fixation by walking insects. Nature 255:330–332

Computation of Motion and Position in the Visual System of the Fly (*Musca*). Experiments with Uniform Stimulation

H. BÜLTHOFF and C. WEHRHAHN[1]

Positional information is required in solving orientation tasks. In open loop experiments it was found that a resting stripe does not influence the time averaged (over 120 s) yaw torque signal of a tethered flying fly irrespective of the position at which it is presented. An open loop response was elicited when either the stripe was moved or the head of the fly was free to move (Reichardt 1973, Wehrhahn 1980). Thus light flux changes are necessary for the extraction of positional information. It is known that flies are able to perceive motion and a mechanism responsible for the computation of motion proposed earlier (Hassenstein and Reichardt 1956), was found to apply also for the fly visual system (Fermi and Reichardt 1963). This mechanism selectively extracts the phase relations between the signals of neighbouring photoreceptors and thus enables the fly to follow a moving object. The direction of the response is inverted when this phase relation is inverted. The existence of a mechanism selectively extracting the position of an object can be tested by stimulating a fly with a resting object, whose luminosity is modulated in time. The result of the first of such experiments where the time averaged torque response to a flickered stripe has been determined, was negative. Thus it was concluded that no separate mechanism exists for the computation of position (Reichardt 1973). However, the motion and flicker stimuli were different in these experiments.

In the experiments carried out by Pick (1974), a stimulus was used which gave an identical light modulation to the single photoreceptors. The phase relation between the light fluxes in different ommatidia was the only parameter which varied, and again the time averaged ("stationary") torque response was determined. The averaging procedure (over 60 s) was started 30 s after the beginning of the stimulus, and a response to flicker was found which strongly depended on the position of the flickered object. It was concluded that a separate channel for the computation of motion exists. These experiments were later repeated with only slight variations and were essentially confirmed (Reichardt 1979). Experiments in which the fixation and tracking behaviour of free flying houseflies was studied, revealed that the time in which a response develops is very short ($\approx$ 200 ms) compared to 60 s because of the fast moving targets (Wehrhahn 1979). Therefore experiments with short time stimulation (2 s) were carried out which essentially showed no significant contribution of a position sensitive system to the flight torque response (Wehrhahn and Hausen 1980, Wehrhahn 1981). However, the stimulus in these experiments was different from that used by Pick

---

[1] MPI für biol. Kybernetik, Spemannstraße 38, 7400 Tübingen, FRG

Localization and Orientation in Biology and Engineering
ed. by Varjú/Schnitzler
© Springer Verlag Berlin Heidelberg 1984

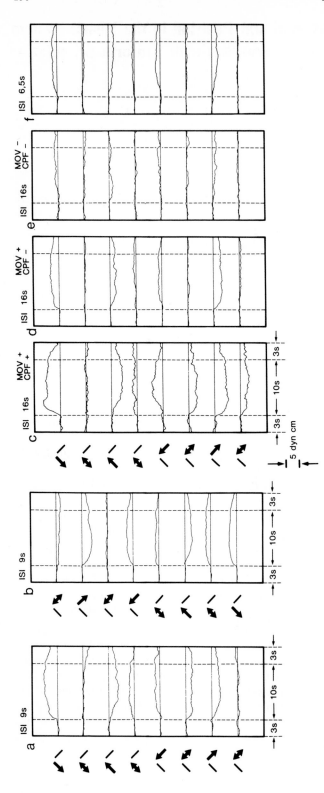

Fig. 1a–f. Average optomotor response of 4–15 female *Musca d.* flies to horizontal movements *(oblique arrows)* and flicker *(double arrows)* of a periodic sinusoidal grating (λ = 13.3°; m = 42%; w/λ = 3 Hz; H = 42 cd/m² oriented vertically behind a 6° wide slit at a position of ±30°. *Lines without arrowheads* indicate a uniformly illuminated visual field (60 cd/m²). *Upward deflection* denotes a flight torque response to the left side. The traces have to be read following each other in time with an interstimulus interval *(ISI)* indicated above

(1974) in several respects and this led to some controversy. Strebel (1982) was able to show that by using a small (6° wide) slit and moving an edge very fast behind it, a stimulus which lasted only for 20 ms, a flight torque response can be measured which is independent of the direction of motion, but dependent on the position.

In the experiments described here, an attempt was made to generate a stimulus condition very similar to those used by Pick (1974), but only for short time stimulation. The stimulus was a 6° wide window on a bright (60 cd/m^2) oscilloscopic screen, the position of the window being either 30° to the left or to the right side of the fly in the equatorial plane. The response was recorded in real time for successive stimuli of 10 s duration.

The intensity of the screen within the window was changed sinusoidally in such a way that each photoreceptor was modulated with a frequency of 3 Hz. The phase relation between neighbouring points was such that either motion from front to back, counterphase flicker (CPF) (phase 0°) or motion from back to front was seen by the fly. Figure 1a shows the result of a sequence of experiments with an inter-stimulus interval (ISI) of 9 s. It can be seen that a response to CPF is generated only after a previous motion stimulus from back to front, but not after motion from front to back. In a second series of experiments the sequence of stimulation was changed to test whether front to back motion on one side of the animal also suppresses the response to a flicker stimulus on the contralateral side. Figure 1b shows that this is not the case. Possibly, then, once the fly generated a response in one direction, a second stimulus in the same direction became ineffective due to response saturation. However, after the response to flicker generated in the seventh trace of Fig. 1b, an even larger response to motion from front to back on the same side was found. Thus we can conclude that the response to flicker is suppressed by preceding ipsilateral motion from front to back.

In the plots of Fig. 1c, d, and e, the ISI was 16 s and the animals, whose responses are shown, were classified according to their responsiveness to flicker and motion (c), no response to flicker but to motion (d) or no response at all (e). The variability in the response and particularly in that to flicker is very high from one fly to another and also in one fly at different times. Finally Fig. 1f shows that for short ISI the response to flicker is always very small.

Conclusion

The existence of a separate mechanism evaluating position in the flight torque response of tethered flying houseflies was confirmed. The responses indicating this are reliably suppressed by previous motion from front to back for short ISI (6.5 s) but not for long ISI. Variations in the strength of the response in different flies or in one fly at different times are described.

References

Fermi G, Reichardt W (1963) Optomotorische Reaktionen der Fliege *Musca d.* Abhängigkeit der Reaktion von der Wellenlänge, der Geschwindigkeit, dem Kontrast und der mittleren Leuchtdichte bewegter periodischer Muster. Kybernetik 2:15–28

Hassenstein B, Reichardt W (1956) Systemtheoretische Analyse der Zeit-, Reihenfolgen- und Vorzeichenauswertung bei der Bewegungsperzeption des Rüsselkäfers *Chlorophanus.* Z Naturforsch [C] 11b:513–524

Pick B (1974) Visual flicker induces orientation behaviour in the fly *Musca.* Z Naturforsch [C] 29c:310–312

Reichardt W (1973) Musterinduzierte Flugorientierung. Verhaltensversuche an der Fliege *Musca domestica.* Naturwissenschaften 60:122–138

Reichardt W (1979) Functional characterization of neural interactions through an analysis of behavior. In: Schmitt FO, Worden FG (eds) The Neurosciences. Fourth Study Program, pp 81–103. MIT Press, Cambridge

Strebel J (1982) Eigenschaften der visuell induzierten Drehmomenten-Reaktion von fixiert fliegenden Stubenfliegen *Musca d.* und *Fannia c.* Dissertation, Eberhard-Karl-Universität, Tübingen

Wehrhahn C (1979) Sex-specific differences in the chasing behaviour of houseflies *(Musca).* Biol Cybern 32:239–241

Wehrhahn C (1980) Visual fixation and tracking in flies. In: Segel LA (ed) Mathematical models in molecular and cellular biology. Cambridge, University Press, pp 568–603

Wehrhahn C (1981) Fast and slow flight torque responses in flies and their possible role in visual orientation behaviour. Biol Cybern 40:213–221

Wehrhahn C, Hausen K (1980) How is tracking and fixation accomplished in the nervous system of the fly? A behavioural analysis based on short time stimulation. Biol Cybern 38:179–186

On the Alignment of Movement Detectors Mediating the Landing Response in the Blowfly, *Calliphora erythrocephala*

H. ECKERT[1]

The flight speed of freely flying blowflies is in the order of 3 m/s. A small object is reached within 50 ms after detection. This short "time to collision" necessitates a fast reflex mechanism, the landing response (LR). The LR depends on the detection of a moving object or a moving surround, which in turn requires movement detectors. The orientation of movement detectors relative to the raster axes of the fly's compound eye was studied by presenting moving single (angular width 1.95 deg; velocity 37.6 deg) and multiple stripes (period 12 deg to 22 deg; contrast 0.66; velocity 80 deg/s) (Eckert 1980). A most conspicuous phase of the complex landing behaviour is the upwards and forwards throw of both foreleg tibiae obeying an "all-or-none" rule (Goodman 1960).

Displacements of the stimulus field center along the horizontal plane are denoted as ψ [deg] (ψ = 0 deg lies in the longitudinal axis of the fly); displacements from the horizontal plane to dorsal and ventral eye regions are denoted as $+\rho$ [deg] and $-\rho$ [deg], respectively (details see Eckert 1983).

Figure 1 exemplifies the dependence of the LR (no. of LR per 20 stimulus representations) on the direction of a periodic striped pattern. If stimulation occurs in frontal dorsal eye regions, the LR depends in a cosine fashion on the angle of pattern inclination, with a maximum (preference direction: PD) for upwards motion (●, Fig. 1). Only directly above the insect (ρ = 90 deg, e.g. an approaching predator), no direction of pattern motion induces the LR. In ventral eye regions, PDs correspond to downwards motion (■, Fig. 1). The "switching point" from upwards to downwards motion occurs at ψ = 0 deg; ρ = 0 deg (Eckert 1983). Oppositely directed PD's are also found along the equatorial plane of the eye. In the contralateral part of the right eye's visual field, PD's correspond to right-to-left motion (Fig. 1; the left eye was covered with black paint eliminating visual input from the covered eye), whereas in the ipsilateral part towards lateral eye regions they are given by left-to-right motion (▲, Fig. 1; Eckert 1980, 1983, Wehrhahn et al. 1981). PD's along the horizontal plane and along the median plane thus form a "cross" with an origin at the eye equator. This cross-like arrangement of PD's led to the hypothesis that, actually, PD's form a "flow-field with a pole in the path of flight" (Wehrhahn et al. 1981). This hypothesis requires PD's inclined obliquely against the horizontal plane, as has been confirmed experimentally: in male *Calliphorae*, PD's radiate from a common origin forming such a flow-field (Eckert 1983; cf. Fig. 2). Consequently, a fly approaching a possible landing

[1] Universität Marburg, FB 17 Zoologie, Postfach 1929, D-3550 Marburg; and Universität Bochum, Tierphysiologie, Postfach 102148, 4630 Bochum, FRG

Localization and Orientation in Biology and Engineering
ed. by Varjú/Schnitzler
© Springer Verlag Berlin Heidelberg 1984

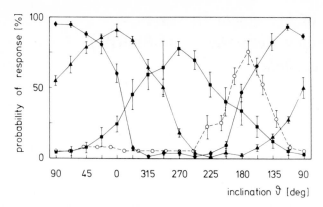

Fig. 1. Dependence of the landing response on the direction ϑ of motion of a moving periodic grating; ϑ denotes the angle between the horizontal plane and the directions of motion (ϑ = 90 deg upwards; ϑ = 0 deg front-to-back; ϑ = 270 deg downwards; ϑ = 180 deg back-to-front). Stimulation occurred 45 deg above (●) and 45 deg below (■) the horizontal plane at ψ = 0 deg; as well as 45 deg lateral to the median plane along the horizontal plane (▲). The *dashed curve* gives the response of insects whose left eye was covered with black paint (○). *Calliphora* ♂

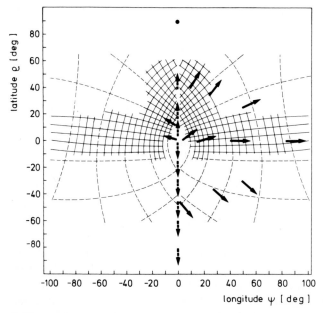

Fig. 2. Preference directions (PD) are super-imposed on a grid of eye axes: "horizontal" z-axes are crossed by "vertical" v-axes. *Dashed lines* denote extrapolated z- and v-axes (Franceschini after Hausen 1981). *Black arrows* denote PDs measured in "monocular" flies and in monocular regions of the eye outside the area of binocular overlap. *Dashed arrows* show vertical PD's arising from interactions between responses from both eyes, such as they are found in intact animal. *Dots* denote parts of the visual field where no direction of motion elicits any LR at all. All PD's are given for the right visual field only. Note that PD's are aligned quite accurately with the z-axes of the eye, which at ρ = 0 deg are almost horizontal; with increasing (decreasing) latitude the z-axes show a local deviation from the horizontal plane, pointing obliquely upwards (downwards). *Calliphora* ♂

object maximizes the "landing stimulus" if it aligns the origin of PD's with the path of flight. The angle between body long axis and direction of flight depends on the flight speed (Wagner, pers. commun): the faster the animal flies the smaller is this angle. A fast-flying animal will thus have to slow down in order to align the origin of PD's with the path of flight. Free-flying houseflies, *Musca*, slow down during landing (Wagner 1982). Contrarily, an object moving away from the fly (simulated by two stripes moving towards one another) provides a stimulus "contraction". Such a stimulus induces inhibitory response components for all directions of motion (Eckert 1980, 1983). Thus, an object moving away from the origin of PD's induces maximum inhibition. Therefore, movement detectors mediating the LR appear to be functionally bi-directional responding with excitation in one, and with inhibition in the opposite direction of motion.

Stimulus fields, 30 deg and 50 deg wide, stimulate large numbers of "elementary" movement detectors (EMD) and PD's will reflect the summed output of all EMD's in the stimulus area. Thus, alignment of subsequently described EMD's with the raster axes of the eye is subject to this restriction.

The compound eye of flies is characterized by eye axes described as "horizontal" z-axes, vertical v-axes, and oblique axes termed x and y (cf. review Hausen 1981). Figure 2 shows that all PD's are aligned with the z-axes but for one eye region: along the median plane in front of the animal vertical PD's are found, however, only in intact "binocular" flies. If PD's are measured in flies with one eye covered with black paint, they are also aligned with the z-axes: thus, vertical PD's measured in intact animals are caused by interaction between both eyes (Eckert 1980, Wehrhahn et al. 1981, Eckert and Hamdorf 1983). The radial distribution of PD's in the fly's compound eye is, therefore, based on two principles: (1) on "bending" of the horizontal z-axes to oblique inclinations against the horizontal plane (Fig. 2), and (2) on binocular interactions.

Acknowledgements. Supported by a Heisenberg stipend (Ec 56/3 and 56/5-2) and grant Ec 56/4-5 by the Deutsche Forschungsgemeinschaft

References

Eckert H (1980) Orientation sensitivity of the visual movement detection system activating the landing response of the blow-flies, *Calliphora* and *Phaenicia*. Biol Cybern 37:235-247

Eckert H (1983) On the landing response of the blowfly, *Calliphora erythrocephala*. Biol Cybern 47:119-130

Eckert H, Hamdorf K (1983) Does of homogeneous population of elementary movement detectors activate the landing response of blowflies, *Calliphora erythrocephala*? Biol Cybern 48:11-18

Goodman L (1960) The landing responses of insects. J Exp Biol 37:854-878

Hausen K (1981) Monokulare und binokulare Bewegungsauswertung in der Lobula Platte von Fliegen. Verh Dtsch Zool Ges (W Rathmayer ed.). Gustav Fischer, Stuttgart New York, pp 49-70

Wagner H (1982) Flow-field variables trigger landing in flies. Nature 297:147-148

Wehrhahn C, Hausen K, Zanker J (1981) Is the landing response of the housefly *(Musca)* driven by motion of a flow-field? Biol Cybern 41:91-99

The Stimulus Efficiency of Intensity Contrast, Spectral Contrast and Polarization Contrast in the Optomotorics of *Pachnoda marginata* (Coleoptera: Cetoniinae) [1]

U. MISCHKE [2]

1 Introduction

The optical environment is composed of radiation of varying intensities, wave lengths, and polarization. From all these contrasts, organisms can form spatial and temporal patterns for their orientation. The optical contrasts which produce a stimulus efficiency in the optomotorics of *Pachnoda* are determined in these studies. The variety of possible combinations of contrasts of the natural light climate has been taken into account in the stimulus program.

2 Materials and Methods

The African rose chafers of the genus *Pachnoda* are, in the wild, diurnal, rapidly flying scarab beetles which feed mainly on Rosaceae. – The double cone apparatus used here (Kaiser 1968, Mischke 1982) permits the reproduction of the light climate contrasts between the diffuse reflecting outer and inner cone (Fig. 1). The rotatable inner cone is composed of trapezoid-shaped stripes and spaces. Upon rotation of the inner cone, the fixed, flying chafers reacts with a head movement in the same direction. The angle of this horizontal head movement is evaluated. The lateral, middle ommatidia are stimulated with a contrast frequency of 6 Hz. The retina in the stimulated eye area is studied with the help of transmission electron microscopy.

3 Results

Monochromatic and heterochromatic contrasts with E-vector contrast values of $0°/90°$ in comparison with $90°/90°$, and $45°/135°$ $(45°/45°)$ from the spectral sensitivity maxima of beetle eyes are offered. The reactions are (1) independent of

---

[1] This paper is a pre-publication of parts of a doctoral thesis completed in the Department of Biology, Freie Universität Berlin

[2] Institut für Allg. Zoologie FU Berlin, Königin-Luise-Straße 1–3, 1000 Berlin 33, FRG

Localization and Orientation in Biology and Engineering
ed. by Varjú/Schnitzler
© Springer Verlag Berlin Heidelberg 1984

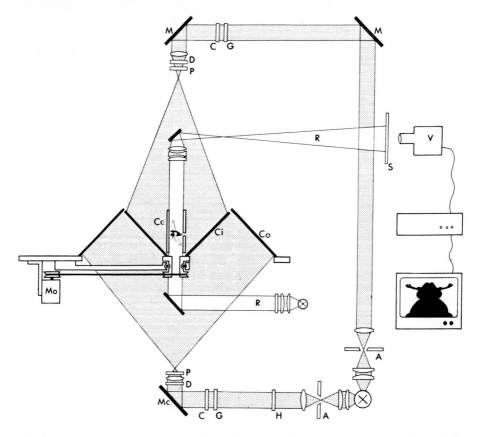

Fig. 1. Schematic drawing of the experimental apparatus for optomotoric analysis. A constant, regulated xenon arc (900 W) with two condensors was used to generate optical contrasts. *A* aperture; *C* color filter; *Cc* central cylinder; *Ci* rotatable inner cone; *Co* outer cone; *D* depolarizer; *G* gray filter; *H* heat-reflection mirror; *M* aluminium-surfaced mirror; *Mc* mirror combination UV/cold light; *Mo* motor; *P* polarizer; *R* red light (RG 695) projection system; *S* screen, *V* video camera

spectral contrasts, (2) independent of E-vector contrasts, (3) independent of combined spectral and E-vector contrasts, and (4) independent of combined intensity and E-vector contrasts. Intensity contrast is the sole stimulus of Pachnoda's optomotoric reactions (Figs. 2 and 3). The morphological organization of the *Pachnoda* eye corresponds to the superposition type. In six of the eight retinula cells of an ommatidium, the microvilli of each cell are mostly arranged parallel to one another.

4 Discussion

Pachnoda exhibits a sensitive reaction to intensity contrasts, but is not stimulated by polarization and spectral contrasts (Figs. 2 and 3). The independence from color of

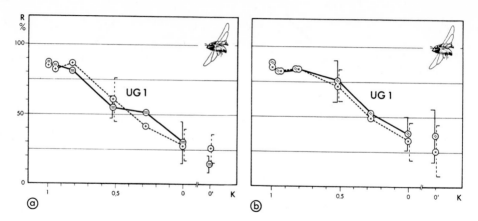

Fig. 2a,b. Stimulus efficiency of monochromatic UV-contrasts *(UG 1)* in connection with identical and varying E-vector inclinations. The optomotor response of *Pachnoda* depends on the intensity contrasts, but independent of the E-vector (Wilcoxon test). Green light (Nal 510), not shown, has a corresponding stimulus efficiency. *K* contrast {K = [(a–b)/(a+b)]; a, b: intensities of the cones}: *0'* control (only RG 695); *R* reaction (average value, standard deviation); **a** E-vectors 0°/90° between the stripes ⊕ and 90°/90° ⊖, **b** E-vectors 45°/135° ⊗ and 45°/45° ⊘

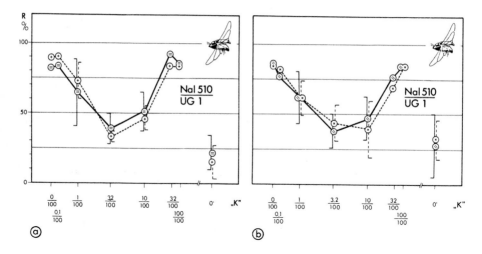

Fig. 3a,b. Stimulus efficiency of heterochromatic UV/green-contrasts with and without E-vector contrasts. All heterochromatic contrast functions generated a uniform graph with a minimum corresponding to the level of the adjusted monochromatic contrasts (K = 0) of Fig. 2. Since the heterochromatic contrasts are not readily defined, the specific grey filter transmission is given in percent ("K"); numerator: *Nal 510*; denominator: *UG 1*; for other symbols see the legend for Fig. 2

the optomotoric reactions is common to insects (Kaiser 1968, Kaiser and Liske 1974, Mischke 1982). The similar reaction of the various insect taxa is independent of the specific retinal and laminal organization. In contrast, the reaction to E-vector contrasts varies in the optomotorics of the insects. *Pachnoda* is insensitive to polarization contrasts, unlike *Musca* (Kirschfeld and Reichardt 1970), *Apis* (Kirschfeld 1973), and *Drosophila* (Wolf et al. 1980). The cause of this polarization insensitivity could be a peculiarity of the visual system of *Pachnoda*. The related beetle *Lethrus* (also Scarabaeidae) can orient itself menotactically with the help of polarized light (Frantsevich et al. 1977). One can thus assume that Scarabaeidae also possess dichromatically absorbing photopigments. Since the polarization insensitivity of *Pachnoda* cannot be explained cytologically on the basis of the microvilli orientation, one might expect, instead, an electrical or neuronal linkage of retinula cells with different specific spatial E-vector sensitivities.

References

Frantsevich L, Govardovski V, Gribakin F, Nikolajev G, Pichka V, Polanovsky A, Shevchenko V, Zolotov V (1977) Astroorientation in *Lethrus* (Coleoptera, Scarabaeidae). J Comp Physiol 121:253–271

Kaiser W (1968) Zur Frage des Unterscheidungsvermögens für Spektralfarben: Eine Untersuchung der Optomotorik der königlichen Glanzfliege *Phormia regina* Meig. Z vergl Physiol 61:71–102

Kaiser W, Liske F (1974) Die optomotorische Reaktion von fixiert fliegenden Bienen bei Reizung mit Spektrallichtern. J Comp Physiol 89:391–408

Kirschfeld K (1973) Optomotorische Reaktionen der Biene auf bewegte „Polarisations-Muster". Z Naturforschg 28c:329–338

Kirschfeld K, Reichardt W (1970) Optomotorische Versuche an *Musca* mit linear polarisiertem Licht. Z Naturforschg 25b:228

Mischke U (1982) Spektrale Eigenschaften des visuellen Systems von *Leptinotarsa decemlineata* (Coleoptera: Chrysomelidae). Zool Beiträge 27:319–334

Wolf R, Gebhardt B, Gademann R, Heisenberg M (1980) Polarization sensitivity of course control in *Drosophila melanogaster*. J Comp Physiol 139:177–191

A Possible Neuronal Mechanism of Velocity and Direction Detection

E. MILEV[1]

1 Introduction

Some cortical neurons are selectively sensitive to direction and velocity of stimuli in their receptive fields. A possible explanation of this property is based on an analogy to the so-called pulse selector for decoding of time pattern codes.

2 The Selector as a Time Pattern Recognizer

A pulse selector (temporal feature extractor) consists of 3 parts (Fig. 1B): (a) network of delay lines, polarity inverters and weighting elements, (b) summing circuit, (c) output indicator (threshold discriminator or amplitude-frequency converter). It functions according to a "lock and key"-principle: its summation output $\Sigma(t)$ obtains a sharp maximum only for a certain matched ("key") input time pattern $m(t)$ – a result of the coincidence of the appropriately delayed components of $m(t)$ (optimal signal compression). The output indicator marks the peak of $\Sigma(t)$, producing a sign for the identification of $m(t)$. This is defined by a time inversion of the selector's single pulse response $h(t)$:

$$m(t) = h(-t) \quad \text{resp.} \quad h(t) = m(-t) .$$

In this case $\Sigma(t)$ corresponds to the autocorrelation function R_m of the key-signal $m(t)$, but as a function of the real time t (not of the time shift τ):

$$\Sigma(t) = \int m(\tau) . \, m(\tau - t) \, d\tau = R_m(t) .$$

Another non-matched input signal $n(t)$ cannot produce such a high peak of $\Sigma(t)$, because in this case $\Sigma(t)$ corresponds to the crosscorrelation function $R_{nm}(t)$ between $n(t)$ and $m(t)$:

$$\Sigma(t) = \int n(\tau) . \, m(\tau - t) \, d\tau = R_{nm}(t) .$$

Devices with these properties are named *matched filters* and play an important role for the detection of signals blurred by noise.

---

[1] Carl-Ludwig-Institut für Physiologie, Karl-Marx-Universität Leipzig, Liebigstraße 27, 7010 Leipzig, DDR

Localization and Orientation in Biology and Engineering
ed. by Varjú/Schnitzler
© Springer Verlag Berlin Heidelberg 1984

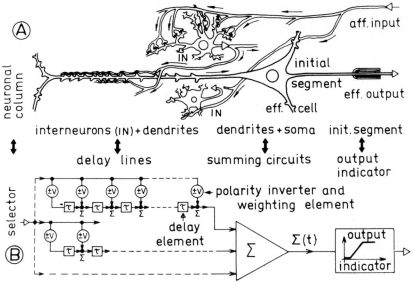

Fig. 1 A,B. The analogy between a neuronal column (**A**) and a pulse selector (**B**)

3 The Selector as a Space-Time Pattern Recognizer

Delay lines with branching points can transform space in time coordinates (Steinbuch and Frank 1961). The spatially distributed inputs must be connected with certain delay line taps. According to the matched filter principle the maximal peak of $\Sigma(t)$ arises only for a certain order of input activations and interactivation intervals, i.e. for a given "key" direction and velocity of activation. It results in a *spatio-temporal* perception based on *temporal* correlation.

4 The Analogy "Neuronal Column – Selector"

The cortical cells may be divided generally in two groups: (a) interneurons – short axon (Golgi II) local circuit cells, (b) efferent cells – long axon (Golgi I) spiking pyramide cells. From the connection of several interneurons with one efferent cell (Fig. 1A) the following functional association to a selector device may be derived: the interneurons and the extended dendritic tree of the efferent cell act as delay lines (by means of PSP[1]-spread in the dendritic "cables" and further delay in interneuronal junctions and axons); the synapses act as polarity inverters and weighting elements; soma and dendrites of the efferent cell as well as contacts within the local circuits act as a summing network; the efferent cell initial segment (trigger zone) acts as an output indicator. This analogy suggests that the same selectional mechanism of matched filtering may operate in the neuronal column. Thus columnar efferent cells

---

[1]   PSP – post synaptic potentials

should be able to respond preferentially (with respect to the greatest somatic potential or to the maximal firing rate) to certain matched input patterns. These key patterns are structurally determined by the interneuronal contacts and may be ascertained by means of a time inversion of the single impulse response of the cell. The receptive field must be connected with a bank of efferent cells with different match; each cell will produce a maximal response only if the stimulus in the receptive field corresponds to its "key" direction and velocity. If the "simple" cells are matched in this way, the higher order cells may be matched by means of convergence too (Hubel and Wiesel 1962). Input-driven plastic changes of the match may explain certain forms of learning and memory.

Several complicating factors limit these model conclusions, e.g. nonlinearity of the PSP-summation, supplementary temporal PSP-summation and spatial correlation. On the other hand they are compatible with several experimental findings (Hubel and Wiesel 1962) and theoretical conceptions of brain functions, such as the theory of the selective neuronal groups (Edelman 1979), the associative holography-like memory (Poggio 1973), the autocorrelative evaluation in the nervous system (Reichardt 1957).

References

Edelman GM (1979) Group selection and phasic reentrant signaling: a theory of higher brain function. In: Schmitt FO, Worden FG (eds) The neurosciences, fourth study program. MIT Press, Cambridge, pp 1115–1139

Hubel DH, Wiesel TN (1962) Receptive fields, binocular interaction and functional architecture in cat's visual cortex. J Physiol 160:106–154

Poggio T (1973) On holographic models of memory. Kybernetik 12:237–238

Reichardt W (1957) Autokorrelations-Auswertung als Funktionsprinzip des Zentralnervensystems. Z Naturforschg 12b:448–457

Steinbuch K, Frank H (1961) Nichtdigitale Lernmatrizen als Perzeptoren. Kybernetik 1:117–124

Optical Monitoring of Ego-Motion for Movement with Respect to a Plane Surface

J.J. KOENDERINK and A.J. van DOORN[1]

Whereas the classical "depth-cues" for statical perspective are inherently ambiguous (they can be used in painting "depth" on a plane surface!), cues from dynamical perspective offer real exterospecific information. These cues can be self-induced (through ego-motion) and thus enable the observer to experiment with the visual world, and thus to objectivate the observations. Gibson (1950) exploited these facts in his well-known treatment of the perception of spatial lay-out through motion parallax. Gibson's analysis has justly drawn a lot of attention, but less well known is the fact that several of the Gibsonean entities, e.g. the famous "focus of expansion", are rather problematic concepts. The more well-known Gibsonean concepts apply to the movement of an observer with respect to a plane surface, and we concentrate on this simple case here.

Consider the movement of a vantage point relative to a point P in ambient space. Let V be the velocity of the observer, $\mathbf{r}$ a unit vector in the visual direction of P, θ the angle between $\mathbf{r}$ and V, ρ the distance from the vantage point to P. The component of V perpendicular to $\mathbf{r}$ is the "veer velocity", it induces a "skid" $-\frac{1}{\rho}(V - (V.r)r) = -V_{veer}/\rho$ of the projection of P in the optic array. The magnitude of this apparent velocity is $\frac{V \sin \theta}{\rho}$, and it assumes constant values for the family of "Vieth-Müller tori" (see Fig. 1a).

If u_∞ is the apparent velocity at which blurring starts, then $\rho < V \sin \theta / u_\infty$ describes the "blur-zone", an important concept in aviatics (Whiteside and Samuel 1970). If δu is the lowest apparent velocity difference that can be resolved, then $\delta \rho = \rho^2 \delta u / V \sin \theta$ is the radial resolving power. If ϵ_0 denotes the smallest angular difference that can be resolved, then $\delta l = \rho \epsilon_0$ is the transverse resolving power. At a distance $\rho = \rho_*$ $\sin \theta$, with $\rho_* = \frac{\epsilon_0 V}{\delta u}$, these resolving powers balance. If you measure distances in terms of just noticeable differences (j.n.d.'s), then the metric for motion parallax is:

$$ds^2 = (\frac{d\rho}{\delta\rho})^2 + (\frac{\rho \, d\theta}{\delta l})^2 = \frac{1}{\epsilon_0^2} [\sin^2\theta \, (d \frac{\rho_*}{\rho})^2 + d\theta^2] .$$

[1] Dept. Medical and Physiological Physics, Physics Laboratory, State University of Utrecht, Princetonplein 5, 3584 CC Utrecht, The Netherlands

Localization and Orientation in Biology and Engineering
ed. by Varjú/Schnitzler
© Springer Verlag Berlin Heidelberg 1984

For a pedestrian

$$\rho_* = 140 \text{ cm. } (\epsilon_0 \approx 1', \text{V} \approx 5 \text{ km h}^{-1}, \delta u \approx 1's^{-1}, u_\infty \approx 100° \text{ s}^{-1})$$

This is the (Riemann) metric of the elliptic plane. From the blur zone to the "vault of heaven" ($\rho = \infty$) you measure a total of $u_\infty / \delta u$ j.n.d.'s. For the human observer this is about 6.10^3 j.n.d.'s, that is about as many j.n.d.'s as go in a (transverse) arc of $100°$. Thus the dynamic visual field is almost as deep as it is wide.

When you move with respect to a plane surface, it can be shown (Koenderink and van Doorn 1981) that the apparent velocity field induced by the movement (the "optical flow" of Gibson) can be derived from two potentials ϕ, ψ:

$$\mathbf{u} = \nabla \phi + i \nabla \psi$$

(the operator i denotes a turn over $\pi/2$ in the positive direction).

Using vector analysis you find that div $\mathbf{u} = \Delta \phi$, curl $\mathbf{u} = \Delta \psi$. Thus the potentials can be thought of as due to the "sources" div $\mathbf{u}$ and curl $\mathbf{u}$, that is the expansion and the vorticitiy. These entities are true scalars with a coordinate independent meaning. They can be obtained locally with isotropic receptive fields. For instance, the divergence merely measures the relative rate of change of a small solid angle. In terms of the parameters of motion, the potentials are (see Fig. 1b):

$$\phi = \frac{1}{2} \frac{\mathbf{V}}{h} \cdot (\mathbf{r} \times \mathbf{e}) \quad \psi = \frac{1}{2} (\mathbf{r} \times \frac{\mathbf{V}}{h}) \cdot (\mathbf{r} \times \mathbf{e}) .$$

The vector $\mathbf{r} \times \mathbf{e}$ specifies the *tilt* of the plane (its direction) as well as its *slant* (its magnitude). Both the skid ($-\mathbf{V}_{veer}/h$) and the orientation of the plane can be found from $\phi(\mathbf{r})$, moreover $\dot{\phi}/\phi = - \dot{h}/h = (\mathbf{V}.\mathbf{e})/h$: thus you can also derive the other component of $\mathbf{V}/h$. (You cannot hope to obtain $\mathbf{V}$ and h separately of course: you measure only angular relations.)

If the motion contains a rotational component, the solenoidal field ($i\nabla\psi$) is affected, but the lamellar field ($\nabla \phi$) is unaffected. In fact you can cancel the solenoidal field with an eye-movement: it is completely propriospecific. Only the lamellar field $\nabla \phi$ carries exterospecific information. For a movement parallel to the plane (e.g. walking, driving a car, etc.) this field has two nodes: a source at a distance h in front of the observer, and a sink at distance h behind the observer. Remarkably enough the Gibsonean "focus of expansion" is absent: the "focus" is not an extremum of the expansion and is not even invariant to eye-movements! (See Fig. 1c,d.)

It remains to be seen whether the human observer is able to detect gradients in movement parallax fields with sufficient sensitivity to profit from these relationships. We ran an extensive set of experiments to find this out (van Doorn and Koenderink 1982), and found that two velocities near a border (a transient of the parallax field) can be distinguished if

$$|\mathbf{v}_1 - \mathbf{v}_2| \geqslant W \frac{|\mathbf{v}_1 + \mathbf{v}_2|}{2} \quad \text{with a "Weber fraction" W near to unity.}$$

This appears insufficient to extract the information of the flow field! These experiments were performed with passive observers, however, and with focal vision. Perhaps self-induced movement in ambient vision is detected with different systems.

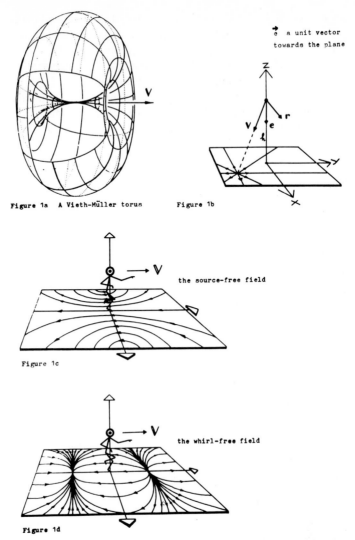

Figure 1a A Vieth-Müller torus   Figure 1b

Figure 1c

Figure 1d

Fig. 1a–d. a Vieth-Müller torus; **b** *e* unit vector towards the plane; **c** source-free field; **d** whirl-free field

There are, however, other possibilities. Instead of monitoring movement gradients, one can also monitor the *geometrical distortions* induced by the flow. For instance, the expansion might be extracted. An especially favourable method appears to be the following: Suppose you measure the orientations (φ) of small, local image details, or, better still, the relative time change ($\delta\dot{\varphi}/\delta\varphi$) of orientation differences ($\delta\varphi$). This can be done without explicit coordinate systems, without even a "reference orientation". It can be shown that

$$\frac{\delta\dot{\varphi}}{\delta\varphi} = A\cos 2\left(\varphi - \mu\right),$$

where μ is the orientation of the bisectrix of the skid and the tilt, whereas A equals the product of V/h and the slant. The result is independent of any rotational component. ($\delta\dot\varphi/\delta\varphi$ depends purely on ϕ.) It appears likely that the simple cells of cortical area 17 of mammals are involved in such analysis: it seems an ideal substrate for the detection of surface slant and tilt as well as ego-motion. The effect is certainly robust enough: for a pedestrian (5 km/h) and a slant of $45°$, at a distance h (~ 1.5 m) the maximum relative change $\delta\dot\varphi/\delta\varphi$ is about 100% s^{-1}. Much smaller values are easily detected by humans.

References

Doorn AJ van, Koenderink JJ (1982) Visibility of movement gradients. Biol Cybern 44:167–175

Gibson JJ (1950) The perception of the visual world. Houghton Mifflin, Boston, Mass.

Koenderink JJ, Doorn AJ van (1981) Exterospecific component of the motion parallax field. J Opt Soc Am 71:953–957

Whiteside TCD, Samuel GD (1970) Blur zone. Nature 225:94–95

Perception of Differences Between Visually and Mechanically Presented Motion Information

H. DISTELMAIER and G. DÖRFEL[1]

1 Problem

Because of increasing use of computer-based simulations of modern man-machine systems (MMS), mathematical models of human perception and behaviour need to be included in these simulations. A long-term goal of the work presented is a mathematical model of the interaction between visual and vestibular motion perception by man especially in vehicle control. Early research in this field investigated human difference perception between visually and mechanically presented motion information.

2 Experiment

2.1 Experimental Setup

For research in the field of motion perception a special motion system was built, (Fig. 1) in which inside a closed compartment, subjects can be moved up to 7 m along a horizontal track with maximum acceleration up to 5 m/s². The visual system consists of a moving belt display with vertical black and white bars on the belt giving the visual impression to subjects of an outside fence (Fig. 2).

2.2 Procedure

The forcing functions for movements of the compartment and the moving belt consisted of modulated sine waves (Fig. 3a) which were equal or different in amplitudes by six factors between 0.4 and 1.6 (Fig. 3b). Subjects inside the compartment had to actuate a button whenever they got the impression that movement of the compartment and relative movement of the bar-pattern were not indentical, i.e., when the moving belt did not accurately represent movements along a motionless outside object (Fig. 3c).

---

[1] Research Institute for Human Engineering (FAT), Königstraße 2, 5307 Wachtberg, FRG

Localization and Orientation in Biology and Engineering
ed. by Varjú/Schnitzler
© Springer Verlag Berlin Heidelberg 1984

Fig. 1. Motion system with track and compartment

Fig. 2. Compartment with visual system inside

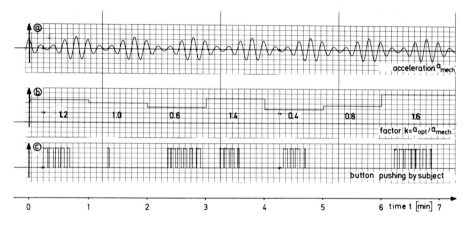

Fig. 3. Record of one session. **a** acceleration a_{mech}, **b** factor steps $k = a_{opt}/a_{mech}$, **c** button pushing by subject

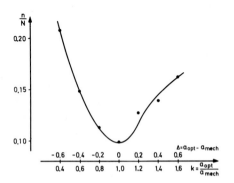

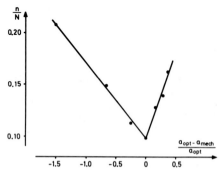

Fig. 4. Relative pushing time n/N vs difference Δ or factor k

Fig. 5. Relative pushing time n/N vs subjective difference estimation

2.3 Results

Duration time of button actuation was taken as a measure for subjects' ability to perceive differences between visually and mechanically presented motion information. To eliminate individual actuation characteristics, button pushing time (n) for each factor was related to total actuation time (N) for all seven steps in one session. Figure 4 shows experimental results. The mean relative actuation duration (n/N) is drawn vs the difference (Δ) between visual and mechanical acceleration values. Directly related to this difference scale there is a second scale in Fig. 4 which gives the ratio (k) of the visual to the mechanical motions. A number of different functional relationships were tried to see whether a linear relationship could be found between the variables in Fig. 4, since such relationships have been found in other studies of human sensory perception. The scale in Fig. 5 which turns out to be the ratio of the difference between visual and mechanical motion values to visual motion values provides a linear relationship. By this calculation the experimental data very well fit in straight line curves. The point in Fig. 5, at motion variables ratio of 0.0, where lines change direction represents the case where visual and mechanical motion values are indentical.

3 Discussion

The results show that an increase in differences between visual and mechanical motion information is accompanied by an increase in subjects' ability to perceive such differences. It can be demonstrated that the duration of motion differences perceived is directly proportional to the difference between visual and mechanical motion values and inversely proportional to visual motion values. That means, to judge differences, subjects related their estimations of these differences to the visual motion information. There are some other studies which also indicate that in visual-vestibular conflict situations subjects tend to rely more on visual information. The question of whether acceleration or velocity variables are used by subjects, separately, or in combination to each other cannot be answered from the experimental results. Statements by the subjects, however, indicate that they used different variables in different situations. At least with the visual information subjects concentrated on acceleration with slow motion presentations but used velocity with fast motion presentations. Two significant details are noticeable from the results shown in Fig. 5. First, there is a difference in the slope of the two straight lines representing the perception of motion differences. This slope might be interpreted as "perceptual sensitivity", showing that there is a difference in sensitivity within the perceptual range investigated. The reasons for this difference can only be speculated about at this time.

The second noticeable detail from Fig. 5 is the value for difference perception in the case where visual and mechanical motion information were indentical. This value does not come out to be zero indicating that subjects perceived differences when there really were none. According to signal detection theory this might be interpreted as "perceptual noise". This possible characteristic of the perceptual system may then always occur so that all measurements are affected and the straight lines are shifted up by this offset value.

Paired Tread Wheels as a Tool for the Analysis of Orientation Behaviour in Arthropods

G. WENDLER, H.-G. HEINZEL, and H. SCHARSTEIN[1]

Most investigations of directional orientation in animals start with a quantitative description of the movements of the animal as a whole. Any further analysis of the underlying system requires measurements of additional variables within the system that are relevant for the orientation process, such as movements of individual legs or groups of legs or neuronal activity. Following an idea of Franceschini, we developed paired tread wheels which allow measurements of such variables in walking arthropods.

We investigated the orientation towards directional optical, acoustical, and wind stimuli and the correcting behaviour in different species of arthropods. As a consequence of their bilateral symmetry, these animals receive most stimuli with paired sense organs, situated on the left and on the right body side. Subsequently, an evaluation is made of those stimuli relevant to orientation. The resulting signals control the leg motor output. In such a system the sensory input from one side of the body may selectively control either the legs of the ipsilateral or the contralateral body side or influence both sides simultaneously, but in a different way.

These different control mechanisms can be investigated by using paired tread wheels which allow the animal to walk at different speeds with its right and left legs and thereby express the momentary turning tendency. The construction of the wheels restricts the legs to moving parallel to the body, while free-walking animals turn by moving their legs with superimposed lateral components. We found, however, that most species readily accept the restricted situation and no change in orientation behaviour can be detected in the tread wheel situation.

The most obvious advantages of this experimental situation are:

1. The animal is tethered which allows us to analyze neuronal and muscular activity while the animal is orientating (Fig. 2a).
2. The mechanical situation for the animal is identical for open loop (pure measurement of the wheel movements) and for closed loop analysis (the difference in the movements of the wheels controls the stimulus situation).

In the following three figures we describe the paired tread wheels and present three application examples.

Acknowledgements. Supported by Deutsche Forschungsgemeinschaft (We 719, He 1118, Scha 264)

---

[1]   Zoologisches Institut der Universität zu Köln, Tierphysiologie, Weyertal 119, 5000 Köln 41, FRG

Localization and Orientation in Biology and Engineering
ed. by Varjú/Schnitzler
© Springer Verlag Berlin Heidelberg 1984

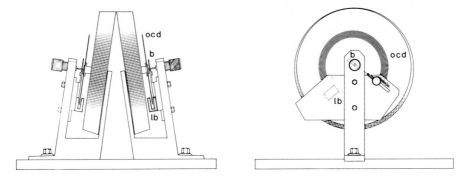

Fig. 1. Schematic drawings of the paired tread wheel. The wheels (ϕ 14 cm) consist of ROHACELL (Röhm). They are mounted with an angle of 18° to each other in order to leave space for the bearings. On the side of each wheel, an optical code disc (*ocd*) (80 mm ϕ, 500 lines) is mounted onto the same axis. The movements of this disc are monitored by a dual channel light barrier (*lb*) which allows to measure speed and to distinguish between forward and backward movements. Friction can be changed by bifurcated spring steel brakes (*b*), touching the axes.

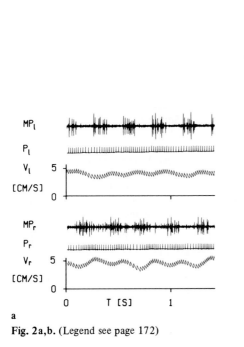

Fig. 2a,b. (Legend see page 172)

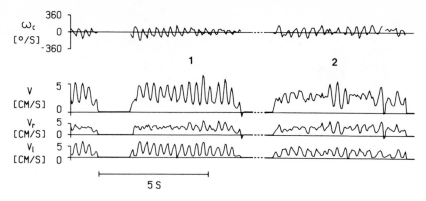

Fig. 3. Unoriented walk of a house cricket *(Acheta domesticus)* on a paired tread wheel (top). As in freely walking crickets, in both situations the walk shows bouts of progression, interrupted by short stops. The oscillations in the bouts are caused by the tripod gait. Left and right side are often in phase *(1)*, but sometimes show antiphase or even different frequencies *(2)*. (Experiment: J. Stabel)

Fig. 2a,b. a Normal walk of a carrion beetle *(Necrophorus humator* F.). The signals of the dual channel light barrier are transformed into unit pulses of the left (P_l) and right (P_r) wheel. They are stored on magnetic tape and later processed with a digital computer (PDP 11/40). Preliminary data (velocities V_l, V_r) can be obtained by filtering (low pass) these pulses. The difference $\Delta V = V_l - V_r$, simultaneously monitored by a highspeed recorder, indicates the momentary turning tendency. In this case, both wheels rotate at a similar speed. Therefore the animal would walk approximately in straight course. The superimposed oscillations (3–4/s) of both wheels demonstrate the high resolution of the system (one pulse for 0.5 mm of walked path). They represent the rhythmic thrust of the legs, which is the consequence of the insect tripod gait. This interpretation is supported by the correlation with the myograms from the left (MP_l) and the right (MP_r) front legs. **b.** Correcting behaviour in woodlice. Reaction of a woodlouse *(Oniscus asellus)* to mechanical stimulation of the left antenna by touching an obstacle *(black bar)*. For details of the stimulation see Schoenemann this Volume. The stimulus causes a velocity reduction on the contralateral side (right turn). After the end of stimulus, the animal turns to the opposite side by reducing the walking speed on the ipsilateral (left) side. This second turn corresponds to the correcting turn in freely walking woodlice when they leave an obstacle. The experiment (B. Schoenemann) shows that antennal input alone determines initial turn and correcting turn

Wind Orientation
in Walking Carrion Beetles (*Necrophorus humator* F.)
Under Closed and Open Loop Conditions

H.G. HEINZEL and H. BÖHM[1]

1 Introduction

To analyze the feedback control system underlying the wind orientation of walking insects (Linsenmair 1973), it is necessary to measure the angular velocity of the insects's turning as a function of wind direction (characteristic curve). These measurements were performed by means of two different experimental setups under closed and open loop conditions.

2 Walk on a Locomotion Compensator Under Closed Loop Conditions

The beetles are free to walk on the apex of the ball of a locomotion-compensator (Kramer sphere, which continuously compensates the translatory movements, but not the walking direction of the beetle; Kramer 1976).

In high-velocity air currents the beetles run nearly straight courses relative to the wind (Fig. 1, 1.5 m/s). The frequency distribution of the walking directions shows a distinct peak (Fig. 1). Correspondingly, the characteristic curve of the system given by the angular velocity ω as a function of walking direction a, is obtained only within a small range (Fig. 1, right middle diagram). The curve shows that in this case the beetle regulates its course to the direction $a = 0°$ (stable direction). Any deviation from the stable direction results in a turn into the opposite direction, thus stabilizing the walking direction at $a = 0°$.

As in the gravity orientation analysis of the grain weevil, *Calandra granaria* (Wendler 1975), it is possible to evaluate the whole characteristic curve. Without air current, beetles often run unorientated in circles (Fig. 1, 0 m/s). When the beetle is exposed to a weak air current, the circles become cycloids (Fig. 1, 0.1 m/s). The peak in the distribution of a shows that the walk is now wind-orientated. But during its cycloidal walk the beetle runs successively in each direction, and thus produces the characteristic curve for the total range, resulting in a sinusoidal curve, which is shifted by a superimposed constant ω value. At an air current velocity of 1.0 m/s (Fig. 1), this animal shows the transition from cycloidal track to straight track and displays a similar

---

[1] Zoologisches Institut der Universität zu Köln, Lehrstuhl Tierphysiologie, Weyertal 119, 5000 Köln 41, FRG

Localization and Orientation in Biology and Engineering
ed. by Varjú/Schnitzler
© Springer Verlag Berlin Heidelberg 1984

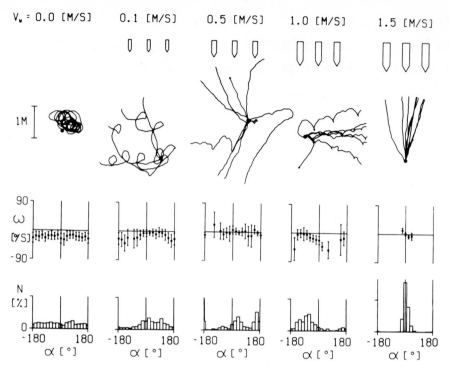

Fig. 1. Walking tracks (*top row*), angular velocity ω as a function of walking direction a (*middle row*), and frequency distribution of a (*lower row*) of one beetle walking at air currents with different velocity V_w. Positive values mean deviation to the left with respect to wind source (a) or rotation to the left (ω). *Vertical lines* three times standard error of mean values. *Black spot* starting point. When the animal moves more than 2 m from this point, the pen returns to it and starts again

sinusoidal curve. At 0.5 m/s velocity the beetle displays a particular feature of the walk. The beetles can change their preferred direction, as demonstrated by the a-distribution with three peaks. Therefore, we obtain no unimodal characteristic curve.

3 Walk on a Paired Tread Wheel Under Open Loop Conditions

The beetles, put on a double joint balance, walk on a paired tread wheel (Wendler et al., Chap. II.12, this Volume), which allows the calculation of the walking speed V and the angular velocity ω_c, i.e. the turning response of the animal under open loop conditions.

Changing the wind direction several times from $a_w = 15°$ (wind from the right side) to $a_w = -15°$ results in clearly alternating walking reactions of the beetles, as demonstrated by the change of the calculated angular velocity ω_c and of a_c, which is the integral of ω_c (Fig. 2, left).

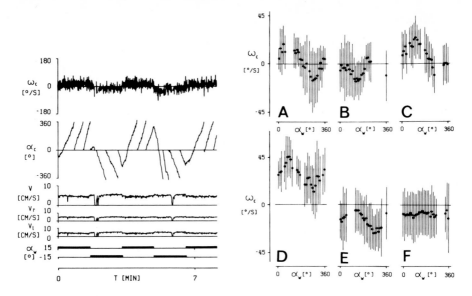

Fig. 2. *Left* parameters of a walk with several changes of wind direction a_w. V_l, V_r speed of left, right tread wheel, V mean speed, ω_c calculated angular velocity of the beetle, a_c integral of ω_c. *Right* A–E Characteristic curves of 5 different animals: calculated mean angular velocity ω_c, and standard deviation as function of wind direction a_w. F Control experiment

The characteristic curves obtained in experiments, where the wind direction is changed, are sine functions shifted by a super-imposed constant ω_c value (Fig. 2, right, A–E). The zero crossing with negative slope (stable direction) is always at a_w values which correspond to a downwind run ($220°$, Fig. 2, right, A). But in many cases the curves do not show a zero crossing (Fig. 2, right, D, E), because there is a high constant ω_c value which can be measured separately by turning the tread wheel together with the animal without air current stimulation (Fig. 2, right, F).

Summing up the results, the experiments lead to similar sinusoidal characteristic curves under closed and open loop conditions. These curves, however, are different in one important aspect, namely with regard to the stable walking direction. Free-walking beetles can maintain stable tracks in all directions relative to the wind, whereas they walk only downwind under open loop conditions. This is similar to the behaviour of *Drosophila* during tethered walk on a ball. The animals show positive reactions to odours which in free walk are strongly repellent (Borst and Heisenberg 1982).

Therefore, the question arises whether the opening of the feedback control system changes the time structure of the external stimulus. Fixing the beetle above the paired tread wheel not only opens the control system with respect to the regulation of the wind direction, but also eliminates the rhythmicity of the beetles body propulsion which may be an important input for the antennae during wind measurement.

Acknowledgements. Supported by Deutsche Forschungsgemeinschaft, He 1118, We 719.

References

Borst A, Heisenberg M (1982) Osmotropotaxis in Drosophila melanogaster. J Comp Physiol 147: 479–484

Kramer E (1976) The orientation of walking honey bees in odour fields with small concentration gradients. Physiol Entomol 1:27–37

Linsenmair KE (1973) Die Windorientierung laufender Insekten. Fortsch Zool 21:59–79

Wendler G (1975) Physiology and systems analysis of gravity orientation in two insect species (Carausius morosus, Calandra granaria). Fortsch Zool 23:33–48

Correcting Behaviour –
a Mechanism to Support Course Stability?

B. SCHOENEMANN[1]

If an insect, millepede or isopod encounters a high obstacle, it turns and runs along the margin. At the end of the obstacle it immediately performs an opposite turn and resumes approximately its previous walking direction. This "correcting behaviour" is very common in animals. It could enable the animal to maintain a nearly straight course although barriers are in its way, even when external orientational cues are missing (Mittelstaedt 1977).

This hypothesis was tested in free-running isopods *(Oniscus asellus)* on a locomotion compensator (Kramer 1976), which automatically measures and compensates all translatory movements. An obstacle of 4 cm length was placed with various angles (Fig. 1b) into the isopod's way once every 60–90 s (3 animals, 10 runs of 45 min duration each).

The reconstruction of the walking paths reveals that the woodlice do not maintain a straight course even when no obstacle is present (see sections between the obstacles in Fig. 1a). Like other unorientated arthropods, these isopods change their walking directions with angular velocities up to $10°/s$. Therefore, the basic approximate straight courses that were to be maintained by the correcting behaviour do not exist in woodlice. This does not mean that they do not show "correcting behaviour". At first glance, however, there seems to be only a very weak correlation between the compensatory turn and the initial turn (see Fig. 2a). Much higher correlations ($r = 0.77$ as against $r = 0.58$) are achieved by separating the data into classes of mean angular velocities which the animals showed before reaching the obstacles. It turns out that the effect of the momentary angular velocity is only to shift the regression lines (Fig. 2b). Since all regression lines are in parallel, the "correcting behaviour" is independent of the momentary angular velocity.

Which sense organs do control the initial turn and the counter-turn besides the proposed intersegmental proprioceptors (Burger 1971) in woodlice (Schäfer 1982)? In order to apply precise stimuli, tethered woodlice were investigated. They walked on top of a paired tread wheel (Wendler et al., Chap. II.12, this Volume) in the dark. The legs of each bodyside drive one wheel. The movements of the wheels are measured with a resolution of $0.5° \triangleq 0.03$ cm. "Obstacles" which could be touched by one antenna were presented at varying angles and durations from one side.

During the antennal contact the contralateral legs slow down (Fig. 3a). When the "obstacle" is removed they resume their previous speed and the ipsilateral legs slow

[1] Zoologisches Institut der Universität zu Köln, Lehrstuhl: Tierphysiologie, Weyertal 119, 5000 Köln 41, FRG

Localization and Orientation in Biology and Engineering
ed. by Varjú/Schnitzler
© Springer Verlag Berlin Heidelberg 1984

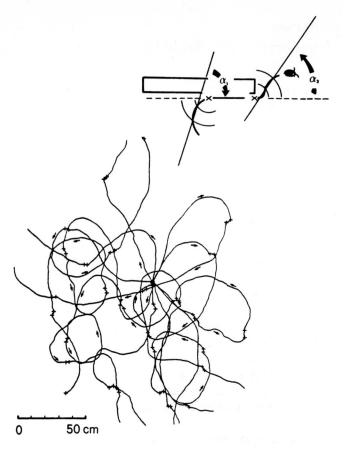

Fig. 1a,b. Track of a woodlouse, running on the locomotion compensator under red light with covered eyes. ● starting point. The pen is reset to this point when the distance exceeds 1 m.
Two crosses mark the points where the animal first touches and then leaves the obstacle.
b The angles are defined as the deviations from the previous walking directions (*left* +). The angles of approach (a_1) and of leaving the obstacle (a_2) are determined by drawing two circles (radius 1.5 cm, 3 cm) around the respective cross and connecting the intersections with the path of the woodlouse by a straight line

down (Fig. 3a). From the wheel's difference in rotating speed an "angular velocity" of the isopod resulting from this was calculated. Often, as in Fig. 3b, the areas under the curve during and after the contact are almost identical with opposite signs. This shows that in woodlice initial turning purely caused by antennal stimuli can be compensated for by counterturning to a great extent.

Acknowledgements. Supported by the Studienstiftung des deutschen Volkes and the Deutsche Forschungsgemeinschaft (We 719).

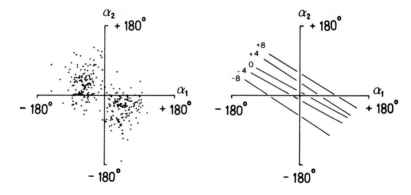

Fig. 2a,b. Dependence of the angle of leaving the obstacle (a_2) on the angles of approach (a_1) n = 349, 1 animal.
b The regression lines in classes of angular velocities ω which the animals showed before reaching the obstacle. Classwidth: $4°/s$. *Numbers* indicate the middle of the class $[°/s]$. Values were only used when ω_1 and ω_2 differ less than $2°/s$

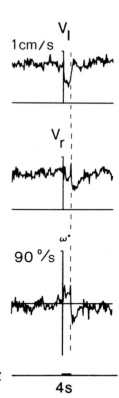

Fig. 3a,b. Velocities of the *left* (v_l) and the *right* (v_r) legs (tread wheel). **b** "Angular velocity" of the animal.
The *traces* show an average of 20 experiments. The 4 s contact to the "obstacle", is marked by a *black bar* and the *vertical lines*. The "obstacle" stands at $30°$ oblique to the longitudinal axis of the animal

References

Burger M-L (1971) Zum Mechanismus der Gegenwendung nach mechanisch aufgezwungener Richtungsänderung bei Schizophyllum sabulosum. Z vgl Physiol 71:219–254

Kramer E (1976) The orientation of walking honeybees in odour fields with small concentration gradients. Physiol Entomol 1:27–37

Mittelstaedt H (1977) Kybernetische Analyse von Orientierungsleistungen. In: Kybernetik, Oldenbourg, München Wien, pp 144–195

Schäfer MW (1982) Ein idiothetischer Mechanismus im Gegendrehverhalten der Assel Oniscus asellus L. Zool Jb Physiol 86:193–208

Acoustical Pattern Recognition in Crickets

TH. WEBER[1]

1 Introduction

Crickets use their songs for intraspecific communication in different behavioral contexts. The European field cricket, *Gryllus campestris* L. has three types of songs. The rivalry song and the courtship song are given after antennal contact with conspecifics; stimulation from sensory inputs (e.g. mechanical, optical) may modulate these songs. Compared to these, the calling song is a long distance communication signal that serves to attract females which are ready to copulate. The calling-song chirp contains four syllables with 5 kHz carrier frequency repeated about every 30 ms (at 20 °C) and the chirp period is 350 ms. The orientated walk to the male by the attracted females is called phonotaxis. This innate behavior is the basis for experiments to clarify which parameters females use to identify the conspecific calling song. The behavioral experiments are also helpful in designing paradigms for electrophysiological studies of pattern recognition in the auditory pathway.

2 Experimental Design

Adult female crickets were placed on top of a sphere (diameter 50 cm) centered in an anechoic chamber, which compensated for the movement of the animal in any direction (designed by E. Kramer, Seewiesen). Two loudspeakers were placed 2 m from the sphere and were separated by an angle of 135°. Acoustical stimuli were presented either sequentially or simultaneously from the two loudspeakers. The response of the female was recorded in terms of velocity and walking direction with respect to the speakers. All experiments were done at room temperature (ca. 20 °C) in the dark (for details see Weber et al. 1981).

[1]   Max-Planck-Institut für Verhaltensphysiologie, Abt. Huber, 8131 Seewiesen, FRG

Localization and Orientation in Biology and Engineering
ed. by Varjú/Schnitzler
© Springer Verlag Berlin Heidelberg 1984

3 Acoustical Signal Properties Essential for Recognition and Localization of Conspecific Song

To determine the necessary and sufficient properties of the calling song for phonotaxis, 2-min paradigms were used, switching between speakers at one minute intervals. Criteria for "attractiveness" of the signal are the duration of characteristic meandering about the required loudspeaker position in each 2-min test and that the speaker switch was tracked correctly. Phonotactically tracking females typically walked in short bouts that started with correcting turns (cf. Weber et al. 1981).

Females respond neither to a continuous tone of 5 kHz carrier frequency, nor to a tone of chirp duration and rhythm, therefore separation of the chirp into syllables is required. The duration of the syllables is not important as long as they are separated by a pause of at least 4–5 ms. Fine details of the males calling song, such as increasing intensity from syllable to syllable and short frequency sweeps at the beginning and the end of the syllable, also play a minor role: the females track taped songs played forwards and backwards with the same precision.

Male *Gryllus campestris* normally call with four syllables per chirp; and rarely with five syllables per chirp. The syllable number per chirp had been thought to be a distinct parameter for recognition (Popov and Shuvalov 1977). However, when one-syllable chirps were played back, females failed to track on the Kramer treadmill. When the chirp consisted of two syllables, the females started tracking but inconsistently; clear tracking was evident when the chirp contained three or more syllables. Under our conditions females also track trill, in which there was no pause between chirps. But some females prefer tracking chirps over trill (Thorson et al. 1982).

The influence of the carrier frequency on phonotaxis was tested by using an attractive chirp pattern. With increasing carrier frequency above 7–8 kHz and above 80 dB intensities, females track with a systematically increasing erroneous angle with respect to the sound source. At 15 kHz the angle goes beyond 90° and therefore leads the female into a direction where she never reaches the sound source. The erroneous angle is shown on both sides of the speaker position and may change from one side to the other during one test program. The typical meandering is maintained in this "anomalous phonotaxis" (Thorson et al. 1982).

These results suggest that in *G. campestris* the syllable interval is the most important acoustic property releasing phonotaxis. As long as the syllable interval is in the range 22–55 ms the females start tracking. Outside this range they may move in the direction of the speaker, but they do not show the typical walking mode and usually change direction or stop. Even females with only one ear show phonotactic response, but walk in cycloids that tend more or less to be at an angle to the speaker position (cf. Zaretsky 1972; Wendler et al. 1980). Therefore, recognition of the conspecific song, expressed in release of phonotaxis, seems to depend on the ca. 30-Hz modulation frequency. In addition to an attractive pattern, both ears are necessary and a carrier frequency close to 5 kHz is required for successful localization.

Pollack (1982) has shown that the same simplicity (e.g. using only a certain range of syllable intervals) for recognition may be true in crickets that produce more complex songs. He has shown in choice experiments with tethered flying *Teleo-*

gryllus oceanicus that females prefer the chirp part of the natural male calling song, whereas males, not as reliably as the females, prefer the trill part. An exception to using only one modulation rate for recognition has been reported by Zaretsky (1972), who used the arena for phonotactic experiments with the African cricket *Scapsipedus marginatus*. The females of this species did not respond with phonotaxis unless both of the two intervals of the conspecific calling song were present.

4 Interaction of Recognition and Localization

As already mentioned, female crickets with only one intact ear show by phonotactic tracking that they can recognize an attractive signal. The question is: Is it possible to discriminate between recognition of the song and localization of the sound source experimentally? How do animals with both ears intact respond to a pattern whose syllable interval is attractive but whose syllables are played back alternating from the two loudspeakers? In this case the attractive pattern results only when both sound sources are active. How do the female crickets respond to two attractive signals, with one on each speaker, whose syllables are interleaved? Do they discriminate between the two attractive patterns or fuse them, resulting in an unattractive pattern?

In both situations one has to pay attention to the fact that in choice experiments on the Kramer treadmill the animal is held in a fixed position of the acoustic field. Even if the intensities of the attractive and less attractive songs are equally adjusted, many females prefer one of the speakers independent of the pattern presented. To resolve preference for one of the patterns the intensity of the speaker not tracked has to be increased, until tracking is consistent with respect to the pattern, and independent of the speaker switches. This compensation was different in intensity (range 1-4.5 dB) for each female and constant in daily tests during 2 weeks. Probably the asymmetry was caused by the loss of hearing sensitivity of one ear.

Each female was tested initially for the range of syllable intervals that released phonotaxis and the females which showed asymmetry in choice experiments were compensated. An attractive song was then emitted by the two speakers (135° separated) in such a way that the syllables alternated between them. This resulted in presenting half of the syllables with double intervals on each speaker, for example, to create a pattern of four syllables repeated at 40 ms interval, two syllables with 80 ms interval were played back from each speaker. Thus the signal of each speaker alone was not attractive due to the longer syllable interval. The composed signal of both loudspeakers released phonotactic tracking of the females. They meandered about in the direction centered between speakers as if they tracked only one attractive signal coming from that direction. They rarely came towards the direction of one speaker, but walked then in a mode typical for an unattractive signal and returned soon to the center between speakers.

Because phonotactic tracking is released, the females recognize the pattern as attractive, although it is separated in space. Considering the directional characteristics of *G. campestris* ears, the first syllable of each composed chirp would stimulate for example the left ear more than the right one during the walk centered between speakers. As the next syllable would arrive from the speaker on the other side of the female at the same angle, the right ear would be stimulated more than the left one. Therefore, the stimulation of both ears would be of equal amount but side-inverted. The information about each syllable and its direction is represented separately in the acoustic sensory pathway. Therefore, both sound sources could possibly be localized, but the tracking centered between the speakers demonstrates that for recognition of the attractive pattern to take place, not the direction of every syllable is taken into account. The information about the different directions of the syllables seems to be integrated over the duration of the whole chirp, indicating that the identified pattern provides the target in space to which the walk is directed.

How do females respond to two attractive signals intermingled so that an unattractive signal results? Attractive chirps containing four syllables repeated at 35-ms intervals were played back from each of the two speakers. The chirps from different speakers were delayed in such a way that the syllables of one chirp were interleaved with the syllables of the other chirp. This resulted in an unattractive pattern with eight syllables repeated at 17.5 ms, if one does not consider the different directions. In this situation, phonotaxis was released, but the females walk was not centered between the speakers, but towards one speaker for about 10–20 s and after changing direction, for about the same time to the other speaker. This resulted in a zig-zag course between the two speaker directions and if the intensity of one speaker was increased, the females only tracked the speaker with the louder signal.

The females apparently recognize that two attractive patterns are present and that they come from different directions. As in the previous experiment, the tracking is directed so that they try to get symmetrical input from the attractive pattern via both ears. This is achieved when the female walks directly to one of the attractive songs, but the frequent changes from one speaker position to the other, demonstrate that the other signal is recognized as well. The signal from the other speaker, which is $135°$ away, certainly causes asymmetrical inputs to the ears, which are in total, probably less than the inputs from the signal directly ahead. When the female is confronted with two attractive signals, and one of higher intensity, she decides to track the more intensive one. Therefore, she seems to be able to relate the signal presented at $135°$ to the one directly ahead, in regard to the intensity.

These results indicate that recognition of the song and localization of the sound source operate with different parameters of the song. Recognition is based on temporal information of the syllables (ca. 30 Hz modulation rate) and the integrated directional information of both ears over attractive chirps; whereas the localization in terms of the acoustic sensory information available uses the carrier frequency and the differential input of both ears.

Acknowledgements. I thank Profs. F. Huber and A.V. Popov for helpful discussions and Drs. H.U. Kleindienst, J. Doherty, and Ulla Klein for criticism of the manuscript, in particular J. Doherty for help with the translation.

References

Pollack GS (1982) Sexual differences in cricket calling song recognition. J Comp Physiol 146: 217–221

Popov AV, Shuvalov VF (1977) Phonotactic behavior of crickets. J Comp Physiol 119:111–126

Thorson J, Weber T, Huber F (1982) Auditory behavior of the cricket. II. Simplicity of calling-song recognition in *Gryllus*, and anomalous phonotaxis at abnormal frequencies. J Comp Physiol 146:361–378

Weber, T, Thorson J, Huber F (1981) Auditory behavior of the cricket. I. Dynamics of compensated walking and discrimination paradigms on the Kramer treadmill. J Comp Physiol 141: 215–232

Wendler G, Dambach M, Schmitz B, Scharstein H (1980) Analysis of the acoustic orientation behaviour in crickets. Naturwissenschaften 67:99–100

Zaretsky MD (1972) Specificity of the calling song and short term changes in the phonotactic response by female crickets *Scapsipedus marginatus* (Gryllidae). J Comp Physiol 79:153–172

Neural Enhancement of Directionality
in Sensory Pathways of Two Arthropods, the Cricket
and the Crayfish

K. WIESE[1]

Signal source localization, not only in invertebrate animals, is supported at the level of the ascending sensory pathways by a neural circuit producing inhibition between the subchannels specific to each body side. This serves to increase activity contrast between subchannels opposed and adjacent to the signal source, and thus directs the next movements of the signal receiver.

In the auditory system of the cricket *Gryllus bimaculatus* the respective bilateral inhibitor neurons are the two giant, complementary omega neurons in the prothoracic ganglion, named in allusion of the axon arrangement (Fig. 1A). Input to an omega cell are the ramifications at the side of the cell soma (Wohlers and Huber 1978). The output of the cell is characterized by the directionality measurement shown in Fig. 1B which was obtained by simultaneously recording from both omega neurons. If a sound source is moved around the preparation, the activity of both omega cells changes in response to the different stimulus directions.

The two complementary cells respond sharply selective to their individual input side only. Observed death of one cell after penetration shows a considerable loss of specificity of the response to stimuli from the contralateral side (see the dotted curve in Fig. 1B).

This is the level of directionality of the tympanal organ, which is a pressure gradient receiver with a sound-conducting tube leading from the prothoracic spiracle to the inner side of the tympanum (Larsen and Michelsen 1978).

If this sound conducting tube is destroyed or blocked, the directionality of the surviving omega cell (Fig. 1B) becomes circular with respect to the origin of the plot, signaling complete loss of directionality in this auditory interneuron.

Two conclusions can be drawn from this observation: the omega cell is a principal mediator of bilateral inhibition and the two inhibitor neurons inhibit each other reciprocally; the latter has been confirmed by Selverston (pers. comm.) by means of simultaneous penetration of both omega neurons. Thereby the type of inhibitory circuit used in the auditory pathway of the cricket is characterized as a system of recurrent inhibition.

In contrast to this situation a different type of lateral inhibitory circuit is used in a sensory pathway of the crayfish. This pathway monitors fluid displacements around the animal by a large population of hair-type sensilla distributed over the body surface. The directionally of the sensillum is prescribed by the articulation of the

---

[1]   Zoologisches Institut, Abteilung Tierphysiologie der Technischen Universität, Pockelstr. 10a, 3300 Braunschweig, FRG

Localization and Orientation in Biology and Engineering
ed. by Varjú/Schnitzler
© Springer Verlag Berlin Heidelberg 1984

Fig. 1 A,B. Photograph of cobalt-stained, intensified omega neuron in whole-mount prothoracic ganglion of *Gryllus bimaculatus*. Ramifications at the side of the soma are inputs, contralateral branchings are outputs mediating postsynaptic inhibition to complementary cell and other auditory interneurons; conduction velocity is 1.2 m/s, the cell is spiking. **B** Directionalities of the both inhibitory omega neurons in the auditory pathway of *Gryllus bimaculatus* (□, ○) normal situation; (– – ○ – –) directionality of omega neuron after killing the complementary cell (each ganglion contains a pair of omega cells); (—●—) directionality of same cell after additional destruction of receptor directionality by cutting tracheal sound conducting pathways

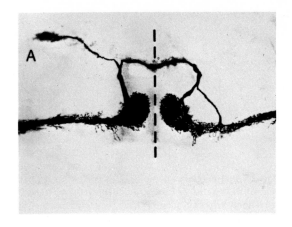

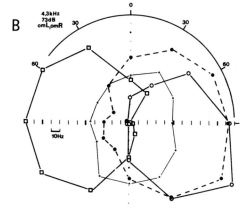

hair shaft, which, in the case of the receptors on the telson, enforces caudo-rostral movements of the shaft. Each direction is monitored selectively by one of the two sensory cells; selectivity of response to movement direction is preserved also at the level of interneurons (Wiese 1976). Only a few cells, belonging to the trigger circuit for escape swimming, respond to both movement directions.

This sensory pathway contains a distinct primary level of interneurons which are not subject to inhibitory influence and contrast enhancement. Evidence for this fact has been brought forward by reversible block of parts of the crayfish's body surface by means of a plastic foil during directionality measurements in interneurons (Wiese and Wollnik 1983). Only in a secondary level of interneurons distinct effects of inhibitory inputs from contralateral and laterally adjacent receptive fields are observed. Figure 2B left shows the directionality of an interneuron not subject to inhibition, still directional due to small receptive field (xxx); the two directionalities shown in Fig. 2B right, belong to interneurons with large receptive fields (xxx); the distinct directional response of one of them is destroyed if receptors on the claws are blocked by a cover of PVC foil.

Reichert et al. (1983) have identified local interneurons (*lateral inhibitory directional selective*) LDS (Fig. 2A) in the sixth abdominal ganglion of *Procambarus*, seeming-

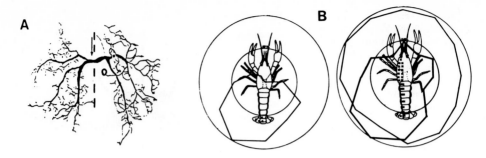

Fig. 2A,B. Functional homologue of cricket omega neurons in the displacement sensitive pathway of the crayfish *Procambarus*. Identified by Reichert et al. (1983), the LDS (*lateral inhibitory directional selective*) neurons have their input at the side of the soma, outputs are contralateral; they are *non-spiking*. **B.** Directionalities of displacement sensitive interneurons recorded from the circumoesophageal connectives of *Procambarus clarkii*; 8 Hz surface waves were used; maximum response observed is set to 100% (*outer circle*); the field of hair-type sensilla on the body surface driving the interneuron is marked by xxx. *Left* distinct directionality due to small receptive field; no inhibitory action was observed to influence this interneuron; *right* two neurons with large receptive fields, one with almost no directionality, the other with distinct directional response due to inhibition powered by receptors on the claw

ly true functional homologues of the omega neurons of crickets however, they provide bilateral inhibition only to a limited number of interneurons in the pathway and do not inhibit each other. They thus represent a system of precurrent inhibition. LDS cells are non-spiking, whereas omega cells regularly produce action potentials.

Figures 3A and 3B describe measurements of inhibitory effectivity (related to coupling factor) in some representative interneurons of the two pathways. The measurements generally involve selective stimulation of the two side specific inputs by using either earphones in the cricket (Kleindienst et al. 1981) or a split-in-two isolated tail fan preparation in the crayfish (Wiese and Schultz 1982). One side of the preparation receives a stimulus of standard strength, and the response of the connected interneurons is monitored. From the contralateral side an inhibitory stimulus is applied, which is increased stepwise in the case of the cricket, and in the case of the crayfish is shifted in phase with respect to the excitatory stimulus, both stimuli consisting of a 3.3 Hz wave motion in the two compartments of the experimental dish.

As a consequence of the inhibition, the responses of the omega neuron and that of a high sensitivity neuron of the anterior connective (LF1) (Rheinländer et al. 1976) gradually decrease. The relation activity decrease produced versus activity increase seen in the inhibitor omega neuron (see Fig. 3A) furnishes an effectivity factor of 0.4 in the omega-omega-interaction and of 0.5 in the omega-LF1 inhibition. In contrast to this procedure, effectivity factors in the crayfish pathway were estimated by the maximum amount of inhibition observed at close to in-phase stimulation of both halves of the tail fan. In this situation 20%–50%, in some special neurons even 80% to 100% of the response to the excitatory stimulus only were removed by inhibition, corresponding roughly to effectivity factors of 0.2–0.5 in many of the interneurons of this displacement sensitive pathway.

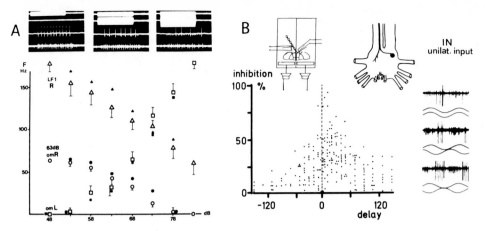

Fig. 3A, B. Measurement of inhibitory effectivity (coupling factor) of two bilateral inhibitory circuits: **A** in the auditory pathway of the cricket; by earphone a standard stimulus is delivered to one ear and the response of the omega neuron (○, ●) and of the interneuron LF 1 (△, ▲) of the anterior connective is recorded (see *inset*). An inhibitory stimulus is delivered to the contralateral ear and has increased stepwise in strength; activity in the inhibitor-omega is seen as small *stipples* in the *insets*, omega-omega inhibition shows a factor of 0.4, omega-LF 1 interaction a factor of 0.5. **B** Effectivity of bilateral inhibition in the displacement sensitive pathway of *Procambarus* (LDS neurons): Dippers produce carefully balanced 3.3 Hz wave motion in the two compartments of the experimental dish (*inset top left*). Motion either in synchrony or with phase lag between compartments; 100% inhibition corresponds to complete suppression of a response to the excitatory stimulus only; delay in ms. Reduction of channel sensitivity in the common mode situation versus full sensitivity in the alternate mode of wave motion (*right*)

Theoretically the principal differences between precurrent and recurrent systems of inhibition are as follows precurrent inhibition shows a common-mode-gain of 1 at cf of 0.0 and of 0 at cf values of 1.0, resulting in high common-mode-rejection; alternate-mode-gain is 2.0 at maximum at cf values of 1.0. In *recurrent* systems the alternate-mode-gain is indefinitely large at cf values of 1.0, in the common-mode only 0.5 at the same cf value, resulting in low common-mode rejection. Recurrent systems, because of the feedback loop involved (reciprocal inhibition between inhibitors), have an inherent tendency to oscillate.

It suits the need to suppress noise in a multisensor array as used in the crayfish, that the crayfish pathway uses precurrent inhibition; on the other hand it suits the acoustic communication of the cricket to keep common-mode rejection low in order not to lose contact to sound sources in front of the listening animal.

The tendency to oscillate as found in the recurrent systems, and close association of the species-specific repetition frequency in the syllable pattern of the song in calling song recognition by listening females (Thorson et al. 1982), has led to growing interest in the dynamic properties of the omega circuit.

In one of the experiments (Fig. 4) one ear of the cricket receives by earphone a pattern of three sound syllables of large intensity, whereas the other ear receives an identical pattern of low intensity, delayed by exactly half the period in the pattern. This leads to alternation of stimuli at the two ears (23, 30, and 40 ms periods of the

4.5 kHz 60|75 dB

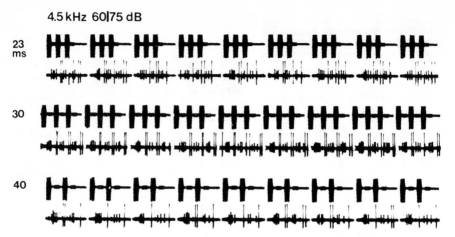

Fig. 4. Resonance-type effect in the response of both omega neurons to alternating sound pulse input to both ears. Each ear receives a pulse pattern of three, on one side the pattern is delayed by half a period. At 30 ms of syllable repetition (pulse) the omega responses appear amplified and generally stabilized; 30 ms corresponds to the period in the conspecific song pattern of *G. bimaculatus*

pattern are used); the responses of the simultaneously recorded omega neurons are amplified and show general stabilization in the 30 ms situation only. The basis of this effect is presumably the post-inhibitory rebound depolarization coinciding in this temporal configuration with the response to the subsequent sound syllable in the stimulus pattern.

There is also a time lag in the circuit visible in omega responses to longer sound pulses and with both omega action potentials occurring side by side. Intervals of about 3–15 ms duration simply do not occur, the reason being that a cell hyperpolarized by its complementary is less likely to build an action potential itself at the peak of hyperpolarization.

Resonance phenomena and time lag are indicators of frequency dependence of the action of the lateral circuit. Accordingly, the response of an omega neuron to an ipsilateral stimulus is maximally reduced by a contralaterally applied sound syllable pattern in cases where sound syllable repetition was adjusted to 25–40 Hz. The inhibition produced is distinctly less, if syllable repetitions outside this range are used (Fig. 5; Wiese 1983).

This frequency band of most effective contrast enhancement coincides with the range of syllable repetition frequencies shown to attract females of *Gryllus bimaculatus* most vigorously (Thorson et al. 1982; see contribution by T. Weber in this Volume). From the phonotaxis point of view sound sources producing syllable rates matching the resonance frequency of the inhibitory circuit – for instance by the conspecific male – leave the listening female with a more precise impression about its position than all other sound sources. The expectation is that the omega neurons, assumed to modulate a large number of the auditory interneurons, form the central oscillator of the receiver system to tune the auditory pathway to the syllable pattern of the conspecific call.

Fig. 5. Frequency dependence of the neuronal directionality enhancement circuit (omega neurons) in the auditory pathway of the cricket; the omega cell is shown to respond to a sound pulse of 3 s duration; the inhibitory stimulus (earphones) consists of a sequence of sound pulses with continuously varying repetition frequency; the strongest inhibitory effect is seen if the lateral inhibitory network is activated at frequencies between 25 and 40 Hz

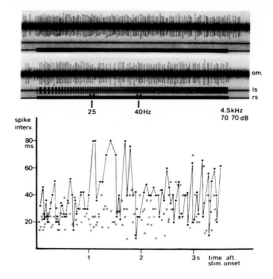

References

Kleindienst HU, Koch U, Wohlers D (1981) Analysis of the cricket auditory system: acoustic stimulation using a closed sound field. J comp Physiol 141:283–296

Larsen ON, Michelsen A (1978) Biophysics of the ensiferan ear III. The cricket ear as a four input system. J comp Physiol 123:217–227

Reichert H, Plummer MR, Wine JJ (to be published 1983) Identified nonspiking local interneurons mediate nonrecurrent lateral inhibition of crayfish mechanosensory interneurons. J comp Phys

Rheinlaender J, Kalmring K, Popov AV, Rehbein HG (1976) Brain projections and information processing of biologically significant sounds by two large ventral cord neurons of *Gryllus bimaculatus*. J comp Physiol 110:251–269

Thorson J, Weber T, Huber F (1982) Auditory behavior of the cricket II. Simplicity of calling song recognition in *Gryllus* and anomalous phonotaxis at abnormal carrier frequencies. J comp Physiol 146:361–378

Wiese K (1976) Mechanoreceptors for near field water displacements in crayfish. J Neurophysiol 39:816–833

Wiese K, Schultz R (1982) Intrasegmental inhibition of the displacement sensitive pathway in the crayfish *Procambarus*. J comp Physiol 147:447–454

Wiese K, Wollnik F (to be published 1983) Directionality of displacement sensitive interneurons in the ventral cord of *Procambarus*. Zool Jb Physiol 87

Wiese K (in preparation 1983) Dynamic properties of bilateral inhibitory network form matched filter in the auditory pathway of the cricket. J comp Physiol

Wohlers DW, Huber F (1978) Intracellular recording and staining of cricket auditory interneurons *(Gryllus camp., Gryllus bimac.)*. J comp Physiol 127:11–28

Wohlers DW, Huber F (1982) Processing of sound signals by six types of neurons in the prothoracic ganglion of the cricket *G. campestris*. J comp Physiol 146:161–173

Homing by Systematic Search

G. HOFFMANN[1]

1 Introduction

The navigational achievements of birds, during migration between wintering areas and breeding grounds, are particularly astonishing because of the distances they cover. But when we consider the orientation mechanisms underlying homing, we find that invertebrates are also capable of remarkable feats. The problem an arthropod has in returning to a certain point is different predominantly in its (absolute) spatial dimensions, and at least some invertebrates have found solutions comparable to those used by birds. To cite but one example, compensation for the sun's movement in determining the direction of return flight was first found in bees (v. Frisch 1948) and shortly thereafter, in birds (Kramer 1950).

In addition to the best known homing methods – navigation and compass orientation – Griffin (as early as 1944) discussed the possibility that simple forms of systematic search might help to explain the homing abilities of birds. But the first quantitative demonstration that an animal is actually capable of such a complex form of spatial orientation, employed an invertebrate, the desert woodlouse *Hemilepistus reaumuri* (Hoffmann 1978). Since then, similar search forms have been described in desert ants of the genus *Cataglyphis* (Wehner and Srinivasan 1981).

2 Results

The isopod *H. reaumuri* inhabits desert regions of North Africa and the Near East. It shelters from the extreme heat and dryness of its habitat in a burrow, ca. 40–90 cm deep, which is occupied in each case by a single family (Linsenmair 1972). If it is to survive, a desert woodlouse must return to its burrow after each excursion. The entrance to the burrow is a small opening, ca. 12 mm in diameter.

To find the food it needs, a woodlouse must often venture a considerable distance away from its burrow. In a foraging excursion it covers 2.6 m, on average, changing direction continually. Having found food, it returns to the burrow by the shortest route, covering an average of ca. 1.1 m. (These averages were calculated from the

---

[1] Zoologisches Institut III der Universität, Röntgenring 10, 8700 Würzburg, FRG

Localization and Orientation in Biology and Engineering
ed. by Varjú/Schnitzler
© Springer Verlag Berlin Heidelberg 1984

Fig. 1. Cumulative frequency distribution F $(r \leqslant r_0)$, giving the frequency (in %) that the "homing-run" error r is less than or equal to the abscissa value r_0. The error r is the distance from the burrow at which a desert woodlouse, returning from a foraging excursion, begins a search pattern. This point can be identified by a distinct change in walking direction. Results of field observations described in more detail in the legend of Fig. 2

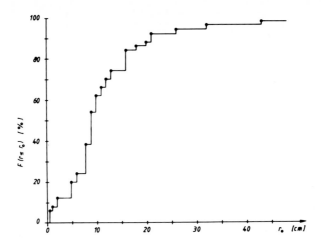

same data as Fig. 1 and Fig. 2.) The longest foraging path of a desert woodlouse yet observed amounted to more than 20 m. At the end of it, the animal returned directly to the burrow from a distance of 6 m. On what mechanism of orientation is this remarkable performance based?

To study this question, one can observe a foraging woodlouse and, at the moment it is about to return home, displace it by several meters without rotating it. It then runs straight to the position where its burrow would be if its entire surroundings had shifted along with the animal, and begins to search for the entrance in this area (Fig. 4a). Obviously, the return to the burrow is directed neither by landmarks nor by a chemically labelled trail. Rather, the animal has stored all the necessary information about the current position of the burrow relative to its own position. It can obtain this information by calculations involving its own active changes of location.

On the return journey, up to the moment it begins searching for the entrance, *H. reaumuri* departs from the true homing direction by only $3.7°$ (average of the *absolute* angle deviation; circular standard deviation $0.07°$; sample size 50). And not only is the direction of the burrow accurately calculated, but its position as well. The distance between the point at which it begins its search and the entrance to the burrow averages 11.4 cm.

But sometimes the returning woodlouse runs to an entirely wrong site. If *H. reaumuri* were able to use visual landmarks to identify the vicinity of its burrow, it could easily compensate for such errors, which exceed 30 cm only in 4% of all cases (Fig. 1). But (as with other terrestrial isopods) the simple structure of its compound eye, which consists of only 29 ommatidia, makes an adequate resolution of the optical surroundings impossible.

H. reaumuri must explore the ground with its antennae until at least one of them has reached the burrow entrance. Not until then can it identify the entrance, by means of contact chemoreceptors (Linsenmair 1972). The maximal distance from which a woodlouse can identify the burrow entrance is therefore only 20 mm (Hoffmann 1983a). The distance from which a foraging woodlouse must find its way back to the burrow is up to 300 times as great. This ratio places enormous demands on the

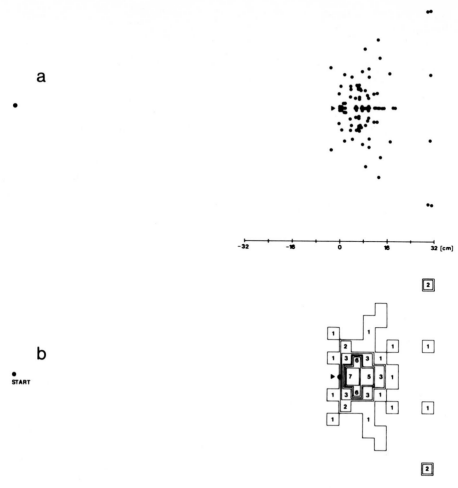

Fig. 2a,b. a Spatial distribution of the burrow-entrance sites in relation to the point (marked by ►) at which a woodlouse returning from a foraging excursion begins to search for it. Results of field observations in which the burrow surroundings were unaltered. Left is the starting point of the return trip (START). The position of the burrow entrance was measured during the observation of 50 isopods foraging in a habitat with a medium density of plants. Each point measured in this way is shown both to the left and to the right of the line connecting the starting point of the return trip and the starting point of the search, because its exact position could not be determined on the basis of the recorded data. b The associated map with contour lines. The spatial density given to a unit area of 16 cm²

homing ability of the desert woodlouse, despite the small absolute length of the return trip.

Without landmarks, how can *H. reaumuri* proceed in finding its burrow when it has reached the point toward which its return journey has been directed, but the burrow is not there? On the assumption (Mittelstaedt 1978), which has been generally accepted, that at this point its data store for the return trip contains the value zero and that it has no further information about the position of the burrow, all it could do would be to wander about at random like a particle in Brownian motion. There are optimal forms of such a Brownian search, which are quite successful (Hoffmann 1983c). But *H. reaumuri* has at most 20 h in which to carry out its search, for by this time at the latest it will have died of water loss. If it erred by only 5 cm in its estimate of burrow position, there would be a 17% probability that such a form of search would mean its death (Hoffmann 1983c).

Therefore *H. reaumuri* must use another searching procedure. Where does it obtain the information about the position of the burrow that it needs in order to employ more effective methods than a Brownian search? The answer is surprisingly simple. In its search *H. reaumuri* takes account of the probability of certain errors of orientation.

Figure 2a shows the spatial distribution of the points at which the burrow is actually located, relative to the point at which the returning woodlouse begins to search for it. The spatial density of these points represents the frequency with which the burrow is located in particular regions. It can best be expressed diagrammatically in the form of contour lines on a map (Fig. 2b). The height of any point on the "mountain" described by the contour lines, the so-called target distribution density, gives the probability density that the burrow is located at that point.

In the case of a slightly different search problem it has been possible to demonstrate quantitatively that *H. reaumuri* uses a systematic search procedure based on the information contained in such a map. If woodlice are displaced directly from their burrows over a certain distance in any arbitrary direction, then at a given later time just as many animals have found their way back to the burrow as would be expected if they had been searching optimally by making use of the map (Hoffmann 1983b).

A woodlouse displaced from its burrow, first concentrates its search within a circular region directly about from the starting point; according to the map, the burrow is most likely to be in this region, because the animal has not actively removed itself from the burrow. If it does not find the burrow in this region, it expands the circle and again searches uniformly through the entire region (Hoffmann 1983b). This apparent overexertion is advantageous because there is a 12% probability that the woodlouse will "overlook" the burrow entrance even if it has been reached by the antennae (Hoffmann, in preparation).

Such a form of systematic search could in principle be carried out by means of a relatively simple and inflexible behavioral program. The information about the spatial distribution of errors in returning to the burrow represented in the map could be rigidly integrated into this behavioral program. Then a neuronal calculation of the information about the animal's location, accumulated during the outward journey, could be restricted to finding one point, the most likely position of the burrow. If the burrow is not found at this point, the subsequent search behavior occurs in the preordained manner.

But the following observations cannot be explained in such a simple way. If a woodlouse is running along a branch on its way home, and the branch is turned into a new position by a gust of wind, the animal turns in the opposite direction and continues to run straight toward the burrow. The angle through which it was passively turned is compensated for by an exactly equal and opposite turn (Fig. 3).

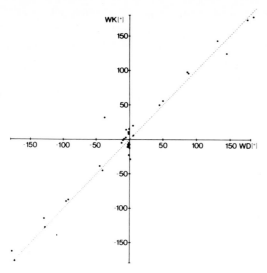

Fig. 3. Compensatory turning. When a woodlouse is passively turned over the angle WD it actively turns in the opposite direction over the angle WK. As they returned from foraging, the animals, walking on a piece of cardboard, were first displaced by ca. 10 m without turning and then rotated over the angle WD

Under comparable conditions, the funnel spider *Agelena labyrinthica* deviates distinctly from the homing direction (Dornfeldt 1975, Görner 1966). The course adopted by the spider is interpreted as a compromise between the directions in which the spider would be sent by two different orientation mechanisms. The allothetic orientation mechanisms, in which direction is determined on the basis of external stimuli, after passive turning, tell the spider to go in a direction different from that indicated by the idiothetic mechanisms, which take account only of the active turning of the animal.

The fact that *H. reaumuri* runs straight back to its burrow even after it has been turned passively, can be reconciled with theoretical ideas about homing mechanisms as developed by Mittelstaedt (1978) on the basis of the experiments on *Agelena*. But to do so one must assume that *H. reaumuri* after passive turning uses only external stimuli to determine its direction. In this case the idiothetic information about the return direction would be lost.

This entirely plausible hypothesis is ruled out by a behavior pattern of *H. reaumuri* that has not previously been demonstrated in any other animal. If, after rotation and compensation, the desert woodlouse does not find its burrow at the point to which it has initially returned, it searches increasingly in the direction in which the burrow would lie if the surroundings, including the burrow, had been rotated along with the passive turning of the animal (Fig. 4b). This behavior can be observed only if the animal has covered a distance far greater than the normal return distance, so that the data stored for the return should be used up. The asymmetry of the search is not very conspicuous, but it is statistically significant (Hoffmann, in prep.).

Therefore, the information contained in the idiothetic orientation mechanisms is not completely lost when compensatory turning follows passive turning, it is merely given less weight than the allothetic information up to a certain time. Its influence on the animal's orientation behavior is not discernible until a period of unsuccessful searching for the entrance has made the probability that the burrow is in the region

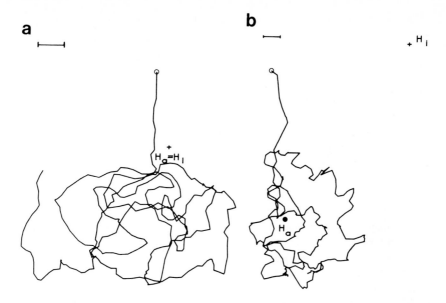

Fig. 4 a,b. Comparison of the search behavior of *H. reaumuri* following displacement without turning **a** and the behavior following displacement with subsequent turning by 96° **b**. H_i, position of home site as estimated by idiothetic information; H_a, position as estimated by allothetic information. Scale, 10 cm

indicated by the allothetic mechanisms just as low as the probability that the burrow is in a region closer to the idiothetically determined direction.

3 Discussion

In what respects do these observations of the homing behavior of the desert wood-louse *H. reaumuri* change our picture of homing performance? For one thing, a wood-louse on the way back to its burrow still retains information about the burrow's position when it has reached the point at which the return-data store should, according to the customary concept, be empty. Moreover, the calculation of the burrow position involving the information from various sources – as demonstrated here with the example of allothetic and idiothetic information sources – seems to be very much more complicated than was previously thought.

  As a way of representing the many items of information a desert woodlouse uses in finding its burrow, and in particular to describe the complex form in which they are linked to one another, the map concept described here has proved especially suitable. The ability of the desert woodlice to adapt their search behavior flexibly to various situations makes it seem likely, at least, that they alter the information described by the map, which they use in their search, by neuronal processing of information about their successive locations during the excursion.

The rapid and accurate homing of a desert woodlouse displaced experimentally, and the explanation of this by an optimal search procedure also cast new light on the homing achievements of other animal species. Decades ago, Griffin (1944) discussed systematic search procedures as a possible explanation of such achievements. The forms of systematic search he considered were necessarily simple ones, for the mathematical methods applied to optimal search procedures, like those used by the desert woodlice, have only recently been worked out (Koopman 1946, Stone 1975). The inconsistency between the flight routes found for pigeons and the simple systematic search procedures suggested by Griffin, as well as the difficulty of following single birds in flight, probably account for the fact that subsequent research on pigeon navigation, with few exceptions, was focussed on monitoring the birds' direction of departure and explaining it by mechanisms of bi-coordinate navigation (summaries by Papi and Walraff 1982, Schmidt-Koenig 1979).

The map concept for homing ability described here offers new points of departure for an attempt to learn more about the orientation processes operating further along in the return trip, the significance of which has repeatedly been emphasized (see, e.g., Schmidt-Koenig 1979, p. 129). It does not, of course, provide any new idea as to how a pigeon might obtain the information described by the map. In this regard it goes no further than Kramer's (1953) concept of map and compass. But it does make possible quantitative inferences as to how an animal can take optimal advantage of this information, especially when it is not sufficient for a direct approach to the target. This may well be the rule in the special case of displacement.

The complicated search behavior of the desert woodlice after passive turning, and its explanation by a correspondingly complicated map, indicate the special advantages of the modified map concept in describing the utilization and evaluation of various information sources. Because pigeons returning to their loft make use of highly diverse information (Papi and Wallraff 1982, Schmidt-Koenig 1979), their behavior may be easier to understand on this basis.

Young pigeons in particular (Wiltschko and Wiltschko, Chap. IV.7, this Volume), as well as other birds (Mittelstaedt and Mittelstaedt 1982, v. Saint-Paul 1982), in navigating after a displacement, resort to mechanisms comparable to those used by *H. reaumuri* to find the way home after foraging. The implication is that the navigational ability of pigeons might have evolved from effective search behavior that serves to compensate for navigational errors in the homing process.

Acknowledgements. Supported by the DFG (Ho 831/1−1).

References

Dornfeldt K (1975) Eine Elementaranalyse des Wirkungsgefüges des Heimfindevermögens der Trichterspinne Agelena labyrinthica (Cl.). Z Tierpsychol 38:267−293
Frisch K v (1948) Gelöste und ungelöste Rätsel der Bienensprache. Naturwiss 35:12−23, 38−43
Goerner P (1966) Über die Koppelung der optischen und kinaesthetischen Orientierung bei der Trichterspinne *Agelena labyrinthica* C L. Z vergl Physiol 53:253−276
Griffin DR (1944) The sensory basis of bird navigation. Quart Rev Biol 19:15−31

Hoffmann G (1978) Experimentelle und theoretische Analyse eines adaptiven Orientierungsverhaltens: Die „optimale" Suche der Wüstenassel *Hemilepistus reaumuri*, Audouin und Savigny (Crustacea, Isopoda, Oniscoidea) nach ihref Höhle. Ph D thesis, Regensburg

Hoffmann G (1983a) The random elements in the systematic search behavior of the desert isopod *Hemilepistus reaumuri*. Behav Ecol Sociobiol 13:81—92

Hoffmann G (1983b) The search behavior of the desert isopod *Hemilepistus reaumuri* as compared with a systematic search. Behav Ecol Sociobiol 13:93—106

Hoffmann G (1983c) Optimization of brownian search strategies. Biol Cybernetics, in press

Koopman BO (1946) Search and screening. Operations Evaluation Group Rep 56, Washington, DC

Kramer G (1950) Weitere Analyse der Faktoren, welche die Zugaktivität des gekäfigten Vogels orientieren. Naturwiss 37:377—378

Kramer G (1953) Wird die Sonnenhöhe bei der Heimfindeorientierung verwertet? J Ornithol 94:201—219

Linsenmair KE (1972) Die Bedeutung familienspezifischer Abzeichen für den Familienzusammenhalt bei der sozialen Wüstenassel Hemilepistus reaumuri Audouin u. Savigny (Crustacea, Isopoda, Oniscoidea). Z Tierpsych 31:131—162

Mittelstaedt H (1978) Kybernetische Analyse von Orientierungsleistungen. In: Hauske G, Butenand E (eds) Kybernetik 1977. R Oldenbourg Verlag, München Wien, pp 144—195

Mittelstaedt H, Mittelstaedt ML (1982) Homing by path integration. In: Papi F, Wallraff HG (eds) Avian navigation. Springer Verlag, Berlin Heidelberg New York, pp 290—297

Papi F, Wallraff HG (eds) (1982) Avian navigation. Springer Verlag, Berlin Heidelberg New York

Saint-Paul U v (1982) Do geese use path integration for walking home? In: Papi F, Wallraff HG (eds) Avian navigation. Springer Verlag, Berlin Heidelberg New York, pp 298—307

Schmidt-Koenig K (1979) Avian orientation and navigation. Academic Press, London

Stone LD (1975) Theory of optimal search. Academic Press, New York San Francisco London

Wehner R, Srinivasan MV (1981) Searching behavior of desert ants, Genus Cataglyphis (Formicidae, Hymenoptera). J comp Physiol 142:315—338

Chapter III Localization and Identification of Targets by Active Systems

The Superlaminated Torus
in the Electric Fish *Eigenmannia*,
a Multimodal Space Integrator

H. SCHEICH[1]

South-American weakly electric fish are known to use a self-generated electric field in conjunction with electroreceptors in the skin, to localize prey and to communicate. In a way similar to echolocation in other animals electrolocation in these fish relies on an active sensory system which is provided with its own energy source, various feedback mechanisms, and a functional match of components for signal processing (see reviews by Scheich and Bullock 1974, Heiligenberg 1977, Bullock 1982). In the case of the genus *Eigenmannia* the discharge of the electric organ, which is driven by a pacemaker in the brainstem, has a maximum intensity of a few mV and a fundamental frequency of between 200 and 400 Hz. Each fish has a stable private frequency to which electroreceptors are tuned. The geometry of the electric field is that of a distorted dipole with more resolution of field lines in the anterior half of the fish (Fig. 1).

Lines of force leave the tail which carries the electric organ and re-enter the fish in the head and body region where the skin is densely covered with electroreceptors. Objects with a conductivity different from the water change the dipole field locally and thus produce different spatial intensity distributions over the receptor map. To simplify the matter, one may say that an "electric shadow" of plants, stones, prey, and enemies, caught in the dipole field, is projected onto the receptor map. A large body of evidence has been gathered since the pioneering studies of Lissman (1958) that electric fish can distinguish between small and large objects with various conductivities and that they estimate distance and speed of objects. Since they are most active during the night, the small distance over which electrolocation is effective (cm range) is still of considerable behavioral advantage over non-electric species.

A somewhat different situation arises from close encounters of several electric fish. The dipole fields are then superimposed on each other leading to overall distortions of each field. Corresponding with theoretical considerations on field geometry it was shown that electroreceptors respond most strongly to stimulation with dipoles oriented at right angle to the axis of the fish (Scheich and Maler 1976). This corresponds to other fish oriented head-on or tail-on. Effective distance for the perception of other electric fields are in the range of meters away. Several electric signals have been identified which rely on the perception of electric fields from other fish (electrocommunication).

One of the most interesting communicative behaviors which involves spatial orientation is the Jamming Avoidance Response (JAR) (Watanabe and Takeda 1963,

---

[1] Zoological Institute, Technical University Darmstadt, FRG

Localization and Orientation in Biology and Engineering
ed. by Varjú/Schnitzler
© Springer Verlag Berlin Heidelberg 1984

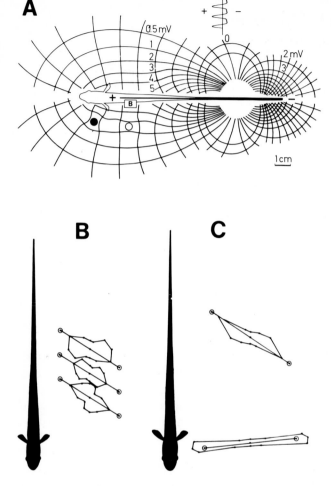

Fig. 1 A–C. A Electric field
of *Eigenmannia* and the
waveshape of the electric
organ discharge. The distortions of the field due to an
object of higher conductivity than water (*filled circle*)
and of lower conductivity
(*open circle*) are also
shown.
B Responses of a torus
P unit to different positions
and orientations of an electric dipole which emitted a
signal with a small ΔF. The
response is plotted in relative polar co-ordinates and
for each position the op-
timal orientation of the di-
polse is shown, which is
the same. C Responses of
two other P units with extreme selectivity for orientation in only those positions along the axis of
Eigenmannia (A from
Scheich 1982, B from
Scheich and Maler 1976)

Bullock 1969, Scheich and Bullock 1974). When two fish with a similar frequency
come within reach of each other their fields are superimposed and each fish will sense
the beating of the two combined frequencies. Consequently, they will shift their
private frequencies away from each other. The resulting larger frequency difference
(larger ΔF with a faster beat frequency) is less detrimental to environmental perception by the receptors, similar to flicker fusion in the retina. The discrimination of
frequency difference also involves distinguishing whether the other fish is lower
($-\Delta F$) or higher in frequency ($+\Delta F$), in order that both fish shift their private frequency in the right direction. The neurophysiological analysis of this behavior has
provided evidence that the key parameters for the discrimination reside in the
amplitude envelopes and the phase shifts of waves in the beat (Scheich and Bullock
1974). As shown also, some analysis of the beat involves convergence of information
from different parts of the receptor map in the skin (Heiligenberg and Bastian 1980).

Two types of so-called tuberous electroreceptors are found in *Eigenmannia* and other weakly electric fish. P units respond to amplitude changes of the field either due to objects or to other electric fish. T units monitor the timing of each electric organ pulse. As a consequence of superposition with another electric field, the resulting phase of the combined field changes with respect to the fish's pulse. Patterns of phase shifts, as perceived and transmitted by T units, depend on ΔF, sign of the ΔF (+ or -), and harmonic content of the two electric frequencies. Different phase shifts are also perceived when objects with capacitative and ohmic impedances are placed in the field which may enable the fish to tell live matter from dead. P and T units appear to have a similar spatial distribution on the skin of *Eigenmannia*. Central P and T unit neurons can be distinguished physiologically and morphologically in the first central nucleus of the electrosensory pathway, the posterior lateral line lobe of the brainstem, and in the second nucleus, the torus semicircularis of the midbrain. Central P units are found in various sub-types which prefer high or low conductivity objects or certain speeds, distances, and shapes. Units in the cerebellum respond to both objects in the water and to the bending of the fish's tail: the latter obviously due to somatosensory input from stretch receptors (Bastian 1976). Since tail bending will change the geometry of the dipole field such units may be involved in a perceptual process which focusses on objects around which the tail is bent.

The torus semicircularis in the midbrain has been identified as the key structure for the control of the JAR. This has been shown by lesion experiments, by electrical brain stimulation which changes the electric organ frequency, and by the finding of feature detecting neurons there (Scheich 1974). The impressive features of the torus in *Eigenmannia* and other electric fish, visible in any type of histological material (Fig. 2), are its enormous size and lamination. Homologous structures in non-electric fish, in amphibians, reptiles, birds, and mammals (inferior colliculus) are much less conspicuous and not obviously laminated (Scheich 1979). These latter toruses are known to receive mainly acoustico-lateralis and some visual and somatosonsory input. Thus, large size and lamination most likely are related to the incorporation of electrosensory neuronal system in the torus. It is noteworthy that high-frequency electric fish like *Eigenmannia*, the electric signal of which are periodic and wave-like, show the largest numbers of distinguishable laminae. This is among other reasons due to the incorporation of central T units in a separate layer. The torus of these electric fish is the brain structure with the largest number of distinguishable cell and fiber laminae known, and is called here *superlaminated torus*. Some reasons for this organization emerge from the anatomical and physiological analysis presented here.

The connectivity pattern of the torus laminae in *Eigenmannia* has been analyzed with a modified HRP (horseradish peroxydase) method (Ebbesson et al. 1981, Scheich and Ebbesson 1981, Scheich and Ebbesson 1983). When HRP distributes in brain structures by micropipette injection, it is absorbed by axon terminals and transported backwards, filling soma and dendrites of the corresponding neurons. It is also absorbed by the somata at the injection site and transported forwards, filling the arborizations of axon terminals. The presence of transported HRP in structures after a certain survival time is verified in histological sections by a dye reaction. HRP injections into the posterior lateral line lobe (LLLP) of the brain stem produce HRP labelling of terminals in layers 5, 6, 7, 9, and 11 in the contralateral torus and also

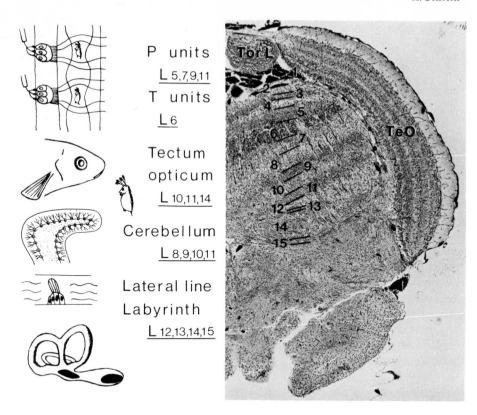

Fig. 2. Lamination of the torus semicirculairs of *Eigenmannia* and representation of sensory modalities in the different layers. From dorsal to ventral P and T unit electroreceptive information, proprioceptive information through the cerebellum, visual information through the tectum opticum and acousticolateralis information all having access to several layers with a certain amount of overlap. The *right side* of a transverse section through the midbrain of *Eigenmannia* is shown in a Bodian fiber stain. *TeO* tectum opticum; *TorL* torus longitudinalis

layer 5 in the ipsilateral torus (Fig. 3). Conversely HRP injections into these layers fill somata and dendrites of electro-receptive neurons in the LLLP. Very small injections substantiate the view that layer 6 of the torus receives input only from second order T units which are morphologically distinct and arranged in one layer of LLLP. Microelectrode recordings have shown that T units are only found in layer 6 and that P unit activity is most reliably recorded in layer 8. This, latter layer mostly contains cells which extend their dendrites into layers 7 and 9, the laminae of massive input from LLLP.

Systematic recordings have shown that the organization of neurons in layers 6 and 8 is in the form of spatial maps of the fish's electroreceptive surface. Congruent spatial maps of P and T units are superimposed vertically. The third order T units in layer 6 have no dendrites but their axons ramify throughout the layer and probably make contacts with other T units. This interaction of T units which receive phase information from various parts of the body surface may be the basis of the temporal com-

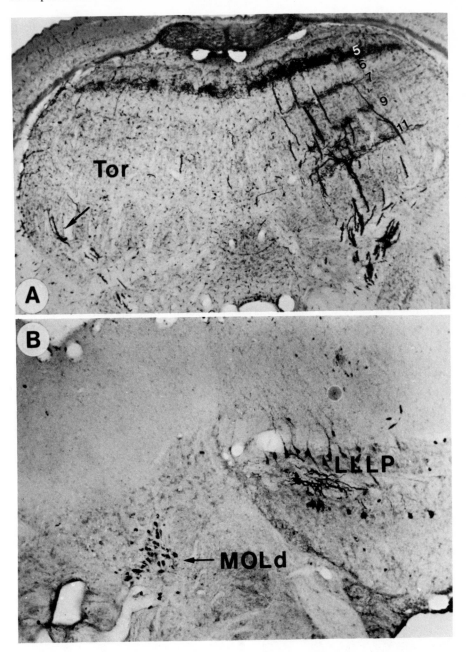

Fig. 3A,B. A shows anterograde HRP labelling of ascending axons and branching terminals in different layers of the contralateral torus after injections in the posterior lateral line lobe (*LLLP*). Arrow points to ascending axons in the ipsilateral torus (*Tor*). **B** shows retrograde labelling of dorsal pyramidal cells (P units) and ventral spherical cells (T units) in LLLP and acousticolateralis cells in *MOLd* after a massive injection which involves all layers of the torus

parator which is used for the control of the JAR, as shown by Bastian and Heiligenberg (1980). P and T units in the torus commonly show preferences for particular orientations and positions of electrical dipoles in the water which mimic other electric fish (Fig. 1). Higher order P units have also been discovered which are excited by a $+\Delta F$ and suppressed by a $-\Delta F$ stimulus: neuronal properties which could be used to control the electric organ pacemaker during the JAR (Scheich 1974). This ΔF decoder neuron obviously receives convergent input from other types of P units which analyze the mirror image asymmetries characteristic of the beat envelopes of + versus $- \Delta F$ situations.

Injections of HRP into deeper layers of the torus and similarly injections into the cerebellum have shown that there is a cerebellar input to torus layers 8, 9, 10, and 11 (Fig. 4). Thus there is a certain overlap of electroreceptive and cerebellar input. The latter probably carries information about the change of electric field geometry due to bending of the tail (see above).

Injections into these deeper torus layers have also provided evidence that the optic tectum of the midbrain projects into layers 10, 11, and 14 and that several layers of the tectum receive input from these torus laminae. This reciprocal connection has a precise topography. Since the tectum is known to incorporate a topographic map of the retina it is likely that this map is projected further onto the corresponding torus layers (Fig. 4).

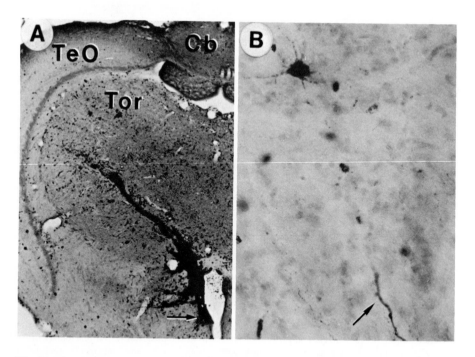

Fig. 4 A,B. A illustrates massive terminal labelling in layer 11 of the torus after HRP injection in the cerebellum; **B** shows retrogradely filled cells and a descending axon (*arrow*) in layers 10 and 11 of the torus after HRP injection in the optic tectum

Still deeper layers of the torus (L 12, 13, 14, 15) receive input from several brain-stem nuclei. They are considered first order targets of the ordinary lateral line system with local information about water movements and targets of the auditory maculae in the labyrinth. Consequently it is assumed that the deepest torus layers analyze information from these modalities.

Even though direct physiological evidence about a corresponding topographic representation of space in the torus is available only from P and T units of the electrosensory system, the remaining inputs to the torus suggest a similar spatial organization. Firstly, the deeper parts of the torus are laminated similar to the superficial electroreceptive parts. Secondly, there is an extensive vertical connectivity among the torus layers which could help to co-ordinate input from the various modalities.

The view which emerges from this type of organization is that multimodal information from corresponding sectors of space around the fish is integrated into the torus which may be used to control orientation responses and to communicate. By comparison it is interesting to note that the incorporation of the electroreceptive system into the torus of these fish represents a strong evolutionary pressure for laminar representation of the other modalities which are not obviously represented in laminae in the torus of other animals.

Acknowledgements. Supported by the German Science Foundation, SFB 45.

References

Bastian J (1976) The range of electrolocation: A comparison of electroreceptor responses and responses of neurons in a gymnotid fish. J Comp Physiol Psychol 108:193–210

Bastian J, Heiligenberg W (1980) Neural correlates of the jamming avoidance response of *Eigenmannia*. J Comp Physiol Psychol 136:135–152

Bullock TH (1969) Species differences in effect of electroreceptor input on electric organ pacemakers and other aspects of behavior in electric fish. Brain Behav Evol 2:85–118

Bullock TH (1982) Electroreception. Ann Rev Neurosci 5:121–170

Ebbesson SOE, Hansel M, Scheich H (1981) An "on the slide" modification of the de Olmos-Heimer horseradish peroxydase method. Neurosci Lett 22:1–4

Heiligenberg W (1977) Principles of electrolocation and jamming avoidance in electric fish. In: Braitenberg V (ed) Studies in brain function. Springer, Berlin Heidelberg New York, p 85

Heiligenberg W, Bastian J (1980) The control of *Eigenmannia's* pacemaker by distributed evolution of electroreceptor afferences. J Comp Physiol Psychol 136:113–133

Lissmann HW (1958) On the function and evolution of electric organs in fish. J Exp Biol 35:156–191

Scheich H (1974) Neuronal analysis of wave from in the time domain: Midbrain units in electric fish during social behavior. Science 185:365–367

Scheich H (1979) Common principles of organization in the central auditory pathway of vertebrates. Verh Dtsch Zool Ges, Gustav Fischer, Stuttgart, pp 155–166

Scheich H (1982) Biophysik der Elektrorezeption. In: Hoppe W, Lohmann W, Markl H, Ziegler H (eds) Biophysik. Springer, Berlin Heidelberg New York, pp 791–805

Scheich H, Bullock TH, Hamstra RH (1973) Coding properties of two classes of afferent nerve fibers: High-frequency electroreceptors in the electric fish *Eigenmannia*. J Neurophysiol 36:39–60

Scheich H, Bullock TH (1974) The role of electroreceptors in the animals life: II. The detection of electric fields from electric organs. In: Fessard A (ed) Handbook of sensory physiology. Springer, Berlin Heidelberg New York, pp 1201–1256

Scheich H, Maler L (1976) Laminar organization of the torus semicircularis related to the input from two types of electroreceptors. Exp Brain Res Suppl I:565–567

Scheich H, Ebbesson SOE (1981) Inputs to the torus semicircularis in the electric fish *Eigenmannia virescens:* A horseradish peroxydase study. Cell Tissue Res 215:531–536

Scheich H, Ebbesson SOE (to be published 1983) The torus semicircularis in the weakly electric fish *(Eigenmannia virescens)*. Advances in Anatomy, Embryology, and Cell Biology, Vol 82

Watanabe A, Takeda K (1963) The change of discharge frequency by AC stimulus in a weak electric fish. J Exp Biol 40:57–66

The Performance of Bat Sonar Systems

H.-U. SCHNITZLER[1]

1 Introduction

Echolocating bats emit orientation sounds and analyze the returning echoes, in order
to detect the presence, and to characterize the location and nature of the reflecting
target. The operational principles of bat echolocation systems are similar to those
of technical systems, e.g. radar and sonar systems. Like those, biological systems
consist of a transmitter which produces and radiates a particular type of signal, and a
receiver which picks up and analyzes the returning echoes. In order to detect a target,
a bat has to decide whether an echo of its own sound is present or not. The detection
is only useful when combined with further information processing. The distance of a
target is determined by the time delay between signal emission and echo reception.
The direction or angular position of a target is determined by interaural echo differences. The relative velocity between bat and target is encoded in the Doppler-shifted
echoes. It can also be determined by range tracking. Fluttering movements of the
target, e.g. wing movements of insects, produce amplitude and frequency modulations
in the echoes. Further target features such as size, shape and surface properties are
encoded in a complex spectral composition and the temporal structure of the whole
echo field. The detection, localization, and characterization of targets is restricted
in the presence of interference factors such as external and internal noise, clutter
echoes, and spectral echo changes due to the frequency-dependent directionality of the
transmitting and receiving antennae and due to atmospheric influences. An echolocation system with an ideally designed type of signal and receiver for the extraction
of all information with maximal accuracy and for a maximal suppression of all possible
interference is not realizable.

   This is well known by the engineer who designs a technical localization system.
He therefore adapts his system to the special problems he wants to solve. For a precise
determination of range in a radar system he will use a signal and receiver design different from in a system to determine the velocity of targets. For bat sonar we propose
the hypothesis that during evolution the echolocation systems of the more than
800 species of bats have been adapted to the special echolocation tasks of each
species.

---

[1]   Lehrstuhl Zoophysiologie, Institut für Biologie III, Eberhard-Karls-Universität, 7400 Tübingen,
    FRG

Localization and Orientation in Biology and Engineering
ed. by Varjú/Schnitzler
© Springer Verlag Berlin Heidelberg 1984

What are the echolocation tasks which have to be solved by the different species of bats? Apart from problems common to all bats, such as finding their way and avoiding obstacles in the dark, the different species must also solve their specific problems, e.g. when searching for food. Bats hunting insects in the open must have strategies different from those hunting insects which fly close to obstacles or sit on surfaces. Other bats feeding on fruit, or pollen and nectar, or small vertebrates, or surface fishes, or on blood also have to solve a wide variety of different echolocation and interference problems.

Comparative studies show that the transmitters and receivers of the echolocation systems of different species of bats have been differently designed during evolution (Schnitzler and Henson 1980, Pye 1980, Neuweiler 1980, Pollack 1980, Suga and O'Neill 1980).

2 Structure and Patterning of Echolocation Signals

The echolocation signals of different species are characterized by differences in frequency structure, duration and sound pressure level. In most species they consist either of frequency modulated (fm) components alone or combinations of constant frequency (cf) components with fm-components. In a few cases pure cf-components have been observed. The fm-signals are mostly rather short, with durations below 5–10 ms, whereas cf-components from a few milliseconds up to more than 100 ms have been observed. Further variations result from a different harmonic content. Figure 1 gives examples for a bat which produces shallow and steep fm-sweeps (Fig. 1B) and another bat which operates with long cf-short fm sounds (Fig. 1A). This

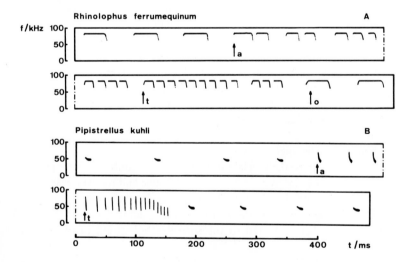

Fig. 1A,B. Sound pattern of a *Rhinolophus ferrumequinum* while passing an obstacle of vertically stretched wires at o (**A**) and of an insect hunting *Pipistrellus kuhli* (**B**). Begin of the approach phase at *a* and of the terminal phase at *t*

example also demonstrates that variation in sound structure are not only found in signals from different species, but also in sounds used by one species in different situations. This can be interpreted as adaptation to changing information-gathering and to interference problems.

In comparable situations different species of bats arrange their sounds in a rather similar way. The three behavioral categories used by Griffin et al. (1960) to describe the pulse patterns of hunting vespertilionid bats can be generalized for all bats.

A bat is considered to be in the free or search flight phase when it does not react to any target with its pulse emission (Fig. 1). In this situation most bats produce a single sound of maximal duration per respiratory cycle and wing beat. Repetition rates between 4–12 pulses/s have been observed. A bat enters the approach phase when it reacts to a target by producing more and shorter sounds. Very often the sounds are arranged in groups, thus indicating that there is still a correlation between sound emission and respiratory cycle and wing beat rate (Fig. 1A). The terminal phase begins when a bat is close to the target. It is characterized by a long group of very short pulses emitted with maximal repetition rate. If bats have caught an insect or passed an obstacle, the emission pattern is switched back to that of the free flight phase.

The echolocation systems of the different species of bats are very efficient in detecting, localizing, and characterizing targets. There is not enough room in this article to describe all the experiments which have been done to measure the performance of the sonar systems of the different species of bats. Detailed reviews have been published by Griffin (1958), Airapetianz and Konstantinov (1970, 1974), Schnitzler (1973b, 1978), Novick (1977), and Schnitzler and Henson (1980).

3 Target Detection

The detection limits of bats have been judged by the observation of insect captures and by measuring the obstacle avoidance skills with vertically and horizontally stretched wires. For different species of bats, sound pressure levels (SPL) of echoes used for detection between 9 and 34 dB SPL (re 0.002 N/m^2) have been estimated. Such detection thresholds still allow the avoidance of wire obstacles with wire diameters between 0.05 and 0.2 mm.

In obstacle avoidance experiments, it was observed that bats collide with new objects placed in an accustomed flight path even when the echo SPL was far above the detection limit. This disregard of strong echoes returning from newly placed obstacles reveals another very important orientation system, the spatial memory. Bats establish a very precise three-dimensional space picture (Neuweiler and Möhres 1967) and rely on this information in a well-known environment. This space memory completely dominates the echolocation system when bats fly on accustomed flight paths. We suggest that this allows bats to concentrate only on targets in which they are interested, e.g. on their insect prey, and to disregard all echoes returning from the environment.

The detection threshold of the fm-bat *Eptesicus fuscus* has been determined in a controlled experiment with a two alternative forced choice procedure (Kick 1982). The bats were able to detect spherical targets with diameters of 4.8 and 19.1 mm at distances of 2.9 and 5.1 m. The calculation of the echo amplitude for the measured detection ranges showed that the bats needed about 0 dB echo SPL for detection, i.e. comparable to the approximate threshold of hearing. This experiment, however, was done under very favorable conditions. The noise level was low, the bats were not in flight, they knew the approximate position of the target, and they could probe the target with several pulse-echo pairs. Therefore it can be assumed that under natural conditions the detection threshold must be higher.

4 Fixation and Tracking of Targets

The high capture efficiency observed in different species which feed on insects, indicates that bats know exactly where their targets are. It was estimated that bats pinpoint their prey in space within about one cubic centimeter (Webster 1963). The speed of action is also astonishing. A complete sequence with detection, fixation, tracking, and capture may only last 300–500 ms. Field observations showed that the fm-bat *Myotis lucifugus* pursued an insect every 3 s and probably caught every second one.

Photographs of bats catching flying insects or mealworms thrown into the air show that immediately after detection the bats turn towards the target and fixate it with their acoustical beam. The tracking of targets, i.e. the ability to determine the pathways of targets and to predict their future position was studied in a few species feeding on insects.

Experiments with ballistic targets reveal that bats are able to evaluate the course of targets and to make predictions regarding their future position. But these experiments contained a strong learning component. Experiments with flying insects which are capable of maneuvering and sometimes of making evasive movements give an unclear picture. Some authors found that bats predict the flight paths of insects, others described how the bats only followed insects when these made flight maneuvers. Sudden changes in flight direction, however, commonly result in successful evasions.

5 Determination of Range

Bats determine the range of targets by measuring the time lag between sound emission and echo reception. Experiments to determine the accuracy of range measurement have been done by Simmons (1973, 1979) and Airapetianz and Konstantinov (1974). In these experiments bats had to decide which of two similar targets offered in different directions, and at different absolute ranges, was closer (Fig. 2A). After the bats had learned this task the discrimination performance was measured as the percentage of correct choices for different range differences (Δr) between the two targets. The discrimination threshold was set at 75% (Fig. 2C).

In different species of bats thresholds between 12 and 25 mm range difference have been measured. These correspond to time differences (Δt) between 70 and 145 μs. In another experiment the two targets were simulated by the bat's echolocation signals which were played back by two loudspeakers with a time difference of Δt. These artificial echoes simulated two phantom targets with a range difference of Δr. The discrimination performance was similar to the thresholds measured with real targets, even when the echoes were presented successively by the two loudspeakers and not simultaneously (Simmons and Lavender 1976). This indicates that the bats are able to compare successive time or range measurements and that the discrimination threshold is a measure for the accuracy of range determination.

Simmons (1973) used his experimental data to reach conclusions about the type of information processing in the auditory receiver of bats. Strother (1961) had already pointed out that the signals of fm-bats are rather similar to the signals used in chirp or pulse compression radar systems. He therefore assumed that echo processing in bats might be similar to that in the matched filter receiver of a pulse compression radar system. Since then various aspects of this receiver model have been discussed by many authors (reviewed by Schnitzler and Henson 1980; see Altes, Chap. III.5, this Volume).

In a matched filter receiver (ideal receiver, optimal filter) the returning echo is cross-correlated with the transmitted signal, by passing the echo through a filter which is matched to the signal and afterwards passed through an envelope detector. Therefore the signal energy at the output is concentrated within a small time interval. This compression increases the signal-to-noise ratio and allows an optimal time measurement. Simmons suggested that the hearing system of bats is matched to the waveform of their signals. Therefore the cross-correlation function (ccf) between echo and emitted signal should be a measure for the accuracy with which the bats are able to determine the arrival time of echoes. In these first experiments he further assumed that with the high frequencies used by bats, phase information is not available in the auditory system. The postulated matched filter would therefore be a semicoherent receiver, i.e. the fine structure of the ccf would not be available and only the envelope of the ccf could be used to make predictions regarding the time or range measurement accuracy. Since the echoes in the discrimination experiment are similar to the emitted sounds, Simmons used the autocorrelation function (acf) instead of the ccf (Fig. 2B). He further took the line which connects the peaks of the acf as envelope and assumed that this envelope predicts the range discrimination performance of the bats. After correction of the prediction curve necessary to compensate for the bats head movement during the experiment, the measured performance curve shows good correspondence with the corrected prediction curve (Fig. 2C). Such a close fit between predicted curve and measured performance was also found in three other species.

These results and some further arguments led Simmons (1973) to the conclusion that there is a neural equivalent of a matched filter in bats. Today we know that his experimental approach does not allow this far-reaching conclusion.

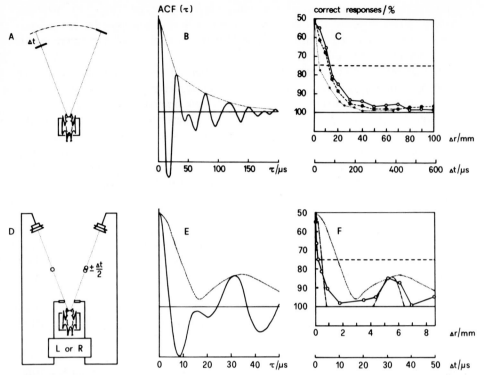

Fig. 2 A–F. Performance and prediction of the accuracy of range determination in the fm-bat *Eptesicus fuscus*. Experimental set-up for the discrimination between real targets differing in range (A) and for the discrimination between a jittering and a stable phantom target (D). Auto-correlation functions of a bat sound (*solid line*) with the derived prediction curves (*dotted line*) (B and E). Comparison of the measured performance curve (*solid line*) with the prediction curves (C and F). In C the dashed prediction curve was obtained after correcting the envelope curve (*dotted line*) for head movements. In F the *dashed line* corresponds to the acf and the *dotted line* to the half-wave rectified acf. (Adapted from Simmons 1973, 1979)

The accuracy of range estimation is always limited by the noise conditions. A prediction of range accuracy is only possible if the signal energy and the noise power per unit band width are known. At high signal-to-noise ratios – as was the case in the experiments described above – nonoptimal receivers may also achieve a very good performance.

In a second approach Simmons (1979) improved his method to measure the accuracy of range determination in the fm-bat *Eptesicus fuscus*. In the new experiment the bats had to discriminate between two phantom targets which were created by playing back echolocation sounds by two loudspeakers. One target was fixed in space at a distance of about 50 cm, which corresponds to a time delay of about 3 ms. The second phantom target was moved towards and away from its position from pulse to pulse by adding or subtracting a delay of $\Delta t/2$ so that it was jittering in time by a time difference of Δt (Fig. 2D). A comparator determined which microphone received the stronger signal and enabled only this channel to return echoes. After the bats had learned this task the discrimination performance was measured for a jitter range of

Δt from 80 down to 0 μs (Fig. 2F). The bats scanned back and forth while examining the targets. They emitted 1-20 signals before making a decision. The pulses were about 1-1.5 ms long and contained 2 to 3 harmonics of a fundamental sweep from 55-33 kHz.

The advantage of this experiment is that the bats could make a decision by scanning only one target, whereas, in the experimental approach described above, they had to compare distances of two targets positioned in different directions. That means that the discrimination performance in the jitter experiments is not so much endangered by head movements changing the target range and obscuring the discrimination performance.

The bats discriminated the jittering echo from the non-jittering one with astonishingly high accuracy (Fig. 2F). In consecutive time measurements, they could detect time differences of 0.5 μs at an absolute time of 3 ms. That corresponds to a range difference of about 0.1 mm at an absolute range of 50 cm. That means that the performance measured with the jitter approach is more than 100 times better than the performance measured in bats which had to decide which of two presented targets was closer. Another result is that at Δt around 30 μs the bats' performance was not as good as at shorter and longer Δt.

In this experiment Simmons (1979) used the same line of argument as in the experiments described above. He also compared the measured data with the acf of the emitted signal. He concluded that the bats perform as though they perceived a half-wave rectified version of the acf. From the reduced performance at about 30 μs, a Δt which corresponds to the average period of the bats' signals, he concluded that the bats perceive small time intervals well enough to sense phase information in sonar signals. He assumed that the acf with its full fine structure is the most appropriate representation of a bat's sonar signal, to be used in discussing the contribution of the signals to perception by echolocation. That means that bats would use a fully coherent optimal receiver. He further suggested that bats using broad-band echolocation sounds use temporal information, not only for ranging, but also for resolution and angular localization (Simmons et al. 1983).

The question of whether this experimental approach allows such far-reaching conclusions is again a matter of conjecture. A range accuracy prediction is only possible for situations where the signal energy and the noise power per unit band width are known. The result, however, that the discrimination performance is reduced at about 30 μs, which corresponds to the average period of the bats' signal, is a strong point for the assumption that bats are able to sense phase information. A possible explanation for this peak in the performance curve could be that the bats perceived a non-jittering scatter echo from the surroundings together with the jittering echoes and that they were able to discriminate the interference patterns between the scatter and the playback echoes originating at different Δt. In this case the bats would not need phase information for the discrimination.

The interpretation of the results of the jitter experiment by Simmons calls for a drastic change in the present view of information processing in the auditory system of bats and maybe also of other mammals. Therefore the questions still unresolved should be cleared up by further experiments as soon as possible.

6 Determination of Angular Direction

For the determination of angles the bats use their two ears, which act as directional antennae. Such a system delivers information on interaural phase, time, and sound pressure differences, as well as monaural cues. It is not yet clear how this information is used by the bats to achieve the angular localization of a target in the horizontal and in the vertical plane.

That bats are able to localize a target in both planes is demonstrated by insect-catching bats which, after detection of an insect, turn their head towards it and keep it in the middle of their acoustical beam. From photographs, the accuracy of the head aim was judged to be ±5 degrees (Webster and Brazier 1965).

In discrimination experiments it was demonstrated that in the horizontal plane the fm-bats *Eptesicus fuscus* and *Phyllostomus hastatus* are able to discriminate between a fixed angle of 19° and a variable angle when the variable angle was 6°-8° larger in *Eptesicus* and 4°-6° larger in *Phyllostomus* (Peff and Simmons 1972). The cf-fm bat *Rhinolophus ferrumequinum* could discriminate two targets separated by at least 4.5° from two targets separated by 0.4° (Airapetianz and Konstantinov 1974).

Monaural echo attenuation, by plugging one ear, reduced the ability to avoid wire obstacles in several species of bats. In *Rhinolophus ferrumequinum* the avoidance score with an obstacle of vertically stretched wires was reduced from 90% to 60% when a monaural plug causing an attenuation of 15-20 dB was inserted. If the second ear was plugged in the same way, the bats reached 90% again (Flieger and Schnitzler 1973). This fact demonstrates that the difference of echo SPL plays an important role in angle determination.

Simmons et al. (1983) report a new experiment with *Eptesicus fuscus* in which the bats were trained to discriminate between pairs of vertical rods separated by horizontal angles of different sizes. They found that the bats were able to discriminate between angles which differed by 1.5°, a much better performance than in the first experiment with *Eptesicus*. They concluded that this performance can be explained on the basis of interaural arrival time differences alone.

Whether time cues alone, or a combination of time and intensity cues together, or mainly intensity cues are used for horizontal angle localization, remains to be investigated further experiments.

The problem of angle determination in the vertical plane is even less understood. In neurophysiological measurements Grinnell and Grinnell (1965) found that the directionality of hearing is frequency-dependent and depends mainly on the structure of the external ear. They demonstrated that every angle in space is uniquely characterized by a family of binaural ratios of echo spectral components of different frequencies.

Simmons et al. (1983) report an experiment with *Eptesicus fuscus* which were able to discriminate between angles differing by 3°. They suggest that the bats use the time delays of secondary echoes from the tragus of the ear to determine vertical angles. In obstacle avoidance experiments with the cf-fm bat *Rhinolophus ferrumequinum* (Schnitzler and Ostwald unpubl.) it was shown that with horizontally stretched wires the avoidance score was lower when the bats' ear movements were stopped by severing the motor innervation of the ears, whereas the bats still reached a normal score in obstacles with vertically stretched wires. That suggests that, in bats with ear movements, at least part of the information for the vertical angle determination is gained by scanning. It may well be that bats with ear movements like *Rhinolophus* determine the vertical angle in a way different from bats without ear movement but a well-developed tragus like *Eptesicus*.

7 Determination of Relative Velocity

The relative velocity between bat and target can be determined either by measuring the decrease in distance between consecutive range measurements, or by evaluating the Doppler shifts in echoes. Echolocation systems characterized by signals with long cf-components and receivers with a high accuracy for the determination of frequency deviations, should be especially suited to determine velocity via Doppler shifts. Such echolocation systems are found in rhinolophid and hipposiderid bats and in the mormoopid bat, *Pteronotus parnellii*. That Doppler shifts play an important role in the echolocation systems of these bats is shown by a behavior which was termed Doppler shift compensation (Schnitzler 1968, 1970a, 1978, Gustafson and Schnitzler 1979). If such a cf-fm bat is in flight, the echo frequency heard by the bat is altered by Doppler shifts. If the target is stationary, the amount of Doppler shift is determined by the flight speed of the bat and the angular position of the target. If the target is moving, however, the amount of Doppler shift is determined by the relative velocity between bat and target.

Bats flying towards stationary targets lower their emission frequency, in order to compensate for Doppler shifts caused by their own flight velocity. In echoes returning from directly ahead, the echo frequency is therefore kept constant at the so-called reference frequency, a frequency about 200 Hz higher than the frequency emitted when in rest. Experiments with horseshoe bats flying in a wind tunnel, or in a helium-oxygen gas mixture, revealed that the bats use a feedback control system to keep the echo frequency constant (Schnitzler 1973a). In resting bats, electronically produced Doppler shifts, simulating target movements towards the bat, are also compensated for, but only at rather high sound pressure levels (Schuller et al. 1974). Insect-catching bats receive echoes both from stationary targets and from moving insects. In the case of the Greater Horseshoe bats, it was shown that they compensate only for the Doppler shifts in the echoes from stationary surroundings and not for the additional Doppler shifts caused by the flight movement of the insects (Trappe and Schnitzler 1982). That means that the frequency of echoes from the surroundings is kept constant, near the reference frequency, whereas the frequencies of echoes from insects are above or below the reference frequency, according to the insects' relative movement towards or away from the bat (Fig. 3).

The difference between emission frequency and the frequency of the echoes from the surroundings encodes the flight speed of the bat. The difference between the frequencies of the emitted signals and insect echoes is a measure for the relative

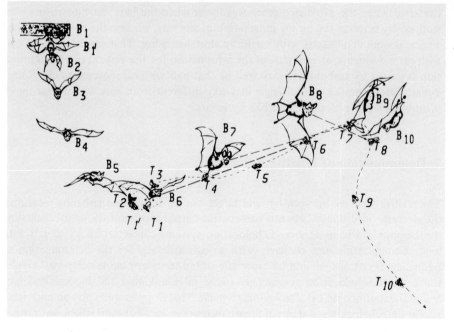

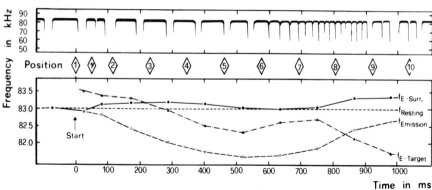

Fig. 3. Multiexposure photograph of a horseshoe bat pursuing a moth which escapes with a sharp turn between exposure 7 and 8 (*upper graph*). Spectrograms of orientation sounds emitted during the pursuit (*middle graph*). Frequencies of the cf-portions of the sounds emitted before take-off ($f_{Resting}$) and during flight ($f_{Emission}$), of the echoes returning from stationary targets in front of the bat ($f_{E\text{-}Surr}$), and from the insect ($f_{E\text{-}Target}$) (*lower graph*). (Trappe and Schnitzler 1982)

velocity between bat and target. The deviation of the frequency of the insect echoes from the reference frequency indicates the flight speed at which the target moves towards or away from the bat. We do not know, however, if all this information is used by the bats.

Doppler-compensating bats are very sensitive to Doppler shifts. Even slow movements of the hand or a large target toward a bat evoke the typical fright reaction, characterized by an increase of scanning movements and a retraction of the body with the hind legs.

The accuracy of frequency determination has only been measured in the Greater Horseshoe bat (Heilmann 1982). In the frequency range of the orientation signals, these bats can discriminate between frequency differences of about 50 Hz in successive signals. That suggests that these bats should be able to detect relative velocities of about 0.1 m/s by evaluating Doppler shifts.

8 Classification of Targets

Besides echo information for the localization of targets and for the determination of their velocity, additional echo cues are available, which can be used to classify targets.

The beating wings of insects cause strong frequency and amplitude modulations in insect echoes. It was suggested that the Doppler shift compensation is an adaptation for the detection of fluttering prey (Schnitzler 1970b). The cf-component of the returning echoes is used as a carrier for the rhythmical modulations caused by the oscillating wings.

This hypothesis was tested in an experiment where horseshoe bats had to discriminate between an oscillating target and a motionless target (Schnitzler 1978). By controlling amplitude and oscillation frequency of the target, frequency modulations of defined frequency swing and modulation frequency were produced so that thresholds of frequency swing at different oscillation frequencies could be determined (Fig. 4). With their 83 kHz signals, bats were able to detect oscillations which created frequency modulations of about 60 Hz depth at oscillation frequencies between 5–40 Hz down to only 12 Hz in a target oscillating with 200 Hz. The analysis of insect echoes (Schnitzler 1978, Schnitzler et al. 1983) showed that frequency modulations caused by the wing movements of insects are far above this threshold. The insect echoes are characterized by rhythmical amplitude peaks and spectral broaden-

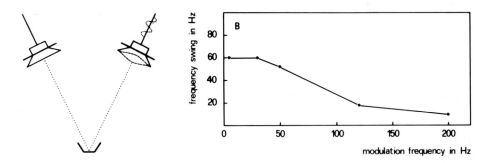

Fig. 4. Experimental set-up and performance of *Rhinolophus ferrumequinum* when discriminating between a target oscillating at different oscillation frequencies and a non-oscillating target, on the basis of Doppler shifts caused by the oscillations. The threshold curve indicates the echo frequency swing (maximal frequency difference in the sinusoidally modulated echoes) necessary to make the discrimination with 75% correct choiches. (Adapted from Schnitzler 1978)

ings encoding the wing beat rate, the angular orientation and species-specific information (see also Menne, Chap. III.3., this Volume).

Behavioral observations show that horseshoe bats use fluttering target information
to distinguish an insect echo from unmodulated background clutter and to hunt
selectively for insect species differing from others in the wing beat rate (Trappe 1982).
Transmitter and receiver of the horseshoe bats' echolocation system are especially
adapted for the processing of this behaviorally relevant information. By Doppler-
shift compensation the bats uncouple the insect echoes from the Doppler shifts
caused by their own flight movement. The insect echoes are therefore kept in an
"expectation window". The receiver is characterized by a corresponding "analysis
window". A highly expanded frequency representation on the basilar membrane of
the cochlea at the expectation window, leads to a strong over-representation of
sharply tuned neurons with special response characteristics throughout the whole
auditory pathway (reviewed in Neuweiler 1980, Pollack 1980, Schnitzler and Henson
1980, Schnitzler and Ostwald 1983).

Other target properties such as size, shape, and surface properties are also encoded
in complex temporal and spectral changes of the echo structure.

Behavioral experiments have shown that bats are able to discriminate between targets differing in size when the echo SPL differed by 1.5–3 dB. Targets differing in
shape and material can be discriminated probably mainly by using differences in the
echo spectra. Bats that have learned to catch mealworms projected into their flight
path can learn to discriminate these from other targets, e.g. from plastic discs,
apparently according to spectral differences in the echoes. Targets producing two
wave fronts can be discriminated at target depth differences larger than 1 mm, also
on the basis of spectral cues.

Under natural conditions this classification mechanism may frequently not be
necessary. For instance in bats hunting insects in the open, it might be a good strategy
to pursue each detected target, since the chance that it is not an insect is remote.
Therefore many bats try to intercept pebbles if they are thrown into their flight
path. If necessary, however, they can learn to discriminate a pebble from a meal-
worm in the laboratory.

As a general rule it can be stated that cf-signals are good for the evaluation of
fluttering target information, whereas broad-band fm-signals are useful for target
classification on the basis of spectral cues.

9 Conclusion

The echolocation systems of the more than 800 species of bats have been adapted
during evolution to the special information-gathering requirements of each species.
This led to differences in the design of the transmitters and receivers and in the
performance of the different echolocation systems. Therefore, it is important to
remember that the echolocation performances described in this paper are only
typical of specific systems and cannot be generalized for all species of bats.

Acknowledgements. I thank Heidi Hackbarth, Dieter Menne, Jo Ostwald, and Dick Altes for critical discussions and again Heidi Hackbarth for technical assistance. Our research on fluttering target detection in horseshoe bats has been supported by the Deutsche Forschungsgemeinschaft, Schn 138/7-15.

References

Airapetianz ES, Konstantinov AI (1970) Echolocation in nature. Nauka, Leningrad (in Russian). English translation: Israel Program of Scientific Translations. Jerusalem 1973

Airapetianz ES, Konstantinov AI (1974) Echolocation in nature. Nauka, Leningrad (in Russian). English translation: Joint Publications Research Service, No 63328. 1000 North Glebe Road, Airlington, Virginia 22201

Flieger E, Schnitzler H-U (1973) Ortungsleistungen der Fledermaus *Rhinolophus ferrumequinum* bei ein- und beidseitiger Ohrverstopfung. J Comp Physiol 82:93–102

Griffin DR (1958) Listening in the dark. Yale University Press, New Haven

Griffin DR, Webster FA, Michael CR (1960) The echolocation of flying insects by bats. Anim Behav 8:141–154

Grinnell AD, Grinnell VS (1965) Neural correlates of vertical localization in echolocating bats. J Physiol 181:830–851

Gustafson Y, Schnitzler H-U (1979) Echolocation and obstacle avoidance in the hipposiderid bat *Asellia tridens.* J Comp Physiol 131:161–167

Heilmann U (1982) Das Frequenzunterscheidungsvermögen der Großen Hufeisennase, *Rhinolophus ferrumequinum.* 56. Hauptvers Dt Ges Säugetierk

Kick SA (1982) Target-detection by the echolocating bat, *Eptesicus fuscus.* J Comp Physiol 145:431–435

Neuweiler G (1980) Auditory processing of echoes: peripheral processing. In: Busnel RG, Fish JF (eds) Animal sonar systems. Plenum, New York, p 519

Neuweiler G, Möhres FP (1967) Die Rolle des Ortungsgedächtnisses bei der Orientierung der Großblatt-Fledermaus *Megaderma lyra.* Z vergl Physiol 57:147–171

Novick A (1977) Acoustic orientation. In: Wimsatt WA (ed) Biology of bats, vol III. Academic, New York, p 73

Peff TC, Simmons JA (1972) Horizontal-angle resolution by echolocating bats. J Acoust Soc Am 51:2063–2065

Pollack GD (1980) Organizational and encoding features of single neurons in the inferior colliculus of bats. In: Busnel RG, Fish JF (eds) Animal sonar systems. Plenum, New York, p 549

Pye JD (1980) Echolocation signals and echoes in air. In: Busnel RG, Fish JF (eds) Animal sonar systems. Plenum, New York, p 309

Schnitzler H-U (1968) Die Ultraschall-Ortungslaute der Hufeisen-Fledermäuse (Chiroptera-Rhinolophidae) in verschiedenen Orientierungssituationen. Z vergl Physiol 57:376–408

Schnitzler H-U (1970a) Echoortung bei der Fledermaus *Chilonycteris rubiginosa.* Z vergl Physiol 68:25–39

Schnitzler H-U (1970b) Comparison of the echolocation behavior in *Rhinolophus ferrumequinum* and *Chilonycteris rubiginosa.* Bijdr Dierk 40:77–80

Schnitzler H-U (1973a) Control of Doppler shift compensation in the greater horseshoe bat, *Rhinolophus ferrumequinum.* J Comp Physiol 82:79–92

Schnitzler H-U (1973b) Die Echoortung der Fledermäuse und ihre hörphysiologischen Grundlagen. Fortschr Zool 21:136–189

Schnitzler H-U (1978) Die Detektion von Bewegungen durch Echoortung bei Fledermäusen. Verh Dtsch Zool Ges. Gustav Fischer, Stuttgart, p 16

Schnitzler H-U, Henson OW Jr (1980) Performance of airborne animal sonar systems: I. Microchiroptera. In: Busnel RG, Fish JF (eds) Animal sonar systems. Plenum, New York, p 109

Schnitzler H-U, Menne D, Kober R, Heblich K (1983) The acoustical image of fluttering insects in echolocating bats. In: Huber F, Markl H (eds) Neuroethology and behavioral physiology: Roots and growing points. Springer, Berlin Heidelberg, p 235–250

Schnitzler H-U, Ostwald J (1983) Adaptations for the detection of fluttering insects by echolocating horseshoe bats. In: Ewert JP, Capranica RR, Ingle DJ (eds) Advances in vertebrate neuroethology. Plenum, New York, p 801

Schuller G, Beuter K, Schnitzler H-U (1974) Response to frequency shifted artificial echoes in the bat *Rhinolophus ferrumequinum*. J Comp Physiol 89:275–286

Simmons JA (1973) The resolution of target range by echolocating bats. J Acoust Soc Am 54: 157–173

Simmons JA (1979) Perception of echo phase information in bat sonar. Science 204:1336–1338

Simmons JA, Kick SA, Lawrence BD (1983) Localization with biosonar signals in bats. In: Ewert JP, Capranica RR, Ingle DJ (eds) Advances in vertebrate neuroethology. Plenum, New York, p 247

Simmons JA, Lavender WA (1976) Representation of target range in the sonar receivers of echolocating bats. J Acoust Soc Am 60 (suppl 1):5

Strother GK (1961) Note on the possible use of ultrasonic pulse compression by bats. J Acoust Soc Am 33:696–697

Suga N, O'Neill WE (1980) Auditory processing of echoes: representation of acoustic information from the environment in the bat cerebral cortex. In: Busnel RG, Fish JF (eds) Animal sonar systems. Plenum, New York, p 589

Trappe M (1982) Verhalten und Echoortung der Großen Hufeisennase *(Rhinolophus ferrumequinum)* beim Insektenfang. Dissertation, University of Tübingen, Germany

Trappe M, Schnitzler H-U (1982) Doppler-shift compensation in insect-catching horseshoe bats. Naturwissenschaften 69:193–194

Webster FA (1963) Active energy radiating systems: the bat and ultrasonic principles II, acoustical control of airborne interception by bats. Proc Int Congr Tech and Blindness AFB, New York 1:49–135

Webster FA, Brazier OB (1965) Experimental studies on target detection, evaluation, and interception by echolocating bats. Aerospace Medical Res Lab, Wright-Patterson Air Force Base, Ohio, AD 628055

Short-Time Spectral Analysis of the Ultrasound Echoes of Flying Insects

D. MENNE[1]

1 Introduction

The Greater Horseshoe Bat, *Rhinolophus ferrumequinum*, hunts insects in free flight using an active sonar system for orientation and pursuit of prey.

Behavioral experiments have shown (Trappe 1982) that these bats hunt fluttering insects exclusively; they do not react to insects sitting motionless on a wall. They also discriminate between species of insect: some noctuids are favoured prey, while the cockchafer of similar size and with the same wingbeat frequency is often avoided.

When the bat shows the first recognizable pursuit action, e.g. starting from a resting place or turning in flight, it emits orientation sounds of 80 ms duration on the average. These consist of a long CF-part at about 83 kHZ (Fig. 2) and a short FM-part (only partly visible in Fig. 2). Echoes from the surroundings are Doppler-shifted in proportion to the relative velocity of the bat and the surroundings. It is well known (Schnitzler 1968) that bats perceive this frequency shift, because they alter the frequency of the sounds emitted to bring the echo frequency back within hearing focus. When a fluttering prey enters the sonar beam, the echoes will also be modulated in frequency and amplitude to the wingbeat period.

It has been shown (Schnitzler et al. 1983) that the fluctuation in amplitude is no smooth function of time but consists of a series of glints showing that the wing of the insect is not a diffuse reflector. These short processes were represented using an autoregressive spectral estimation method.

The present paper will extend this method to the study of the echoes of other insects. Using linear prediction methods and procedures of speech recognition, insect echoes can be described and classified quantitatively. It is possible to select one wingbeat period which is typical for the insect and which may serve as a reference pattern for the recognition of an insect by its echo.

2 Recording and Frequency Analysis of the Echoes

The insect species examined in this study are shown in Table 1.

---

[1] Lehrstuhl Zoophysiologie, Institut für Biologie III, Eberhard-Karls-Universität, 7400 Tübingen, FRG

Localization and Orientation in Biology and Engineering
ed. by Varjú/Schnitzler
© Springer Verlag Berlin Heidelberg 1984

Table 1. Insect species

Insect	Family	Wings	Wingbeat period (ms)
Tipula sp.	Tipulidae	2	14
Ochropleura plecta	Noctuidae	4	21
Chrysopa sp.	Chrysopidae	4	54

Only a single specimen of each species has been examined. Recordings of the echoes were made by Kober (1983).

The insects were suspended at their thorax at a distance of 60 cm from a loud-speaker. A microphone was mounted next to the loudspeaker pointing in the same direction. The aspect angles were from 0 degrees (insect flying towards loudspeaker and microphone) to 180 degrees (insect flying away) in steps of 30 degrees, giving seven angles of incidence of the ultrasound. The loudspeaker emitted a continuous tone of 80 kHz. The echoes were recorded on tape. For sampling, the echo signal (80 ± 2.5 kHz) was transformed to baseband (2.5 ± 2.5 kHz). The echoes tested were stored on disk. For each section of flight the wingbeat period was determined using autocorrelation. Records in which the wingbeat period varied more than 10% from the mean were discarded. A 3-ms Hamming window was applied to segments of the digital data and linear prediction coefficients (LPC) were calculated using the auto-correlation method (Markel and Gray 1976). The analysis window was advanced in steps of 1 ms; the resulting 67% overlap between data segments was chosen to smooth out small fluctuations in wingbeat period which could make the recognition process unreliable.

The short-time spectra can be obtained from the LPC. Time-frequency representations for different orders of the autoregressive model are shown in Fig. 1.

The point density is proportional to the sound pressure level of the returning echo. Frequency is plotted on the ordinate relative to the loudspeaker frequency of 80 kHz. Higher intensities above the center line indicate that the reflecting part of the wing was moving towards the microphone. The black and white figures in this book cannot adequately reproduce the time and frequency resolution of the method that was used.

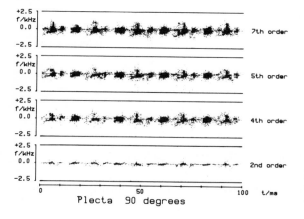

Fig. 1. Time-frequency representations for different orders of the autoregressive model

The seventh order model, which is used for all further investigations, results in a mean prediction residual of less than 0.1%.

Figure 2 shows the results of the autoregressive frequency analysis for three species from 0 degrees (front view). The echo pattern of the two-winged Tipula is especially simple. There is only one short glint per cycle with a negative Doppler shift, showing that the greatest intensity is reflected during the down-back stroke of the wing. After each glint there is a drop-out, where the intensity is even lower than when the animal does not move its wings. This could be the result of some kind of negative interference of sound waves.

One period of the pattern of *Ochropleura plecta* and *Chrysopa* consists of two bursts which could be identified as the upstroke and downstroke of the wings. As with *Tipula*, the mean frequency of the echo shifts to lower values, which is characteristic of all recordings from 0 degrees.

The dependence on aspect angle for one insect is depicted in Fig. 3. The echo pattern from 180 degrees (back view), can be thought of as the mirror image of the pattern at 0 degrees, reflected at the carrier-frequency axis. The frequency of the main

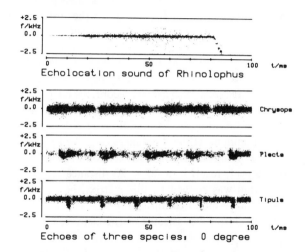

Fig. 2. Results of autoregressive frequency analysis

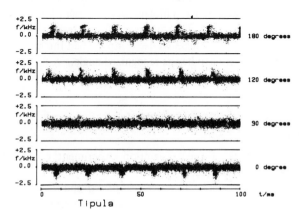

Fig. 3. Dependence on aspect angle for *Tipula*

glint is positively Doppler shifted, but the echo of the upstroke of the wing is also visible. The side view (90 degrees) seems to consist mainly of drop-out glints, but examination of the original figures show that the strokes can still be distinctly separated – an average, however, the positive and the negative Doppler-shifted parts have about equal energy.

3 Finding a Reference Template of Insect Echoes

The procedure used to find a reference template is a slightly modified version of the method described in Rabiner et al. (1979). For each of 7 aspect angles of the three species, 20 wingbeat periods were chosen for the template (for some angles of Chrysopa, only 12 or 15 periods could be used). The measure of similarity between each two of the wingbeats, expressed as a distance, was determined using the Itakura method (Itakura 1975), which gives the distance of two 3-ms frames by the logarithmic ratio of an autoprediction and a cross prediction residual. The distance of two wingbeats is obtained by adding up the distances of all 3-ms frames of these two wingbeat periods. In this paper only wingbeats of the same species are compared with each other; as their periods are selected as having equal length, no dynamic time warping is necessary as would to the case in speech recognition. In contrast to speech signals, the wingbeat signals have to be assumed as being periodic; as one period is cut out randomly from a larger record, the relative phase of the wingbeat is not known. To get a reliable distance, the Itakura method was modified by determining the distance for all cyclic permutations of one period, from the fixed alternative period – the minimum of these distances gives the final distance between two periods.

The modified Itakura distances for all pairs of the 20 wingbeats of one insect at one aspect, result in a 20 × 20 distance matrix. A k-nearest-neighbour clustering algorithm was applied to this matrix; the parameters in the cluster algorithm were interactively modified until, at most, two stable clusters resulted. The center element of each cluster was selected as the element whose maximum distance from all other elements in the cluster was minimal. This center element identified the reference template for the insect at the given angle, the "typical wingbeat". To test the reliability of this reference template, each element was used to predict the aspect angle of this insect echo. The distance of the test element to all reference elements was determined; the angle whose template had the minimum distance to the test element was considered to be the recognized angle.

4 Recognition Experiment

The results of the recognition experiment can be summarized in the confusion matrices (Table 2).

In the cases of *Ochropleura plecta* and *Tipula* at least 18 of 20 examined echo periods from each angle are recognized correctly. All but one confusion occurs be-

Table 2. Results of recognition experiment

Confusion Matrix: Chrysopa spec.

Presented angle (degrees)	Recognized angle (degrees)							n
	0	30	60	90	120	150	180	
0	17			3				20
30	3	7	1	1				12
60		1	13	1				15
90	1			11				12
120		1			10		1	12
150						15		15
180	12	1	4				3	20

Confusion Matrix: Ochropleura plecta

Presented angle (degrees)	Recognized angle (degrees)							n
	0	30	60	90	120	150	180	
0	18	2						20
30		20						20
60		1	19					20
90				20				20
120					20			20
150						20		20
180	1						19	20

Confusion Matrix: Tipula spec.

Presented angle (degrees)	Recognized angle (degrees)							n
	0	30	60	90	120	150	180	
0	19	1						20
30		20						20
60			18	2				20
90				20				20
120				1	19			20
150						20		20
180						1	19	20

tween adjacent angles. For *Chrysopa* the confusion matrix is more complex. Confusions between angles farther away do occur, and can only partly be explained by the fact that in some cases less than 20 patterns have been analyzed, resulting in unstable clusters. The most striking error occurs at 180 degrees (n = 20), which was often confused with 0 degree. Inspection of the relative phase at which the confusion occurred shows that in these cases the upstroke of the front view was taken for the downstroke of the back view.

5 Discussion

To do more research on the cues the bat can extract from moth echoes, it is necessary to isolate typical single wingbeats and to reproduce them at will. This is done in our laboratory by quadrature-sampling one wingbeat and reproducing it by single-side band modulation. The problem of finding one wingbeat cycle typical for the insect at the given aspect, can be solved objectively by choosing the reference template by means of the method described in this paper.

One other objective is to be able to describe and classify quantitatively the information coded in the insect echoes. Does a noctuid and a beetle-class exist? What is the feature overlap between different noctuid species? Which species of insects can be easily recognised by the bat, which are difficult to distinguish?

The results presented in the confusion matrices should be only taken as an indication that such questions could be answered by the speech recognition method. They do not represent a realistic recognition experiment, because the data were only from one animal of each species and the "presented angle" in the matrices were taken from the ensemble from which the reference template was selected. This paper has concentrated on angle recognition, but the bat will hear a prey presenting itself from different angles, and it probably treats the angle as a nuisance parameter when classifying insects. By applying the same methods as described, it should be possible to find echo structures which characterize the insect, irrespective of the aspect angle.

References

Itakura F (1975) Minimum prediction residual principle applied to speech recognition. IEEE Trans Acoust Speech Signal Processing ASSP-23:67−72

Kober R (1983) Analyse des Ultraschallechos von einheimischen Insekten. Staatsexamensarbeit, Universität Tübingen

Levinson SE, Rabiner LR, Rosenberg AE, Wilpon JG (1979) Interactive clustering techniques for selecting speaker-independent reference templates for isolated word recognition. IEEE Trans Acoust Speech Signal Processing ASSP-27:134−141

Markel JD, Gray AH (1976) Linear prediction of speech. Springer, Berlin Heidelberg New York

Schnitzler HU (1968) Die Ultraschall-Ortungslaute der Hufeisen-Fledermäuse (Chiroptera-Rhinolophidae) in verschiedenen Orientierungssituationen. Z vergl Physiol 57:376−408

Schnitzler H-U, Menne D, Kober R, Heblich K (1983) The acoustical image of fluttering insects in echolocating bats. In: Huber F, Markl H (eds) Neuroethology and behavioral physiology: Roots and growing points. Springer, Berlin Heidelberg New York Tokyo

Trappe M (1982) Verhalten und Echoortung der Großen Hufeisennase *(Rhinolophus ferrumequinum)* beim Insektenfang. Thesis, Universität Tübingen

Control of Echolocation Pulses
in the CF-FM-Bat *Rhinolophus rouxi*:
Neurophysiological Investigations of the Function
of the Brain Stem Motor Nucleus Innervating the Larynx:
the Nucleus Ambiguus

R. RÜBSAMEN[1]

Microchiroptera are nocturnal mammals. They orient themselves in space and hunt for insects by echolocating actively. They emit ultrasonic pulses and extract information about environment and prey from the reflected echoes. The echolocation pulses are produced in the larynx which is innervated by two vagal nerves on each side: the superior laryngeal nerves (SLN) and the recurrent laryngeal nerves (RLN). The motorneurons of these nerves have been shown to be located in two completely separate regions of the nucleus ambiguus by the Horseradish peroxidase technique (Schweizer et al. 1981). The motorneurons of SLN are located in the anterior part of the nucleus extending into the caudal pole of the motor nucleus of the facial nerve. The motorneurons associated with RLN are located in the caudal part of the nucleus.

In an effort to understand the vocal control mechanisms in horseshoe bats multi- and single unit recordings were made from different portions of the nucleus ambiguus in the vocalizing bat using a stereotactic approach. The echolocation pulses were picked up by a microphone and played back with a defined positive frequency shift. Neural activity, along with the bat's vocalization and respiration, were recorded by a tape recorder. Of a total of 379 events recorded, 117 could not be correlated with the bat's vocalization, since no vocalization occured during these recordings. Of the remaining 262 recordings, 189 were single unit recordings. The discharge pattern of these units was classified into eight different categories according to the temporary relationship of units firing with the vocalization (Fig. 1). Additionally, units could be classified into three categories on the basis of their relative responses during respiration and vocalization: (1) some units showed no discrimination between respiration with and without vocalization; (2) other units discharged during the respiration cycle but changed the discharge pattern when vocalization occurred; (3) a number of units showed activity which was correlated to vocalization only.

Neurons of type A (14%) showed a tonic discharge activity starting at 30-50 ms prior to the onset of vocalization and ending 15 ms before the termination of the call. The discharge rate is 150-200 impulses/s. If vocalization occurred together with respiration, activity was higher than when respiration was not accompanied by vocalization.

Neurons of type B (9%) showed an activity that was best correlated to the onset of vocalization. Activity started with a burst of 3-4 spikes 10 ms prior to vocalization and then continued at some lower rate up to 15 ms before the end of the pulse. Activity of this type only occurred during vocalization.

---

[1] Zoologisches Institut der Universität München, FRG

Localization and Orientation in Biology and Engineering
ed. by Varjú/Schnitzler
© Springer Verlag Berlin Heidelberg 1984

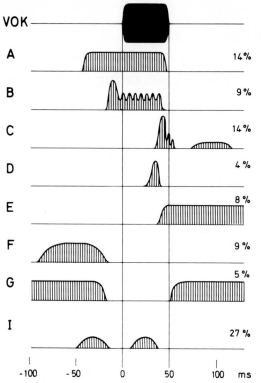

Fig. 1. Scheme of the different discharge patterns of single units recorded in the nucleus ambiguus of the horseshoe bat when correlated to the production of echolocation pulses. *Upmost trace* envelope of the vocalization. *Column* to the *left* different types of units. *Right* percentage of these types is given. (Further explanation see text)

Neurons of type C (14%) showed a phasic discharge activity 5 ms before the termination of the call. These 3–5 spikes we followed by after-discharges at some lower rate up to 100 ms after vocalization. Activity of this type only occurred during vocalization.

Neurons of type D formed only a small group containing 4% of the total. The units showed a prominent activity burst in a temporary relation to vocalization, typical for the individual cell.

Neurons of type E (8%) showed tonic discharges starting at the end of vocalization and continously decreasing after 150 ms from the end of vocalization.

Neurons of type F (9%) showed a broad activity starting 100–50 ms before the onset of vocalization and ending 20 ms before the call. Type E and type F did not change their activity pattern when the bat was not vocalizing during the expiration cycle.

Neurons of type G (5%) formed a small group of units which showed an ongoing tonic discharge activity with pauses if vocalization occurred.

Neurons of type I (27%) formed a heterogenous group, all with a rather low discharge rate. Some showed only one spike per vocalization just before the onset of vocalization: others one spike before the termination of the call. Others showed a discharge activity with 3–5 spikes before the vocalization and again 3–5 spikes during the CF-part of the call. Many of these units were only active during vocalization.

When the bat showed Doppler-shift compensation, units of the types A, B, and I changed either the discharge rate or the discharge pattern during vocalization. Some units type I stopped firing when the bat lowered the frequency of the CF-part of the vocalization. Some units type A showed a correlation between the discharge rate and the frequency of the CF-component of the call.

A comparison of these recordings with the gross activity recorded from the RLN and SLN (Schuller and Rübsamen 1981; Rübsamen and Schuller 1981) shows that type A activity corresponds closely the motor activity found in the SLN. It is thought that these units are responsible for controlling the frequency of the CF-component and they are exclusively found in the anterior part of the nucleus ambiguus. Types B and C units are presumably involved in controlling the time course of the call. Type B can best be correlated to the onset of vocalization and type C to the end or the FM-part of vocalization. Thus the combination of these two elements can produce the gross activity pattern observed from the RLN. These units are located in the caudal part of the nucleus ambiguus. Types E and F units are presumably the motor-control for the valve mechanism of the larynx during respiration and type I units are probably acting as interneurons according to their low discharge rate. These units are scattered all over the nucleus ambiguus.

All elements necessary for the control of vocalization were recorded in the nucleus ambiguus. In addition to the motoneurons controlling the frequency course and the time pattern of the echolocation pulses, neural elements were found that possibly serve the function of co-ordination of activities within the nucleus itself and/or the function of co-ordination between the two homonomous nuclei.

References

Rübsamen R, Schuller G (1981) Laryngeal nerve activity during pulse emission in the CF-FM bat, *Rhinolophus ferrumequinum* II. The recurrent laryngeal nerve. J Comp Physiol 143:323–327 143:323–327

Schuller G, Rübsamen R (1981) Laryngeal nerve activity during pulse emission in the CF-FM bat, *Rhinolophus ferrumequinum*. I. Superior laryngeal nerve (external motor branch). J Comp Physiol 143:317–321

Schweizer H, Rübsamen R, Rühle C (1981) Localization of brainstemm notoneurons innervating the laryngeal muscles in the rufous horseshoe bat, *Rhinolophus rouxi*. Brain Res 230:41–50

Echolocation as Seen from the Viewpoint of Radar/Sonar Theory

R. A. ALTES[1]

1 Introduction

The design and performance of optimum detectors and estimators depend on the echo waveform. The echo, in turn, depends on the target model. Animal echolocation targets are different from targets that concern man-made systems, with the exception of targets that vibrate or have moving parts. With this discrepancy in mind, the discussion focuses on detectors and estimators for a point target model (to compare with some experimental results), a non-random model for a time varying extended target (flying insect or swimming fish), and a model for a random time varying extended target.

2 Point Target Model

2.1 Matched Filter Detection

A point target returns an echo that is a delayed, Doppler distorted, attenuated version of the transmitted signal. If the range rate and range are known, a single correlator or filter can be used to detect the target. To maximize output signal-to-noise ratio, or to implement a likelihood ratio test in additive Gaussian noise, a whiten-and-match filter is used. The output of the filter is sampled at the time of expected maximum response, and the resulting sample value is compared with a threshold. The performance of the filter is completely determined by the output signal to noise ratio (SNR). From Van Trees (1971), page 246,

$$SNR = [(\log P_F)/(\log P_D)] - 1 \qquad (1)$$

where P_F is the probability of false alarm and P_D the probability of detection. By measuring P_F and P_D, i.e., by determining the receiver operating characteristic (ROC) of an animal's performance (Van Trees 1968), output SNR can be inferred from experimental data, if a matched filter detector is assumed.

When no internal noise is added by the receiver itself, output SNR can be predicted from the observed signal and external noise. If the power spectrum of the external

---

[1] Orincon Corporation, 3366 N. Torrey Pines Ct., La Jolla, CA 92037

Localization and Orientation in Biology and Engineering
ed. by Varjú/Schnitzler
© Springer Verlag Berlin Heidelberg 1984

noise is $N(f)$ and the Fourier transform of the echo is $U(f)$, then the filter transfer function is $U^*(f)/N(f)$. At the optimum sampling time, the ratio of output signal power to expected output noise power is

$$\text{SNR} = |\int_{-\infty}^{\infty} U(f)\,[U^*(f)/N(f)]\,df\,|^2 \,/\, [\int_{-\infty}^{\infty} N(f)\,|U^*(f)/N(f)|^2 \,df]$$

$$= \int_{-\infty}^{\infty} [|U(f)|^2/N(f)]\,df\,. \tag{2}$$

In the case of white noise, $N(f)$ equals $N_0/2$, and output SNR equals $2\,E_u/N_0$, where E_u is signal energy. Although detectability in white noise depends upon signal energy and is independent of other signal parameters, detectability in clutter (using a single pulse) tends to be better for signals with good resolution in range and Doppler. Relatively large time-center frequency product, Tf_0, is required for good Doppler resolution if the signal is narrowband. Large time-bandwidth product, TB, is necessary (but not sufficient) for good Doppler resolution if the signal is wideband (Altes 1971; Altes and Titlebaum 1970). A large bandwidth, B, is necessary for good range resolution. Constant frequency (cf) signals with smooth envelopes, typically used at long range by cf-fm bats (Pye 1980), have BT approximately equal to one, but they are highly Doppler resolvent because $f_0 T$ is large. For the chirped signals used at long range by fm bats, BT can be 200 or more, and good range resolution can be obtained because B is large. Although these chirped signals have large BT, some of them are apparently designed to be insensitive to Doppler.

Given a point target model, together with the assumption that optimum processing is used, one can sometimes infer the clutter suppression and parameter estimation capabilities of a sonar system by observing the waveform. This inference process has yielded valuable insights into bat sonar. Other insights, however, can be obtained by relaxing assumptions about point targets and/or matched filters. The relaxation of such assumptions is the main subject of this paper.

Some experiments support a matched filter receiver model (Simmons 1980), while others suggest that different signal processors may be more applicable to echolocation. In one set of experiments cited as evidence against matched filtering in bats (Schnitzler and Henson 1980), it was found that the fm bat *Eptesicus fuscus* has range-independent estimation error for planar (point) targets. To see why this result appears to contradict matched filter predictions, some parameter estimation results must be considered.

2.2 Estimation with Matched Filters

If a parameter is unknown, one can synthesize many detectors, each based on a different value of the unknown parameter. The optimum detector with the largest response corresponds to a maximum likelihood (ML) estimate of the parameter. Matched filter or correlation processing is thus central to estimation and to detection of deterministic signals with unkown parameters. The expected rms error of the ML estimator is given by the Cramér-Rao lower bound (Van Trees 1968). In the case of delay estimation, this lower bound is (Cook and Bernfeld 1967).

$$\Delta\tau = 1/(B\sqrt{\text{SNR}})\,. \tag{3}$$

As delay increases, SNR is expected to decrease because of spreading loss (which decreases signal energy by a factor of range^{-4}) and atmospheric absorption (Pye 1980; Lawrence and Simmons 1982). A matched filter receiver model thus seems to become tenuous if a bat's range uncertainty ΔR does not increase with range R, as reported by Simmons (1973).

There are several reasonable arguments, however, justifying the observed results in terms of a matched filter model. First, both amplitude and duration of FM bat signals tend to increase in the presence of noise (Simmons et al. 1975). The resulting increase in signal energy could offset range-dependent SNR losses. Another argument is the effect of internal noise. SNR at the filter output may be dominated by internal noise when planar targets that are large relative to an insect are presented in a quiet environment. If receiver performance is limited by internal noise, $\Delta \tau$ will be nearly independent of input (external) signal-to-noise ratio.

If internal noise were indeed large relative to external noise in the experiments reported by Simmons (1973), a matched filter model would predict that ΔR should increase with R if the experiments were repeated in a more noisy environment. Range estimation experiments with dolphins (Murchison 1980) in a comparatively noisy environment (Au 1980) indeed indicate an increase in ΔR as target range is increased. Again, these data are complicated by the fact that dolphins can adjust to noise by raising transmitted signal levels, altering signal spectra (Au 1980), and averaging over more echo pulses. It is also possible that internal noise level depends upon external noise level (Green and Swets 1966), as one would expect in a system that uses random point processes to encode stimulus levels (Snyder 1975).

The seemingly contradictory results in Simmons (1973) and Murchison (1980) do not necessarily imply different detection or estimation models in bats and dolphins. The results do suggest the need for more experimental control of important variables. More control could be obtained if electronic echo simulators were used. Transmitted echolocation signals could then be retransmitted to the animals which appropriate delays, and noise could be electronically added to the simulated echoes before retransmission. Such an experiment would allow delay estimation error to be measured as a function of input SNR at each delay. Internal noise could also be estimated from detection performance for reduced signal levels, by using Eq. (1).

Another argument for rejection of a matched filter model is that the peripheral auditory system is conventionally modelled as a bank of bandpass filters followed by envelope detectors (Evans 1975; Siebert 1968). It would seem that such a system would destroy the phase information necessary for pulse compression of a wideband signal. To further consider the validity of this argument, it is necessary to investigate the information destroyed by such a filter bank processor.

2.3 Implementation of a Matched Filter
Using a Spectrogram Representation of Echoes

A bank of filters with transfer functions $\{V(f-f_i)\}_{i=1}^{N}$ followed by envelope detectors creates a two-dimensional (time-frequency) signal representation called a spectrogram. The two-dimensional Fourier transform of a spectrogram is the product of the narrowband ambiguity function of the input signal and the ambiguity function of the filter transfer function $V(f)$ (Altes 1980a). If the filter ambiguity function is non-zero everywhere [e.g., if $V(f)$ is Gaussian], then it is theoretically possible to obtain the input ambiguity function from the spectrogram. Integrating the product of the input ambiguity function and the ambiguity function of a reference signal is equivalent to correlating the input signal with the reference signal, followed by envelope detection (Altes 1980a). It is thus theoretically possible to synthesize a matched filter followed by an envelope detector if only the spectrogram of the input data is given.

Practical limitations to matched-filter-equivalent spectrogram processing are related to the assumption that the filter ambiguity function is nonzero at all relevant parts of the ambiguity plane. If this is not the case, then some information about the signal may be destroyed (Johannesma 1982). For example, suppose the filters are strictly band limited, and the signal consists of two sinusoids, one of which is phase shifted relative to the other (Tukey 1982). This relative phase shift is supposedly measurable, since the signal can theoretically be reconstructed from its spectrogram, aside from a constant, unknown phase shift of the whole signal (Altes 1978 and 1980a). However, envelope detectors at the bandpass filter outputs destroy information about the relative phase shift between two sinusoids separated in frequency by more than the filter bandwidth.

The problem can be related to the fact that the relative phase between two sinusoids at frequency f_1 and f_2 is contained in the signal ambiguity function, $\chi_{uu}(\tau, \phi)$, evaluated at $\phi = \pm(f_2 - f_1)$. In order to recover the relative phase from a spectrogram, the filter ambiguity function, $\chi_{vv}(\tau, \phi)$, must be nonzero at $\phi = \pm(f_2 - f_1)$. This condition does not hold if the filter transfer function $V(f)$ is strictly band limited with a bandwidth less than $|f_2 - f_1|$.

Other examples can be constructed. Consider a signal that is composed of two impulses at different times t_1 and t_2, where one impulse is phase shifted relative to the other (e.g., a positive pulse followed by a negative one). In this case, relative phase information is preserved in a spectrogram representation only if $\chi_{vv}(\tau, \phi)$ is nonzero at $\tau = \pm(t_2 - t_1)$. This condition does not hold if the filter impulse response is strictly time limited with a duration less than $|t_2 - t_1|$.

The examples indicate that the best auditory filters for matched-filter-equivalent spectrogram processing are broad both in time and frequency. The filters have broad critical bandwidth (Zwicker 1961; Green 1960) and a long critical interval (Vel'min and Dubrovskiy 1976). The corresponding neurons have overlapping, broad turning curves as in Pfeiffer and Kim (1975). The neurons should also fire for a long time (the length of an echolocation signal) in reaction to a sound pulse. A bank of such broadly tuned neurons may seem unexciting if one is looking for specific feature detectors, but such a configuration is essential for synthesizing a matched filter with a large time-bandwidth product from a spectrogram.

The above considerations imply that the ability of mammals to perform optimum processing for a point target model may be limited if a specific kind of spectrogram representation is not synthesized in an animal's auditory system. A necessary condition for matched filter processing with spectrograms is the existence of a bank of overlapping filters, broadly tuned in both frequency and time. If such filters do not possess very large bandwidths at high frequencies, harmonics may not be coherently combined. Another limitation is that the equivalent matched filter must be followed by an envelope detector, which corresponds to optimum detection for a signal that is known except for phase. Despite these limitations, it would appear that spectrogram pulse compression is feasible for the chirped signals used in animal echolocation. For example, the required broadband, low-Q neural tuning curves have been found in *Rhinolophus* (Schnitzler and Henson 1980, Möller et al. 1978). Filters that are broad in frequency and have long impulse response can be obtained by using asymmetric transfer functions that rise rapidly in frequency and then decay relatively slowly, as observed by Zwicker and Fastl (1972) with masking experiments on humans.

2.4 The Effects of Unknown Parameters on Initial Detection Performance

If a signal is known except for, e.g., delay, Doppler shift, azimuth and elevation, the outputs of a bank of filters must continuously be observed in order to simultaneously detect the target and to form a joint estimate of the unknown parameters. Consideration of more than one sampling time and more than one filter output increases the chances for noise to cross a set threshold level and trigger a false alarm. This phenomenon can result in substantial degradation of detector performance, i.e., an increase in false alarm probability for a given probability of detection (Nolte and Jaarsma 1967, Urick 1975).

One solution to this problem is to use signals that are parameter tolerant, such that a perturbation of a given parameter does not necessitate a separate filter to detect the perturbed signal. Some fm bat signals, for example, are Doppler tolerant (Cahlander 1966). The envelope detected matched filter responses for such signals are comparatively insensitive to a Doppler-induced signal-filter mismatch in time scale (Altes and Titlebaum 1970, 1975, Altes 1980b). The cf signals used by some other bats are target tolerant (Altes 1971); different matched filters are not required if the signal is passed through any time invariant linear target filter and the correlator response is envelope detected.

3 Non-Random Time-Varying Extended Target Model

3.1 Matched Filter Detection for Echoes from Time-Varying Targets

Although echoes from time varying targets are often modelled as random processes, it is sometimes feasible to describe such echoes in deterministic terms. If a deterministic description is applicable, a receiver designed for random echoes is suboptimum. If the signal and target impulse response are non-random, then the echo is non-random, and a matched filter can be used to detect it.

For a constant frequency transmission, the echo from a flying insect can have surprisingly large time-bandwidth product. Echoes from moving insect wings often have a bandwidth of roughly 2 kHz for an 80 kHz cf *Rhinolophus* signal (Schnitzler and Henson 1980), resulting in an echo time-bandwidth product of 100 for a 50 ms signal duration. This value of TB is the same as that obtained with a 2 ms *Myotis* signal that has 50 kHz bandwidth, or a 3.3 ms *Eptesicus* signal with a 30 kHz bandwidth. The complexity of a matched filter is a function of the echo time-bandwidth product. For example, a finite impulse response (FIR) filter synthesized with a tapped delay line has $2 B_u T_u$ real-valued taps if matched to an echo waveform with time-bandwidth product equal to $B_u T_u$. A matched filter receiver for a wing-beat-modulated cf echo is then comparable in complexity with the processing required for a chirped signal. Although the matched filter must be more complex due to wingbeat modulation, large TB implies that the echo carries more information about the target, since there are more degrees of freedom for specifying different echo waveforms.

3.2 Non-Random Filter Characterization for cf Signals

Can the transfer function of a flying insect really be modelled as a deterministic time varying function? The echo of a cf bat signal from a flying insect is a delayed version of the waveform

$$e(t) = \int_{-\infty}^{\infty} H(t,f) \, U(f) \exp{(j \, 2 \, \pi f t)} \, df \approx H(t,f_0) \, u(t) \qquad (4)$$

where $H(t,f)$ is the insect's time-varying transfer function, and $U(f)$ is the Fourier transform of the transmitted narrowband signal $u(t)$ with center frequency f_0. The approximation in (4) applies when $H(t,f)$ is relatively constant over the bandwidth of the signal. This condition occurs when the signal bandwidth is much less than the Nyquist sampling interval for $H(t,f)$ in the frequency domain. The frequency domain sampling interval is the inverse of twice the duration of the target impulse response. For a target 1.5 cm long, the impulse response duration is 0.1 ms. A 50 ms cf *Rhinolophus* pulse has a bandwidth of approximately 20 Hz, which is indeed small relative to the 5 kHz frequency domain sampling interval.

Although Eq. (4) appears to account only for amplitude modulation, it can also incorporate frequency modulation. Assuming that signal and echo are written as complex analytic functions (Rihaczek 1969), $H(t,f_0)$ can be written as the product of a cross section term, $\sigma(t,f_0)$, and a time-varying phase term $\exp{[j \, 4 \, \pi f_0 \dot{R}(t)/c]}$. The cross section term causes the echo to be amplitude-modulated, and the phase term introduces a time-dependent Doppler shift that causes the echo to be frequency modulated.

Assuming that the envelope of $u(t)$ in Eq. (4) is approximately constant, the echo provides complete information about $H(t,f)$ at $f = f_0$, if the duration of $u(t)$ is long enough to include a complete wingbeat (5–50 ms, depending on wing beat frequency). Having made such measurements on a variety of targets, a cf sonar can in principle implement a matched filter bank receiver for various flying insects at different aspects. Changing the phase of the insect wingbeat motion at the time of reflection is equivalent to introducing a time delay into the modulation function $H(t,f_0)$. Specific hypotheses about the phase of the wingbeat are then unnecessary if the signal duration is sufficiently long and the matched filter output is continuously observed (a technique that implements a continuum of delay hypotheses).

If matched filter processing is used, the large time-bandwidth product of the echo from a flying insect can be fully exploited to differentiate between target types, even though a low time-bandwidth product cf signal is transmitted. Neurons sensitive to wing-beat-induced amplitude and frequency modulation have been identified in the inferior colliculus and cortex of cf-fm bats (Pollak 1980, Suga and O'Neill 1980).

3.3 Non-Random Filter Characterization for Wideband Signals

Characterization of a target with moving parts as a non-random time varying transfer function is more difficult over a broader bandwidth. By using an FIR filter model (tapped delay line with time-varying tap weights), a necessary condition for a deterministic target impulse response representation is easily obtained from the sampling theorem (Kailath 1962). If the maximum bandwidth of all the time-varying tap gains is $\hat{B}$, then each tap must be sampled with at least $2\hat{B}$ real samples per second or $\hat{B}$ complex-valued samples per second. For complex-valued samples, this requirement implies that the sonar pulse repetition frequency (PRF) must be greater than $\hat{B}$ pulses per second. For an fm bat signal with an average frequency of 40 kHz, $\hat{B}$ is approximately 1 kHz, and the sampling requirement is satisfied only at ranges less than 16 cm. For a dolphin sonar, $\hat{B}$ is less than 400 Hz for an average echolocation signal frequency of 70 kHz (Hester 1967). The maximum unambiguous range for non-random target characterization in water is then about 2m.

Although the time varying FIR tap weights cannot be deterministically measured at long range, the target descriptions obtained at short range should be adequate for matched filter design. The resulting matched filters represent specific hypotheses for parameters such as wing or tail beat frequency and target aspect. High PRF transfer function estimation is unnecessary for detection or identification after the matched filters have been implemented. In principle, it would appear that non-random filter characterization and appropriate matched filter design is feasible both for broadband echolocation systems and for narrowband ones. Unfortunately, broadband echoes tend to be very sensitive to small aspect changes. The many aspect hypotheses necessitated by this effect can degrade detector performance, as discussed in Section 2.4.

4 Random, Time-Varying, Extended Target Model

4.1 A Scattering Function Description and a Locally Optimum Detector

Let the impulse response of a time-varying target be $h(t, \tau)$. This function can be interpreted as the FIR filter tap gain at delay τ, evaluated at time t. The cross correlation between two different taps at τ and $\tau + \tau_0$, at two different times t and $t + t_0$, is

$$R_h(t, t + t_0, \tau, \tau + \tau_0) = E\{h(t, \tau)h^*(t + t_0, \tau + \tau_0)\} \tag{5}$$

where $E\{\cdot\}$ denotes expected value. If the tap weights are uncorrelated, then $E\{h(t, \tau) h^*(t + t_0, \tau + \tau_0)\}$ will be zero for $\tau_0 \neq 0$. Assuming for the moment that the tap weights are indeed uncorrelated, and that the time variation of the tap gain at delay τ is a wide-sense stationary process, we have

$$R_h(t, t + t_0, \tau, \tau + \tau_0) = R_h(t_0, \tau) \delta(\tau_0 - \tau). \tag{6}$$

In this case, the target can be described by its scattering function (Kailath 1962)

$$\Gamma_h(\tau,\phi) = \int_{-\infty}^{\infty} R_h(t_0,\tau) \exp(-j2\pi\phi t_0) \, dt_0 \, . \tag{7}$$

The scattering function is the echo power at delay τ and Doppler shift ϕ (Green 1962).

Because of the periodicity of insect wing motion or fish tail movement, the assumption that the tap gains are wide-sense stationary is very reasonable (Yaglom 1962). This same periodicity could also result in high correlation between different tap gains at particular values of t_0, corresponding to predictable wing positions at different times. Such correlations between different taps would invalidate the uncorrelated scatterer assumption in Eq. (6). Experimental results, however, indicate that wing sonar cross section is very small when the wing is not perpendicular to the direction of propagation (Schnitzler and Henson 1980). At delays such that the wing cross sections can be large, the filter tap weights are a sequence of periodic, narrow time pulses. The tap gains are comparatively small at other delays. The uncorrelated scatterer model is thus reasonable except for high correlation between the two tap gains corresponding to maximum echo strength wing positions. Further thought indicates that these two taps, where the wings are occasionally perpendicular to the direction of propagation, are separated by a range difference less than the width of the insect's thorax, and thus lie within the same range resolution cell. This observation implies that there is effectively only one time varying tap, and the wide-sense stationary, uncorrelated scatterer model is applicable. The scattering function thus seems to be an acceptable description for a flying insect, although its general applicability for wideband signals has recently been questioned (Ziomek and Sibul 1982).

The applicability of the scattering function description implies that a spectrogram correlator is a locally optimum target detector, i.e., one optimized for low SNR (Altes 1980a). The echo spectrogram is the convolution in time and frequency of the scattering function, $\Gamma_h(\tau,\phi)$, with the signal spectrogram, $S_{uv}(\tau,\phi)$. The locally optimum detector correlates the echo spectrogram with a template and compares the result with a threshold. As indicated in Section 2.3, such a detector could easily be implemented using mammalian hearing models.

A more general optimum statistic for detecting Gaussian random signals in white Gaussian noise correlates the data spectrogram with a template that depends upon the expected echo spectrogram and the expected noise power (Altes 1984). Specifically, the log-likelihood ratio for testing H_1 (Gaussian random signal + noise) vs. H_0 (noise alone) is

$$\varphi[z(t)] = \int_{T_i}^{T_f} \int_{-\infty}^{\infty} A(t,f) \, S_{zv}(t,f) \, df \, dt \underset{H_0}{\overset{H_1}{\gtrless}} \gamma \tag{8}$$

where $z(t)$ is the data observed for $T_1 \leqslant t \leqslant T_f$, $S_{zv}(t,f)$ is the spectrogram of the data, γ is a threshold, and

$$A(t,f) = \frac{E\{S_{ev}(t,f)\}}{(N_0/2) E_v + E\{S_{ev}(t,f)\}} \, . \tag{9}$$

In Eq. (9), $E\{S_{ev}(t,f)\}$ is the expected echo spectrogram, $N_0/2$ the expected noise power, and E_v the energy of the window function $v(t)$.

For very small SNR, the template $A(t,f)$ becomes the expected echo spectrogram divided by $N_0/2$, and the test becomes identical to the locally optimum detector (Altes 1980a). The spectrogram correlator is similar to Green's energy summation model for human detection of a sum of sinusoids at different frequencies (Green 1958, Green et al. 1959).

4.2 Performance of the Spectrogram Correlator for a Random Target Model

A spectrogram correlator computes a weighted sum of envelope detected samples in the time-frequency plane. The samples can be assumed to be statistically independent if they are $(2B_v)^{-1}$ s apart in time and $(2T_v)^{-1}$ Hz apart in frequency, where B_v and T_v are the approximate (ten dB) bandwidth and duration of $v(t)$, the filter or window function used to construct the spectrogram (Altes 1980). The effective output SNR for substitution into (1) can be obtained by determining the SNR at each statistically independent spectrogram sample, computing a weighted sum of these SNR values over n_s samples, and multiplying this sum by a relative integration-improvement factor, $I_i(n_s)/n_s$. The factor $I_i(n_s)/n_s$ is the ratio of effective SNR improvement with noncoherent addition to SNR improvement with coherent addition, when n_s independent samples are observed.

The n_s relevant spectrogram samples are those given nonzero weight by the spectrogram correlator, i.e., time-frequency points where the expected echo spectrogram is nonzero. The SNR at each spectrogram sample is determined by the same calculation as in Eq. (2), where $U^*(f)/N(f)$ is replaced by $V(f-f_0)\exp(-j2\pi ft_0)$, the transfer function of the filter used to compute the spectrogram at frequency f_0 and time t_0.

The weights for the weighted sum of sample SNR values are determined by the template correlated with the data spectrogram. For low SNR, this template is the expected echo spectrogram divided by $N_0/2$. The integration improvement factor $I_i(n_s)$ is between $\sqrt{n_s}$ and n_s, and is described in Marcum (1960) and Skolnik (1962). For small signal-to-noise ratio and large echo time-bandwidth product, the effective output SNR is approximately $BT(SNR_{in})^2$, where SNR_{in} is the input signal power in B divided by the input noise power in B (Urick 1975, Peterson et al. 1954).

The numer of independent spectrogram samples n_s and $BT(SNR_{in})^2$ are both proportional to echo time-bandwidth product. The use of a cf signal to detect a flying insect at long range can thus be justified in terms of large echo TB product, without resorting to a matched filter model.

5 Summary

Several target models have been reviewed. The optimum detectors for these models have been related to spectrogram representations of echoes. The utility of parameter tolerant processing has been discussed, and some new advantages of transmitting a cf signal have been found. One of these advantages is the possibility of obtaining a non-random description of the time varying transfer function of a flying insect at long range. Another advantage is enhanced echo detectability for a spectrogram correlator receiver. This enhancement, relative to the transmitted signal, is a consequence of the large echo time-bandwidth product induced by wing beat modulation of the cf signal.

References

Altes RA (1980a) Detection, estimation, and classification with spectrograms. J Acoust Soc Am 67:1232

Altes RA (1980b) Models for echolocation. In: Busnel R-G, Fish JF (eds) Animal sonar systems. Plenum, New York London, p 625

Altes RA (1978) Possible reconstruction of auditory signals by the central nervous system. J Acoust Soc Am 64, Supp No 1, p 137

Altes RA (1971) Methods of wideband signal design for radar and sonar systems. Fed Clearinghouse No AD 732—494

Altes RA (1981) Texture analysis with spectrograms. IEEE trans on Sonics and Ultrasonics SU-31 (to be published in July, 1984)

Altes RA, Titlebaum EL (1975) Graphical derivations of radar, sonar, and communication signals. IEEE trans on Aerosp and Electron Sys AES-11, p 38

Altes RA, Titlebaum EL (1970) Bat signals as optimally Doppler tolerant waveforms. J Acoust Soc Am 48:1014

Au WWL (1980) Echolocation signals of the Atlantic bottlenose dolphin *(Tursiops truncatus)* in open waters. In: Animal sonar systems (op. cit.), p 251

Cahlander DA (1966) Echolocation with wideband waveforms: bat sonar signals. Fed Clearinghouse No AD 605—322

Cook CE, Bernfeld M (1967) Radar signals. Academic Press, London New York

Evans EF (1975) Cochlear nerve and cochlear nucleus. In: Keidel WD, Neff WD (eds) Handbook of sensory physiology, vol 5, part II. Springer, Berlin Heidelberg New York, p 1

Green DM (1960) Auditory detection of a noise signal. J Acoust Soc Am 32:121

Green DM (1958) Detection of multiple component signals in noise. J Acoust Soc Amer 30:904

Green DM, McKey MJ, Licklider JCR (1959) Detection of a pulsed sinusoid in noise as a function of frequency. J Acoust Soc Am 31:1446

Green DM, Swets JA (1966) Signal detection theory and psychophysics. Krieger, Huntington, pp 255—257

Green PE (1962) Radar measurement of target characteristics. In: Harrington JV, Evans JV (eds) Radar Astronomy. McGraw-Hill, New York

Hester FJ (1967) Identification of biological sonar targets from body-motion Doppler shifts. In: Tavgola WN (ed) Marine bio-acoustics, vol 2. Pergammon, Oxford, p 59

Kailath T (1962) Measurements on time-variant communication channels. IRE Trans on Info Theory, IT-8, p 229

Lawrence BD, Simmons JA (1982) Measurements of atmospheric attenuation at ultrasonic frequencies and the significance for echolocation by bats. J Acoust Soc Am 71:585

Marcum JI (1960) A statistical theory of target detection by pulsed radar. IRE Trans on Inform Theory, Mathematical Appendix, IT-6, p 145

Möller J, Neuweiler G, Zöller H (1978) Response characteristics of inferior colliculus neurons of the awake CF-FM bat, *Rhinolophus ferrumequinum*. I. Single tone stimulation. J Comp Physiol A 125:217

Murchison AE (1980) Detection range and range resolution of echolocating bottlenose porpoise *(Tursiops truncatus)*. In: Animal sonar systems (op. cit.), p 43

Nolte LW, Jaarsma D (1967) More on the detection of one of M orthogonal signals. J Acoust Soc Am 41:497

Peterson WW, Birdsall TG, Fox WC (1954) The theory of signal detectability. IRE Trans on Inform Theory, IT-4, p 171

Pfeiffer RR, Kim DO ((1975) Cochlear nerve fiber responses: distribution along the cochlear partition. J Acoust Soc Am 58:867

Pollak GD (1980) Organizational and encoding features of single neurons in the inferior colliculus of bats. In: Animal sonar systems (op. cit.), p 549

Pye JD (1980) Echolocation signals and echoes in air. In: Animal sonar systems (op. cit.), p 309

Rihaczek AW (1969) Principles of high-resolution radar, chap 2. McGraw-Hill, New York Toronto London Sydney

Siebert WM (1968) Stimulus transformations in the peripheral auditory system. In: Kolers PA, Eden M (eds) Recognizing patterns. MIT Press, Cambridge, p 104

Schnitzler HU, Henson OW (1980) Performance of airborne animal sonar systems: I microchiroptera. In: Animal sonar systems (op. cit.), p 109

Simmons JA (1980) The processing of sonar echoes by bats. In: Animal sonar systems (op. cit.), p 695

Simmons JA (1973) The resolution of target range by echolocating bats. J Acoust Soc Am 54:157

Simmons JA, Lavender WA, Lavender VA (1975) Adaptation of echolocation to environmental noise by the bat, *Eptesicus fuscus*. F.I.B.R.C. 1975; Kenya National Academy for Advancement of Arts and Science, p 97

Snyder DL (1975) Random point processes. Wiley, New York London Sydney Toronto

Suga N, O'Neill WE (1980) Auditory processing of echoes: representation of acoustic information from the environment in the bat cerebral cortex. In: Animal sonar systems (op. cit.), p 589

Urick RJ (1975) Principles of underwater sound, 2nd edn. McGraw-Hill, New York, pp 349–354

Van Trees HL (1971) Detection, estimation, and modulation theory, part III. Wiley, New York London Sydney Toronto

Van Trees HL (1968) Detection, estimation, and modulation theory, part I. Wiley, New York London Sydney

Vel'min VA, Durbrovskiy NA (1976) The critical interval of active hearing in dolphins. Sov Phys Acoust 22:351

Yaglom AM (1962) Stationary random functions. Prentice-Hall, Englewood Cliffs

Ziomek LJ, Sibul LH (1982) Broadband and narrowband signal-to-interference ratio expressions for a doubly spread target. J Acoust Soc Amer 72:804

Zwicker E (1961) Subdivision of the audible frequency range into critical bands. J Acoust Soc Am 33:248

Zwicker E, Fastl H (1972) On the development of the critical band. J Acoust Soc Am 52:699

Remarks on Recent Developments in Inverse Scattering Theory

K.J. LANGENBERG[1], G. BOLLIG, D. BRÜCK, and M. FISCHER[2]

1 Introduction

Inverse scattering, as applied to such different fields as radar, geophysical probing, medical diagnostics, or non-destructive testing of materials with ultrasound attempts to identify or reconstruct a geometrical obstacle and structure illuminated by an electromagnetic, acoustic or elastic wave from the information contained in the scattered field. As illustrated by the far-field behavior, this information is available through the radiation pattern of the obstacle, which is generally a function of direction and frequency uniquely determined by the scattering geometry. The inversion process, therefore, requires the variation of the mutual situation of the transmitter and receiver and/or the frequency of the exciting wave.

The minimum amount of data is obtained for time harmonic excitation from a prescribed direction, and recording the (scalar) scattered field within an aperture surrounding the scatterer completely. A proposed reconstruction scheme is called Generalized Holography (Porter 1970) or Exact Inverse Scattering (Bojarski 1973) whose results are, unfortunately, not unique (Bleistein and Cohen 1977, Devaney and Sherman 1982). Therefore, one has to change either to multiple experiments, i.e. recording the scattered field for a variety of different angles of incidence of the exciting wave (Prosser 1969, Wolf 1969, Devaney 1982), or choosing an impulsive, i.e. broadband excitation (Fischer and Langenberg 1984); in the latter case, special assumptions about the induced surface distribution or the scattering potential are then necessary (Bleistein and Cohen 1977) and, when explicitly introduced, yield algorithms like POFFIS (Physical Optics Far-Field Inverse Scattering) (Bojarski 1982, Bleistein 1976) and IBA (Inverse Born Approximation) (Rose and Richardson 1983). Alternatively, a heuristic inversion of the direct scattering problem has been formulated, which is closely related to synthetic aperture radar, and, for non-destructive testing applications, has been acronymed SAFT (Synthetic Aperture Focusing Technique) (Ganapathy et al. 1982).

We demonstrate that in the far-field all three above-mentioned reconstruction schemes are identical.

[1] University of Kassel, Dept. Electrical Engineering, FB 16, 3500 Kassel, FRG
[2] Fachbereich 12.2 Elektrotechnik, Universität des Saarlandes, 6600 Saarbrücken, FRG

Localization and Orientation in Biology and Engineering
ed. by Varjú/Schnitzler
© Springer Verlag Berlin Heidelberg 1984

2 Non-Destructive Flaw Detection with Ultrasound

Typical flaws in elastic materials are voids whose shapes range from voluminous, nearly spherical structures to linear or penny-shaped cracks. The computation of ultrasound scattering by such geometries is highly involved because of the occurrence of longitudinal as well as transverse waves and their mutual coupling on the flaw surfaces. Nevertheless, elastic inverse scattering should account for these effects and try to utilize them; the first attempts have been made in connection with IBA, but all other algorithms are essentially scalar, and, if investigated with respect to their basic potential, this assumption would be a very useful one. Computer simulations reveal (Bollig and Langenberg 1983) that at least for longitudinal-longitudinal scattering a scalar, in this case rigid, boundary condition in connection with propagation media supporting no shear waves is a quite satisfactory model. Hence, we restrict ourselves to a scalar formulation.

3 Exact Inverse Scattering

Consider a homogeneous fluid with constant compressibility κ_0 and density ρ_0 where a scattering body with surface S_c and spatially varying and frequency dependent $\kappa(\mathbf{r}, k)$ and $\rho(\mathbf{r}, k)$ is embedded; then the frequency domain (wave number k) total pressure field $\phi(\mathbf{r}, k)$ satisfies the inhomogeneous Helmholtz equation (Morse and Ingard 1968)

$$(\Delta + k^2)\phi(\mathbf{r}, k) = \nabla \cdot \gamma_\rho(\mathbf{r}, k) \nabla \phi(\mathbf{r}, k) - k^2 \gamma_\kappa(\mathbf{r}, k) \phi(\mathbf{r}, k) \tag{1}$$

with $\gamma_\rho(\mathbf{r}, k) = [\rho(\mathbf{r}, k) - \rho_0]/\rho_0$ and $\gamma_\kappa(\mathbf{r}, k) = [\kappa(\mathbf{r}, k) - \kappa_0]/\kappa_0$; the right hand side of Eq. (1) is called the scattering potential and can be interpreted as a source term $-q(\mathbf{r}, k)$. Among other formulations, Exact Inverse Scattering derives the following integral equation for $q(\mathbf{r}, k)$

$$\theta_h(\mathbf{r}, k) = 2j[\gamma(\mathbf{r}) q(\mathbf{r}, k)] * G_i(\mathbf{r}, k) \tag{2}$$

were * denotes a three-dimensional convolution, and $\gamma(\mathbf{r})$ is the characteristic function of the volume V_M bounded by a surface S_M, where $\phi(\mathbf{r}, k)$ and $\nabla\phi(\mathbf{r}, k)$ are known (measured) for all $\mathbf{r}$ and fixed k, where k is prescribed by the frequency of the incident time harmonic wave. The quantity $\theta_h(\mathbf{r}, k)$ is called the holographic reconstruction and is defined through the backward wave propagation argument by

$$\theta_h(\mathbf{r}, k) = \int_{S_M} dS' \cdot \{G^*(\mathbf{r} - \mathbf{r}', k) \nabla'\phi(\mathbf{r}', k) - \phi(\mathbf{r}', k) \nabla'G^*(\mathbf{r} - \mathbf{r}', k)\} . \tag{3}$$

G^* is the complex conjugate of the free space Green's function and G_i its imaginary part. It can be shown that due to the non-uniqueness of the solution of Eq. (2) the holographic reconstruction θ_h is already the best possible information about the source which can be obtained out of Eq. (2) (Fischer and Langenberg 1984). This is the reason for the relative success of a special version of Eq. (3) for planar measurement surfaces S_M, the so-called Rayleigh-Sommerfeld holography, especially for crack-like scatterers (Berger et al. 1981), which reveal very specific source and scattered field distributions (Langenberg et al. 1983a). On the other hand, uniqueness can be enforced for broadband excitation, and, by introducing special assumptions concerning the source distribution, pertinent algorithms like IBA and POFFIS can be derived.

4 Broadband Synthetic Aperture Algorithms

4.1 Inverse Born Approximation (IBA)

The Inverse Born technique starts with an iterative solution of Eq. (1) in terms of

$$\phi_{n+1}(r,k) = \phi_i(r,r_0,k) + \int_{V_c} d^3 r' \frac{e^{jk|r-r'|}}{4\pi|r-r'|} q(r',k)\,\phi_n(r',k) \tag{4}$$

$$n = 0,1,2,3 \dots$$

where the incident field $\phi_i(r',r_0,k)$ from a source point r_0 is inserted as zero-order approximation for the total field $\phi_0(r',k)$ under the integral, which might be sufficiently valid for a weak scatterer. Assuming a point source at r_0, introducing the far-field approximation

$$e^{jk|r-r'|}/4\pi|r-r'| \cong e^{jk(r-\hat{r}\cdot r')}/4\pi r \text{ with } \hat{r} = r/r$$

and manipulating the differential operators in q yields for the first-order scattered field $\phi_s = \phi_1 - \phi_i$

$$\phi_s(r,k) = k^2 \frac{e^{jk(r+r_0)}}{(4\pi r)(4\pi r_0)} \int d^3 r' \, e^{-jk(\hat{r}+\hat{r}_0)\cdot r'} \, \gamma(r')\,p(r',k) \tag{5}$$

with p being a modified source called object function and γ denoting the characteristic function of the scattering volume V_c. Now we recognize the integral in Eq. (5) as a Fourier integral with the Fourier variable $K = k(\hat{r}+\hat{r}_0)$; therefore, the Fourier transformation, $\tilde{p}(K,k)$ of the object function is given by

$$\tilde{p}(K,k) = (4\pi r)(4\pi r_0)\, e^{-jk(r+r_0)} \frac{\phi_s(r,k)}{k^2} \tag{6}$$

and can be experimentally determined within the volume of the "double" Ewald sphere $|K| < 2k$ by variation of $\hat{r}+\hat{r}_0$ for fixed $\hat{r}_0$, i.e. performance of multiple experiments to determine the time harmonic scattered field for various $\hat{r}_0$ yields a data set out of which the object function can be calculated via Fourier transform. Independently moving transmitters and receivers are needed for these experiments and this is not practical. Alternatively, the same data space is spanned by variation of the frequency and $\hat{r}_0$, but with $\hat{r} = \hat{r}_0$, i.e. with a so-called monostatic or impulse echo technique; provided the object function does not depend explicitly on the frequency, Eq. (6) is then a broadband reconstruction scheme based on the far-field Born approximation, called IBA. Incidentally, fixed $\hat{r}_0$ combined with varying k and $\hat{r}$ leads to a sub-set of data space revealing Eq. (6) to be the FIFFIS technique (Frequency Independent Far-Field

Scattering) (Detlefsen 1979), which is well-known to those people connected with radar; it is suspected that the results range between time harmonic holography and IBA.

4.2 Physical Optics Far-Field Inverse Scattering (POFFIS)

The POFFIS identity has been derived primarily under the assumption of a perfect scatterer, be it either rigid or soft. It is obtained by writing down the Kirchhoff integral far-field solution of the homogeneous Helmholtz equation for the scattered field similar to Eq. (3), using G instead of G*, with S_c as integration surface. By inserting the Kirchhoff (Physical Optics) assumption – total field proportional to incident field on the illuminated side of the scatterer and zero in the shadow zone – and presuming a monostatic arrangement $r = r_0$, combining the field for r_0 with the complex conjugate for $-r_0$, and changing from the surface to a volume integral and by introducing the characteristic function of the scatterer the relation

$$\tilde{\gamma}(K) = (4\pi r)^2 \, e^{-2jkr} \, \frac{\phi_s(r,k) + \phi_s^*(-r,k)}{(2k)^2} \tag{7}$$

with $K = 2k\hat{r} = 2\frac{\omega}{c}\hat{r}$ is yielded, which is already very similar to Eq. (6) and becomes nearly identical, if the inverse Fourier transform F_K^{-1} with respect to K is combined with $\phi_s(r,k)$ and $\phi_s^*(-r,k)$ to give

$$\gamma(r') = 2(4\pi r)^2 \, \text{Re} \, F_K^{-1} \left\{ \frac{\phi_s(\hat{r},k)e^{-2jkr}}{(2k)^2} \, ; r' \right\} \tag{8}$$

At first sight this result seems surprising because Eqs. (6) and (8) have been derived under completely different assumptions: weak scattering is valid for low frequencies and Physical Optics for high frequencies. But, ignoring the fact that weak scattering is not meaningful for a perfect scatterer and repeating the procedure of the foregoing section e.g. for a rigid scatterer, setting in Eq. (1) $\gamma_\rho \to \infty$, $\nabla\phi \to 0$, $\kappa(r,k) = 0$, and $\gamma_\kappa(r,k) = -\gamma(r)$, we obtain the identity (6) for the characteristic function itself.

The deeper reason for this coincidence is that both POFFIS and IBA concentrate on the scattering surface, the first procedure by definition and the latter one by assuming an undisturbed incident field inside the scatterer. The rôle of this surface detection is still more emphasized if a POFFIS identity for the singular function, which peaks on the surface, is derived (Cohen and Bleistein 1979). It consists essentially of a multiplication of the right hand side of Eq. (7) by k, thus accentuating high frequencies and therefore being related to edge enhancement procedures.

Equations (8) and (6) respectively become still more intuitive if they are interpreted in the time domain. This is easily done by rewriting Eq. (8) for spherical coordinates in K-space $d^3K = K^2 \, dKd\Omega$

$$\gamma(r') = \frac{16r^2}{c} \, \text{Re} \int d\Omega \, \frac{1}{2\pi} \int_{-\infty}^{\infty} d\omega \phi_s\left(\hat{r},\frac{\omega}{c}\right) e^{-2j\frac{r}{c}\omega} e^{2j\frac{\hat{r}\cdot r'}{c}\omega} U(\omega) \tag{9}$$

where $U(\omega)$ denotes the unit step-function. The ω-integral is an inverse Fourier integral for the time $t = 0$ and therefore

$$\gamma(\mathbf{r}') = \frac{16\,r^2}{c} \operatorname{Re} \int d\Omega \phi_s(\hat{\mathbf{r}}, t) * U(t) \Big|_{t = 2\frac{r}{c} - 2\frac{\hat{\mathbf{r}} \cdot \mathbf{r}'}{c}} \tag{10}$$

with

$$U(t) = \frac{1}{2}\delta(t) + \frac{1}{2\pi} \operatorname{Pf} \frac{1}{jt} \tag{11}$$

and $\phi_s(\hat{\mathbf{r}}, t)$ the scattered time domain impulse response.

Hence

$$\gamma(\mathbf{r}') = \frac{8\,r^2}{c} \int d\Omega \phi_s(\hat{\mathbf{r}}, 2\tfrac{r}{c} - 2\,\frac{\hat{\mathbf{r}} \cdot \mathbf{r}'}{c}) . \tag{12}$$

For fixed reconstruction point $\mathbf{r}'$, Eq. (12) is just the collection of all time domain amplitudes which originate from that point with the correct travel time under the Physical Optics and far-field assumptions. We state that Eq. (8) is useful for processing and Eq. (12) for interpretation.

4.3 Synthetic Aperture Focusing Technique (SAFT)

Originally, SAFT is an intuitive reconstruction procedure closely related to Synthetic Aperture Radar (without time domain-matched filtering) which works directly with broadband transient signals not only in experiment but also for processing. Therefore, it heuristically picks up the data collection idea according to

$$\gamma_{SAFT}(\mathbf{r}') = \int d\Omega F(\hat{\mathbf{r}}, \frac{2}{c}|\mathbf{r} - \mathbf{r}'|) \tag{13}$$

where $F(\hat{\mathbf{r}}, t)$ is the monostatic backscattered field for pulse excitation $F(t)$. An image is then obtained via envelope detection in the reconstruction (pixel) space $\mathbf{r}'$. It is obvious that Eq. (13) represents primarily an extension of Eq. (12) without utilizing the far-field approximation, which is, in a strict mathematical sense, incorrect; on the other hand, at least the physical assumption to justify Eq. (13) in the far-field is revealed comparing it with Eq. (12), namely the Physical Optics assumption. Secondly, Eq. (13) extends Eq. (12) to arbitrary excitations $F(t)$, for example short sine bursts as they are used in ultrasonic material testing. This can be justified (again for the far-field) by defining a "new" characteristic (or singular) function out of Eq. (10)

$$\gamma_{burst}(\mathbf{r}') \cong |\int d\Omega \phi_s(\hat{\mathbf{r}}, t) * U(t) * F(t)|_{t = 2\frac{r}{c} - 2\frac{\hat{\mathbf{r}} \cdot \mathbf{r}'}{c}} \tag{14}$$

where the imaginary part of $U(t)$ yields the Hilbert transform of $F(t)$, and, hence, the absolute value in Eq. (14) accounts for an envelope detection: well understood in data, not in pixel space.

Even though SAFT cannot be justified in terms of rigorous mathematics it works nicely in practice. The reason is that for actual data, IBA and POFFIS reduce to a mere time of flight reconstruction procedure (with noise suppression on behalf of the averaging in Eq. (10) (Langenberg et al. 1983b) and these times are correctly accounted for by SAFT, even in the near-field.

References

Berger M, Brück D, Fischer M, Langenberg KJ, Oberst J, Schmitz V (1981) Potential and limits of holographic reconstruction algorithms. J Nondestr Eval 2:85

Bleistein N (1976) Physical optics far-field inverse scattering in the time domain. J Acoust Soc Am 60:1249

Bleistein N, Cohen JK (1977) Non-uniqueness in the inverse source problem in acoustics and electromagnetics. J Math Phys 18:194

Bojarski NN (1973) Inverse scattering. Naval Air Systems Command, Washington, D.C., Naval Air Systems Command Rep., Contract N 000 19-73-C-0312, Sec 11, p 3

Bojarski NN (1982) A survey of the physical optics inverse scattering. IEEE Trans Ant Propagat AP-30:980

Bollig G, Langenberg KJ (1983) The singularity expansion method as applied to the elastodynamic scattering problem. Wave Motion 5:331

Cohen JK, Bleistein N (1979) The singular function of a surface and physical optics inverse scattering. Wave Motion 1:153

Detlefsen J (1979) Abbildung mit Mikrowellen. Fortschrittberichte der VDI-Zeitschriften, Reihe 10, Nr 5. VDI-Verlag, Düsseldorf

Devaney AJ, Sherman GC (1982) Non-uniqueness in inverse source and scattering problems. IEEE Trans Ant Propagat AP-30:1034

Devaney AJ (1982) An inversion formula for inverse scattering with-in the Born approximation. Optics Letts 7:111

Fischer M, Langenberg KJ (1984) Limitations and defects of exact inverse scattering. IEEE Trans Ant Propagat (accepted for publication)

Ganapathy S, Wu WS, Schmult B (1982) Analysis and design for a real-time system for nondestructive evaluation in the nuclear industry. Ultrasonics 20:249

Langenberg KJ, Brück D, Fischer M (1983a) Inverse scattering algorithms. In: Höller P (ed) Research and development to new procedures in NDT. Springer, Berlin Heidelberg New York

Langenberg KJ, Bollig G, Brück D, Fischer M (1983b) Remarks on recent developments in inverse scattering theory. Proc Int Symp URSI Comm B, Santiago de Compostela, Spain

Morse PM, Ingard KU (1968) Theoretical acoustics. McGraw Hill, New York

Porter RP (1970) Diffraction-limited scalar image formation with holograms of arbitrary shape. J Opt Soc Am 60:1051

Prosser RT (1969) Formal solution of inverse scattering problems. J Math Phys 10:1819

Rose JH, Richardson JM (1983) Time domain Born approximation. J Nondestr Eval 3 Eval

Wolf E (1969) Three-dimensional structure determination of semi-transparent objects from holographic data. Optics Comm 1:153

Signal Extraction and Target Detection for Localization by Radar

G. KÄS[1]

1 Why Use Radar?

Today information is often used to increase the probability of survival. Some thousands of years ago, it was perhaps needed to recognize a beast early enough to escape; today it is needed in many cases for collision avoidance. Information is only useful when it is available in sufficient time for a helpful reaction; that means sufficient to avoid collision. The range R_e to detect an obstacle must be greater than the distance R_m that is needed for collision avoidance

$$R_e \geqslant R_m \ .$$

R_m depends on the velocity v and the time t_r needed for reaction. If we say in a simplified way $R_m = v \, t_r$, then the necessary value of R_e increases with growing velocity v.

$$R_m \sim v \quad \text{(with constant } t_r) \ .$$

In the simplest case of optical target detection R_e is equal to R_{opt}, the range for detection with optical means. Collision cannot be avoided if (1) optical sight is highly reduced (by rain or fog for instance) or (2) velocity is very high because in both cases is

$$R_e = R_{opt} < R_m \ .$$

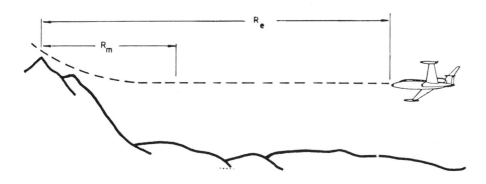

---

[1] Pfaffenhofen a.d. Ilm, FRG

Localization and Orientation in Biology and Engineering
ed. by Varjú/Schnitzler
© Springer Verlag Berlin Heidelberg 1984

To solve this problem it is necessary to increase range by using electromagnetic waves with a wavelength far greater than the wavelength of light. Attenuation by the atmosphere is highly reduced. Therefore the range of radar systems is far greater than the range of optical systems. This means that collision avoidance is achievable, although the velocity is high and the optical sight is reduced,

$$R_e \geqslant R_m \text{ and } R_e > R_{opt} .$$

2 Radar Detection

Electromagnetic waves are reflected if the permittivity or permeability in the medium of propagation changes. A receiver may record the transmitter signal that is reflected by the target.

To locate a target on a (two-dimensions) plane at least two measured values are needed for the two coordinates, usually obtained by measuring range and angle.

The target angle is given by using high gain antennas illuminating a very small sector only. All incoming (echo) signals are attached to this sector. Range is obtained by measuring the time of propagation the signal needs for the distance transmitter-target back and forth. Therefore short microwave pulses are transmitted and the time Δt is measured. The velocity of propagation c is well known and the distance R is calculated as

$$R = c \cdot \Delta t/2 .$$

The signal power P_e of the echo diminishes with increasing distance R (Fig. 3). The maximum range is reached if a signal cannot be detected with sufficient probability within the noise.

$$P_e = \sim \frac{1}{\sqrt[4]{R}} .$$

Target signal extraction from noise is the main problem.

3 Target Signal Processing

To get an optimum in signal-to-noise ratio (SNR) it is necessary to study the structure of signal and noise. Mostly we distinguish between noise and clutter. Internal equip-

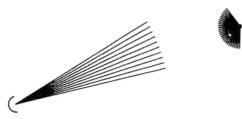

Target

Fig. 2. Target illumination and reflected signal

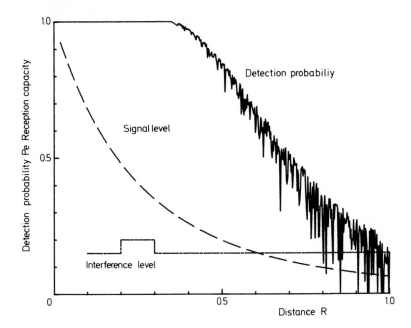

Fig. 3. Probability of detection versus range

ment noise results from changes in voltage, current, temperature etc., in all noted as N, while clutter is a summary of unwanted reflections as echo signals from hills, buildings, clouds an similar objects; when moving in respect to the earth noted as C2, and those from nonmoving objects noted as C1.

To improve the S/N ratio measures concerning receivers are preferred today because transmitters work with the maximum available power and therefore range may be hardly extended. Three examples will be given here in order to emphasize data processing by modern receiver structures: (1) integration of pulses to improve SNR (signal-to-noise ratio); (2) adaptive antenna diagrams (specifically tailored) to improve $SC_1 R$, and (3) Doppler filtering to improve $SC_2 R$.

Integration of pulses offers a SNR improvement of (ideally) 3 dB for twice the number of pulses because the signal is constant in amplitude and polarity while the noise varies statistically in phase and amplitude at the same place from pulse to pulse.

Adaptive antennas improve SCR conditions especially for clutter from nonmoving objects in the simplest case. The socalled clutter may consist of ground clutter, i.e. (unwanted) echos from buildings, hills or other objects situated on the earth with no or respectively small (relative) velocities as created by (the top of) trees or chimneys for instance. All those echos are summarized as ground clutter and noted as C_1 (Clutter of the order 1). To improve $SC_1 R$ we use adaptive antennas. One way is to change object depending signal strength while illuminating. If this is possible, these objects will also answer with signals of different strength. The easiest way of attenuating signal range and azimuth dependency while receiving.

To achieve this effect, the resulting (receiving) antenna diagram is the sum of some different particular diagrams which are partly attenuated depending on range and azimuth. Thus, by attenuating different parts of the antenna diagram by way of attenuating their receiver channels, a unique site-adaptive range-azimuth clutter level may be obtained. Gain control of the different receiving channels effecting the antenna diagrams is done by a range-azimuth generator with programmable range and angle datas.

While adaptive diagrams improve the signal-to-(nonmoving) clutter ratio, SC_1R, the signal-to-(moving)clutter ratio, SC_2R, is improved by doppler filtering.

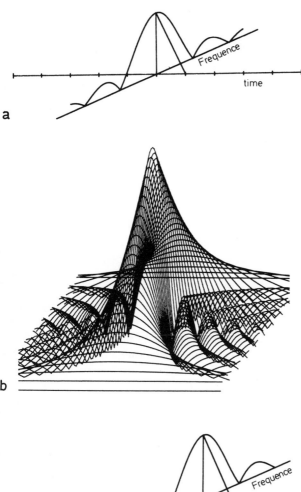

Fig. 4 a–c. Time domain and frequency domain in one signal a, the ambiguity function (plotted) of this signal b, and two signals with different Doppler frequencies depending on the (different) target velocities c

Doppler filtering permits target separation in one range cell, if the targets differ in velocity and if they have therefore different doppler frequency shifts. The resolution in range and azimuth of a radar is limited; in range by the pulse duration and in azimuth (or in angle generally) by the antenna diagram. Those small cells halve the pulse duration (multiplied with the velocity of electromagnetic waves) and about halve the beamwidth of the antenna diagram width are called resolution cells or range cells. If the Doppler shift is not measured, two targets in one single range cell may not be separated, and therefore they may not be detected as two targets, they seem to be only one target.

But if the Doppler shift is measured, the two targets may be separated by their different Doppler frequencies. In this case we get a third degree of freedom and in fact another (additional) way to improve SCR by improving SC_2R besides resolution in range and angle. This improvement can be easily explained with the help of Fig. 4. In this figure **a** shows the power distribution of a signal above the time-frequency plane, **b** shows the complete (three-dimensional) distribution of the signal in **a**, and **c** presents two signals in one range cell separated only by their different target velocity respectively Doppler shift.

Each of these three examples shows the possibility to increase the probability of signal detection in noise and clutter and, while target velocity is a well known distinguishing feature, the possibility of target distinction for localization by radar.

Environment and Factors Influencing
Optimum Multi-Radar Processing in Air Traffic Control

H. VON VILLIEZ[1]

1 Introduction

The more recent air traffic control systems are semi-automated to increase considerably the traffic handling capacity. The basic concept is to assist the controller in his mental data processing workload, and through this to improve the overall system productivity.

The processing of flight plan information is essential for the efficient planning of a conflict-free traffic flow. The processing of radar data is vital as it is the tool by which air traffic control is safely executed. This explains why much effort still goes into the acquisition, processing and display of radar data so as to provide the air traffic controller with the most accurate, instantaneous picture of the actual air situation. The development in using radar information made great progress with the advent of suitable digital processing techniques enabling the transfer of radar data from one or more radar sites, via telephone lines, to places where they could be operationally exploited.

Civil air traffic control systems still use both types of radar, primary, and secondary, simultaneously and will continue to do so for some time to come. Thus, we will continue to base our current air traffic presentation to the controller on both information sources, which will ideally present combined plot data on actual aircraft positions leading to tracks supported by two, three or more radars.

2 Operational Advantages of Multi-Radar Processing

The operational requirement for the provision of radar data in air traffic control calls for reliability in the provision of these data. This has led in most systems to a double radar coverage of the airspace to be controlled. In itself, this does not lead to a multi-radar processing concept; it is primarily the existing density of long range radars in the western part of Europe and in particular their multi-national utilisation as introduced by Eurocontrol. This increases the required availability of radar data and offers considerable advantages in terms of additional functions inherent in an intelligent

---
[1]   Eurocontrol Maastricht U.A.C., 6236 ZH Beek (L), Netherlands

Localization and Orientation in Biology and Engineering
ed. by Varjú/Schnitzler
© Springer Verlag Berlin Heidelberg 1984

exploitation of several powerful radar installations, specifically across national boundaries.

The multi-radar environment, as used by the Eurocontrol Maastricht Upper Area Control Centre is given in Fig. 1. The exploitation of this multiple-radar environment is influenced by the fact that the radars are of different design and come from different manufacturers. However, at the digital extractor level, Eurocontrol succeeded together with its Member States in developing a common specification which guarantees at least the same message format for the build-up and transmission of radar data from the different sites.

3 Selected Functions for System Optimisation

By the wide geographical distribution of these radar stations, we arrive at a situation that any aircraft flying in the upper airspace system is seen simultaneously by on average, 2.5 radars. There is in certain areas a coverage of 4–5 radars offering an enormous excess capacity. Certain selected functions for system optimization deserve our attention.

3.1 Monitoring and Automatic Alignment of All Radars

The radar picture displayed to the air traffic controller is built up from the measurements of different radars. Its quality is highly dependent on the formulae used to fit the different measurements together and the adjustment of the radars themselves.

In the system here under consideration the positions of the radar stations for the system plane are precisely calculated; the radar measurements, however, are calcu-

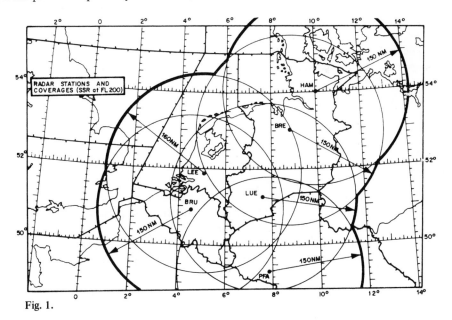

Fig. 1.

lated in local radar planes and then fitted into the system plane by a shift and a turn (north convergence). This results in a mean error of 125 m between true and calculated position in the system plane. The maximum error is 780 m.

Experience has shown that the adjustment of the radars themselves i.e. correct north alignment and range calibration, is the crucial factor for optimum corrections at display level. As there are generally deviations from the optimum adjustment, and as they are time dependent, correction algorithms were developed which ensure optimum adjustment.

Based upon the method of the minimum square deviation they detect and correct adjustment errors of the incoming radar data dynamically, eventually leading to a

- mean distance between aircraft seen by radar pairs: $\bar{d}_{R_1, R_2} = 260 - 390$ m
- mean distance within the overall system: $\bar{d} = 320\ m$.

3.2 Improved Probability of Detection

An important element in the assessment of the useful coverage of a radar the collective coverage of a given number of radars respectively, is the probability of detection (Pd). It is usually given as the percentage of target detections to the number of antenna scans to which the target was exposed. The multi-radar tracking concept as used in the Eurocontrol Maastricht Centre makes it possible to reconstitute an accurate aircraft trajectory, provided that at least one radar or a group of radars provides sufficient measurements to maintain the established system track.

Knowledge of the existence and position of a particular aircraft allows to calculate for every radar antenna the time and predicted position it will next illuminate a particular target. It is thus feasible to measure the ratio between actual target detections and the number of antenna revolutions during which the radar should have observed that target. This principle is especially attractive in areas of weak coverage of a single radar.

Because of the fact that the type of detection, i.e. combined plot, primary or secondary plot is known, the Pd of both PR and SSR may be measured. Using opportunity traffic allows the automatic collection of measurement data leading eventually to statistically significant numbers of observations at practically no extra cost. This outweighs the limited advantages of the much more costly radar checkflight campaigns.

A convincing distinction between primary and secondary radar capability can be obtained from Figs. 2 and 3. Both show the probability of detection (Pd) over the range (NM) with the various height layers as parameter; it starts with layer 1 from FL 00-20 (0– 2,000 ft) in steps of 2,000 ft. Thus, layer 8 shows the Pd value measured between FL 140 and FL 160. The advantage of the secondary surveillance radar in the vicinity of the radar site is clearly visible.

This recently developed method for automatic Pd measurements, with the desired height-layer and/or specific azimuth sectors as parameter, has turned out to be an extremely valuable tool for relevant studies and also for routine performance monitoring of the whole radar system used.

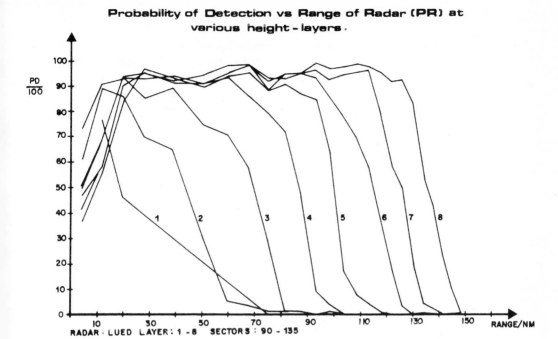

Fig. 2.

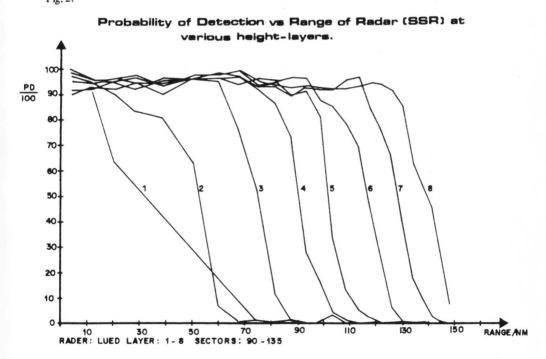

Fig. 3.

3.3 Conflict Alert in the Horizontal and Vertical Plane

The operation of a multi-radar tracking system has in the recent past enabled the implementation of a short term conflict alert function, which is highly appreciated by our air traffic controllers. It is to be seen as a safety net, which warns the controller 2 min in advance about an imminent violation of the separation standards. The flight paths of all aircraft in the system are calculated for the next 128 s and compared to each other every 5 s so as to discover the potential infringement of the fixed separation standards in good time.

The essential expansion was the development of an independent tracking algorithm in the third dimension using successive altitude reporting data (Mode C information of the secondary surveillance radar system) simultaneously from up to three radars.

The basic function creates a single height track as soon as an increase in height is discovered; the tracking up-date cycle is 5 s where altitude information received from the relevant three radars is used. Upon completion of the vertical manoeuvre, the relevant height track will be erased.

A sliding window type selection of measurement points is applied so as to store the recent history covering an interval of up to 2 min. This makes it possible to derive an estimate of the vertical transition rate. As in the majority of tracking systems, it is assumed that the vertical evolution of the aircraft is a uniform straight motion by which the choice of a linear regression method appears to be justified.

This leads to the requirement for a highly sensitive detector of changes in the vertical slope which can be achieved by permanently monitoring the linear correlation coefficient (r) between the respective plot detection time and the reported height values and establishing a theoretical minimum value of $|r|$. Thus we can say: whenever, for a certain sample population, the a posteriori calculated $|r|$-value drops below the a prior calculated statistical minimum, it is highly probable that the aircraft has changed its vertical transition rate. In that case most of the history data are erased and a new value for the rate of climb (or descent) is established as soon as sufficient new "Mode C" data are available.

The next important step is the height prediction. There are two reasons for deviations between the predicted and real height for a particular point of time in the future:

– the errors in the present height and present rate of climb/descent (due to stochastic errors in Mode C data)
– deviations due to frequent changes in the rate of climb/descent as occurring for non-linear height profiles.

Here there are the more serious problems in the design of a conflict alert function because it is difficult for the height tracker to stabilize under such circumstances and consequently difficult to obtain an accurate 2-min prediction. However, the optimization of the height tracking function, as well as the application of adequate tolerance values leading to a proportional height band for the prediction interval, has given satisfactory results.

4 Optimum Display System for Processed Radar Data

The display concept is organized in such a way that all radar data from the different sites feeding the system can construct a coherent radar picture covering the overall area of responsibility. From this large total picture every radar controller selects that portion for his working position/display screen which suits him best in the performance of his individual task regardless of whether two, three or more radars are providing the relevant targets.

The combination of the radar data processing functions, together with carefully selected input facilities and this display concept, guarantee a presentation of all air-craft in the system with a very high quality. This has shown to be the best man-machine interface presently possible in this sensitive profession.

Collision Avoidance in Air Traffic – Conflict Detection and Resolution

W. SCHROER[1]

1 Collision Avoidance: Objective and Fundamentals

Because of the potential danger of serious accidents caused by midair collisions, in recent years great efforts have been made in developing collision avoidance systems (CAS). However no system has been introduced up to now.

The discussion has arisen again since the United States Federal Aviation Administation (FAA) presented their new proposal, TCAS (Threat Alert and Collision Avoidance System), FAA (1982). TCAS is an airborne system using the secondary surveillance radar (SSR) transponder to identify intruders and enable a data-link between conflicting aircrafts.

Each collision avoidance system must work in conjunction with Air Traffic Control (ATC), the final authority on collision prophylaxis. ATC surveys all aircraft, flying with instrument flight rules, (IFR) controlling their separation at aminimum distance of 3 nmi within the Terminal Area or of 5 nmi en route. Compatibility of CAS and ATC can be achieved, if the CAS reacts at a distance not greater than the ATC separation, that is, if ATC has obviously failed.

This paper deals with algorithms of detecting conflicts with airborne systems and discusses collision avoidance by vertical and horizontal maneuvers.

2 The Geometry of Collision and Evasive Maneuvers

2.1 The Geometry of Collision

The conflict geometry can be described by the triangle of collision (Fig. 1a). If both, the own aircraft (velocity v_o) and the intruder (velocity v_i) are flying with constant heading towards the collision point, the relative bearing ϕ of the intruder is constant.

The collision condition is

$$\frac{v_o}{\sin \alpha} = \frac{v_i}{\sin \phi} \, . \tag{1}$$

---

[1] TU Braunschweig, FRG

Localization and Orientation in Biology and Engineering
ed by Varjú/Schnitzler
© Springer Verlag Berlin Heidelberg 1984

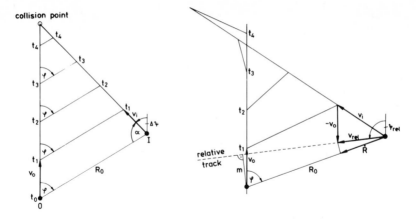

Fig. 1a,b. a Collision triangle. **b** Near miss with geostationary and bordrelative track

The deflection angle

$$a = \arcsin\left(\frac{v_i}{v_o} \sin \phi\right) \tag{2}$$

has one solution for $-\pi \leqslant \phi \leqslant +\pi$ if $v_o < v_i$, and two solutions if $v_o > v_i$, but only for small relative bearings ϕ.

The pilot sees the intruder approaching on the relative track. The relative motion can be evaluated by subtraction of the velocity vectors v_i and v_o (Fig. 1b). The closing velocity $\dot{R}$ depends on the relative velocity v_{rel} and the miss distance m. In near miss situations, it is accelerating

$$\dot{R} = v_{rel} \frac{\sqrt{R^2 - m^2}}{R} . \tag{3}$$

In this paper, a near miss is defined to be a collision, if the miss distance is less than $m_o = 200$ m horizontal or $z_o = 100$ m vertical.

2.2 Resolution Maneuvers

For successful conflict resolution, a minimum alarm distance has to be ensured. Because of strength limitations, the flight mechanical parameters must not exceed any critical threshold. Moreover, evasive maneuvers should only use normal operating accelerations to prevent endangering passengers, when the CAS generates unnecessary alerts.

Figure 2 shows the minimum reaction distance for a vertical maneuver with a climb- or descend-rate of 2,000 ft/min vs. relative bearing of the intruder. The velocity is 200 kts, that of the intruder is shown as parameter.

Figure 3 depicts the same configuration for horizontal maneuvers "turn to" and "turn off" (Form and Schroer 1982). A restriction to only one reaction plane – vertical or horizontal – is unsuitable because of following conflicts by any other air-

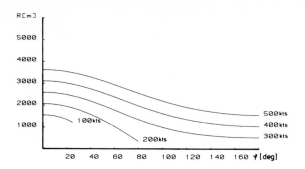

Fig. 2. Vertical maneuver: minimum reaction distance R for a successful conflict resolution vs relative bearing φ. v_0: 200 kts, v_i: parameter; climb/descend rate: 2,000 ft/min

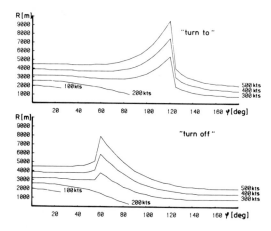

Fig. 3. Horizontal maneuver: minimum reaction distance for a successful conflict resolution vs relative bearing φ v_0: 200 kts; v_i: parameter; banking: 4.5 deg/s, limit 30 deg

craft. However, horizontal maneuvers require the relative bearing measurement on board.

For both maneuver planes, conflict resolution is possible within the ATC separation distance.

3 Conflict Detection and Collision Prediction

3.1 The Distance-Coaltitude-Criterion

The most simple criterion for conflict detection uses a distance threshold at coaltitude. Both the distance and the pressure altitude can be measured and transmitted respectively with an interrogator and a transponder. The conflict decision is positive, if the intruder penetrates the distance threshold.

The probability that two approaching aircrafts will pass each other within a distance $m_0 = 200$ m, depends on the relation $v_i:v_0$. In Fig. 4 it is supposed that the probability of the intruder's heading is uniformly distributed between $0°$ and $360°$; parameter is the distance threshold of decision. This criterion is safe, but suffers from a large number of unnecessary alarms, even if the distance of decision is small.

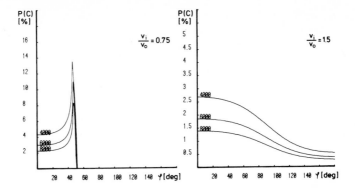

Fig. 4. Probability of collision P(C) vs relative bearing φ uniform heading distribution miss distance threshold m_o = 200 m

3.2 The τ-Criterion

A reduction of false alerts can be achieved by using the τ-criterion (Ata 1971, Zeitlin 1982); the time-to-collision τ can be evaluated by distance measurement and its differentiation. Evasive manoeuvers are to be initiated, if

$$\tau = |\frac{R}{\dot{R}}| \leqslant \tau_o . \tag{4}$$

As the relative bearing is unknown, only vertical maneuvers can be undertaken. τ must be great enough to climb or descend at least 100 m.

The example in Fig. 5 depicts the deflection angle a for collision conditions [see Fig. 1 and Eq. (1)] vs relative bearing at a distance of R = 5,000 m (v_o = 250 kts, v_i = 200 kts). The area of the τ-criterion with $\tau \leqslant \tau_0$ (parameter) is shown as well. In addition, the approach area ($\dot{R} < 0$), which corresponds to the distance criterion is shown. From the relation of these areas, we can evaluate the probability of unnecessary alarms, which is still too high for sufficient operating of the CAS.

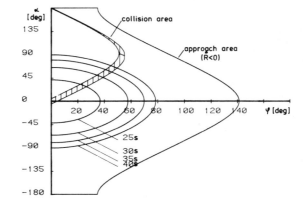

Fig. 5. Deflection angle a vs relative bearing φ
– for collision condition
– for $\tau < \tau_0$
– for (transient) approach
    ($\dot{R} < 0$)

As slowly-arising conflicts will not be detected by the τ-criterion, a combination with the distance-criterion is suitable.

3.3 Tracking of the Intruder

By measuring the relative bearing of an intruder and its distance vs time, it is possible to evaluate the track and to predict the miss distance.

The technology of the relative bearing measurement is still developing (Reed 1982, Welch and Teitelbaum 1982). The azimuth pattern of omnidirectional antenna, mounted on top of a DO 28 Skyservant measured by Bart and Teztlaff (1982) in horizontal and banked attitude gives an impression of the difficulties of the bearing determination (Fig. 6). A precise measurement ($\sigma \leqslant 1°$) seems to be possible with a tolerable economic expence only for air carriers.

The requirements for the track system are defined as: The relative bearing error has a gaussian distribution, and the distance error was found to be of minor influence on the tracking performance. Because of rejection procedures for gross errors and depending on the pulse bandwidth and multipath conditions, we can achieve a spread of about ±50 m. For computing simplification we can regard the error distribution to be uniform.

The estimation of the miss distance is equivalent to the estimate of the angle δ between the line-of-sight and the relative track (Fig. 7). The distribution $F(z)$ of tan $\delta < z$ depends on the cross- and closing speed.

As

$$\tan\delta = \frac{\dot{C}}{\dot{R}} \approx \frac{R \cdot \dot{\phi}}{\dot{R}} = \tau \cdot \dot{\phi} \tag{5}$$

the distribution is

$$F(\tan\delta < z) = \frac{1}{2\,\Delta\tau} \int_{\tau-\Delta\tau}^{\tau+\Delta\tau} \int_{-\infty}^{z/\tau} \frac{1}{\sqrt{2\pi}\,\sigma_{\dot\phi}} \exp\left[-\frac{1}{2}\left(\frac{\dot\phi - \bar{\dot\phi}}{\sigma_{\dot\phi}}\right)^2\right] d\dot\phi\, d\tau , \tag{6}$$

$\bar\tau, \bar{\dot\phi}$ = true value

$\sigma_{\dot\phi}$ = mean error of bearing velocity estimation

$\Delta\tau$ = spread of the τ-error($\pm$)

$$F(z) = \frac{1}{2\,\Delta\tau} \int_{\tau-\Delta\tau}^{\tau+\Delta\tau} \Phi\left(\frac{z/\tau - \bar{\dot\phi}}{\sigma_{\dot\phi}}\right) d\tau , \tag{7}$$

$\Phi(x)$= gauss integral.

The probability of detection $P(D)$ is the integral of $F(z)$ over all true angles $\bar\delta$ which lead to a collision

$$-\theta_0 = -\arcsin\frac{m_0}{R} \leqslant \bar\delta \leqslant +\arcsin\frac{m_0}{R} = +\theta_0 , \tag{8}$$

$$P(D) = \int_{-\theta_0}^{+\theta_0} g_1(\bar\delta) \int_{-\theta^*_{-\bar\delta}}^{\theta^*_{-\bar\delta}} g_2(\Delta\delta)\, d\Delta\delta\, d\bar\delta , \tag{9}$$

$\Delta\delta$ = angle estimation error.

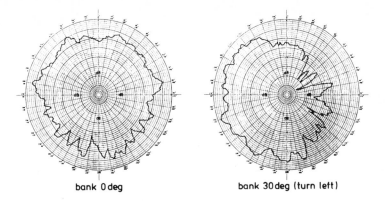

bank 0 deg   bank 30 deg (turn left)

Fig. 6. Horizontal pattern of an omnidirectional L-band antenna mounted on top of a DO 28 Skyservant adopted from: Barth and Tetzlaff (1982)

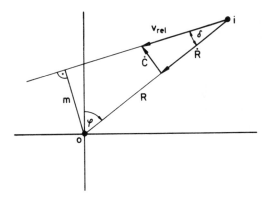

Fig. 7. Relative track with closing- and cross-speed vector

θ^* is the threshold of δ for the detection, that the approach is a conflict. Extension of $\theta^* = \arcsin(m^*/R) \geqslant \theta_0$ enlarges the probability of detection, but also the number of unnecessary alerts (Fig. 8).

The distribution of δ depends on the relation $v_i : v_0$, but is found to be nearly uniform within the small limits of $\pm\theta_0$:

$$P(D) = \frac{1}{2\theta_0} \int_{-\theta_0}^{+\theta_0} \{F[\tan(\theta^* - \bar{\delta})] - F[\tan(-\theta^* - \bar{\delta})]\} \, d\bar{\delta} . \tag{10}$$

Figure 9 shows as numerical evaluation of Eq. (10) the probability of detection vs decision threshold m^* for an approach with a time-to-collision of $\tau = 20$ and 30 s. Figure 10 shows the deflection angle a for conflict conditions ($m \leqslant 200$ m) and conflict decision conditions ($m^* \leqslant 600, 1{,}000$ m). Comparison with the τ-area shows a good decrease in unnecessary alarms, even if $m^* = 1{,}000$ m.

The combination of both τ-criterion and miss distance prediction by tracking with an adaptive m^*-threshold controlled by τ and R seems to be a suitable method for conflict detection. The performance of the tracking depends on the bearing velocity estimation, which may not be worse than $\sigma_{\dot{\phi}} \geqslant 0.1 - 0.3$ deg/s.

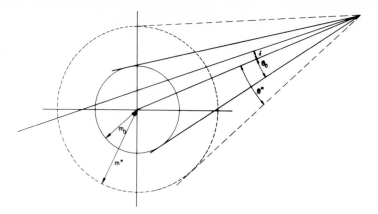

Fig. 8. Relative track of the intruder and threshold for collision decision m[*]

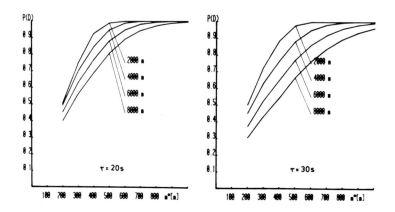

Fig. 9. Probability of detection P(D) vs collision decision threshold m[*] parameter decision distance angle velocity estimation error: $\sigma\varphi = 0.1$ deg/s; τ estimation error $\Delta\tau = \pm 2$ s

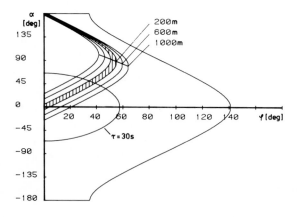

Fig. 10. Deflection angle a vs relative bearing φ
− for collision condition (m = 200 m)
− for collision decision condition (m = 600, 100 m)
− for $\tau < 30$ s

4 Conclusions

To increase air-traffic safety, collision avoidance systems can serve as a back up to ATC. Conflict resolution maneuvers within short distances (to decouple ATC and CAS) are possible. Some algorithms of conflict detection have been analysed.

References

ATA (1971) A.N.T.C. Report No 117 by Air Transport Association Technical Working Group

Barth and Tetzlaff (1982) Flugvermessung der Strahlungsdiagramme von 7 Bordantennen des Flugzeugs DO28 IFZB. Deutsche Forschungs- und Versuchsanstalt für Luft- und Raumfahrt, Braunschweig

FAA (1982) Third TCAS Symposium Federal Aviation Administration. Washington DC

Form and Schroer (1982) Viewpoints on selection of collision avoidance systems. Int Congress of the Institutes of Navigation. Paris

Reed (1982) Enhanced TCAS II. Third TCAS Symposium. Washington DC

Welch and Teitelbaum (1982) TCAS angle-of-arrival engineering results. Second TCAS Symposium. Washington DC

Zeitlin (1982) Minimum TCAS II threat detection and resolution logic status. Third TCAS Symposium. Washington DC

A Ship Encounter – Collision Model

E.M. GOODWIN[1] and J.F. KEMP[2]

1 Aims of Model

In the open sea, ships encounter one another in an unstructured environment. Occasionally, encounters translate into collisions but the translation rate varies with many factors such as the type of encounter, the class of ship, the experience of the navigators and the weather conditions.

In order to reduce the rate at which encounters translate into collisions, it is desirable to devise a model of the translation mechanism so that the effect of changing key variables may be examined. Experiments may then identify measures which may be taken to improve the safety of shipping.

2 Situation to be Modelled

We consider a simple right-angled crossing situation in which ship A has ship B on her own starboard side. Both ships are assumed to have equal speeds and equal lengths. Five different cases for the initial position of ship A with respect to ship B are considered as illustrated in Figs. 1 and 2.

3 Responses of Ships

Reference to the Collision Regulations and experimental evidence suggest that ship A, which has the responsibility to avoid collision, would ideally respond as illustrated in Fig. 2. In practice, due to misperception and other causes, there is likely to be a proportion of inappropriate responses in each case, and experimental evidence suggests that the crossing distances can be expected to be normally distributed with a standard deviation of 600 m.

In order to simplify the model, we assume that the ships in each of the five cases are all initially following the centre lines of the respective path bands, and that the

[1] Polytechnic of North London, UK
[2] City of London Polytechnic, UK

Localization and Orientation in Biology and Engineering
ed. by Varjú/Schnitzler
© Springer Verlag Berlin Heidelberg 1984

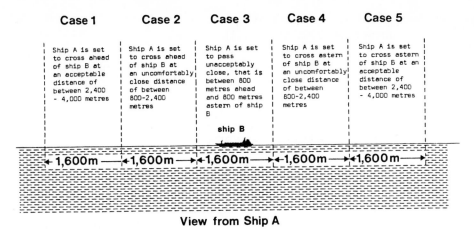

Fig. 1. Cases of initial position. The distances are based on domain theory and traffic surveys. Each of cases 1 to 5 comprises a path band 1,600 m wide for ship A

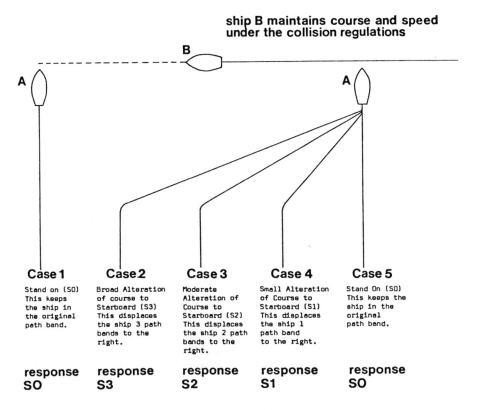

Fig. 2. Ideal manoeuvre. This figure illustrates appropriate action for ship A in each of the 5 initial cases. The effect is that cases 1 and 5 are unchanged and that cases 2, 3, and 4 are each transformed to case 5

distribution of responses made in a particular case is a reflection of the distribution of misperceptions in the initial situation. This allows us to assign probabilities that a particular case may be perceived by the navigator as an adjacent case or even a case once removed. On this basis, we can allocate probabilities to the responses which can be expected for ships in each of the five cases.

4 Example of Distribution of Responses

As an example, we consider ship A in case 2. The tracks of ships perceived are taken to be normally distributed about the centre line of the case 2 path band with a standard deviation of 600 m. The probability of the ship track being correctly perceived as case 2 is 0.81640 since the 1,600-m-wide path band lies within ± 1.33 standard deviations of the centre line. This would lead to a S3 response. The probability of the ship track being perceived as case 3 is 0.09177 since the 1,600-m-wide path band lies between 1.33 and 4.0 standard deviations of the case 2 centre line. This would lead to a S2 response. The probability of the ship track being perceived as case 4 is 0.00003 since the 1,600-m-wide path band lies between 4 and 6.67 standard deviations to the right the case 2 centre line. This would lead to a S1 response. The probability of the ship track being perceived as case 5 is negligible. The probability of it being perceived as case 1 or as a track to the left of case 1 is 0.09180, since these comprise any path bands at a greater distance than 1.33 standard deviations to the left of the case 2 centre line. The response to any of these cases would be S0.

The basis for these probabilities is illustrated in Fig. 3.

Similar probabilities can be assigned to the various responses which might be taken when ship A is faced with cases 1, 3, 4, or 5, and these are summarised in a matrix of transition probabilities, which forms Fig. 4.

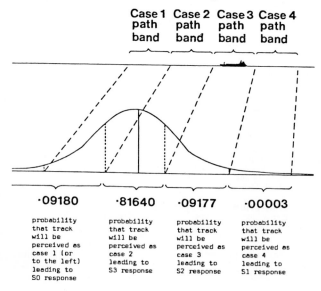

Fig. 3. Misperception probabilities for ship A in case 2 situation

Fig. 4. Matrix of transition probabilities.
The entries in a cell are first the probability
that the response defining the column will
be made given the case which defines the
row and, second, in parenthesis, the new
case which will result from the given response. The notation 5+ signifies that the
ship will take up a new path to the right of
case number 5, i.e. with greater clearance

		RESPONSE			
		S0	S1	S2	S3
CASE	1	·90820(1)	·00000(2)	·00003(3)	·09177(4)
	2	·09180(2)	·00003(3)	·09177(4)	·81640(5)
	3	·00006(3)	·09177(4)	·81640(5)	·09177(5+)
	4	·09180(4)	·81640(5)	·09177(5+)	·00003(5+)
	5	·90820(5)	·09177(5+)	·00003(5+)	·00000(5+)

5 Matrix Results

The main conclusion from the matrix is that the probability of ship A crossing the
path of ship B within the danger zone is 0.00003 when the initial situation is case 1,
0.00003 when the initial situation is case 2, and 0.00006 when the initial situation is
case 3. Given that an encounter occurs, where an encounter comprises any of the
initial cases 1–5, the overall probability that ship A passes through the danger zone
is 0.00012.

At this point, we make a further assumption that, if ship A has made an error
which takes it through the danger zone, then it is purely a matter of chance whether
or not a collision occurs. On this basis, given a crossing of the danger zone, the probability of collision is the ratio of twice the length of the ships involved to the total
width of the zone. This is because there will be a collision if the centre of ship A
crosses the track of ship B within the sector extending from half a length ahead of
ship B to half a length astern of ship B. Taking the length of the ships as 50 m, the
probability of collision, given a crossing of the 1,600-m danger zone is thus 100/
1,600 = 0.0625.

The probability of collision, given an encounter, is the product of the probability
of a danger zone crossing, given an encounter, and the probability of collision, given
a danger zone crossing, that is:

$$0.00012 \times 0.0625 = 7.5 \times 10^{-6} \ .$$

6 Consideration of Results

To test the above result for reasonableness, we note that traffic surveys of the study
area in the southern North Sea suggest that approximately one million crossing encounters would have occurred there over a period of 20 years. The model, therefore,
predicts $7.5 \times 10^{-6} \times 10^{6} = 7.5$ collisions during that period. An investigation by
Cockcroft identified 1 collision resulting from a crossing encounter in the same area
during the 20-year period 1958–1977. This is encouragingly good correspondence in
view of the relatively simple form of the model.

As it stands, the model can be used to evaluate effects such as poor watchkeeping
standards, by assigning an increased probability to the response in each case, or to

evaluate the effect of improvements in instrumentation, by reducing the standard deviation of the distribution of misperceived initial situations.

More important, the model may be extended to take account of possible actions by ship A, and to treat the overall responses as occurring in stages over a period of time rather than as single, instantaneous decisions. The extended model may then be used to investigate the reasons for the different rates at which encounters translate into collisions given different initial situations.

It may also be used to estimate the effect of proposed changes in the Collision Regulations, and to suggest the most effective operational interpretation of some of the existing rules.

The model has been developed as a means to examine the way in which encounters between ships may translate into collisions. It may well have potential for investigating comparable situations in other fields, such as the interaction between predator and prey in a biological context.

Chapter IV   Navigation, Bird Migration and Homing

Pacific Systems of Trade and Navigation

A. D. COUPER[1]

1 Introduction

When the Europeans first explored the Pacific they found that almost every island at which they called was inhabited. The people spoke different languages although these were often clearly related, and had distinctive but related cultures. They frequently demonstrated knowledge of the wider sea region which they inhabited, even of islands and archipelagos far distant from their homelands. They all appear to have grown coconut and breadfruit, planted tuber crops and possessed fowls, pigs, and even dogs. The problem of how they came to be there, and what navigational capabilities they possessed for voyaging, have been the subjects of discussion since the early contact period.

This paper deals with Pacific voyaging and indigenous trade. It reviews what is known about the systems of navigation which evolved during the Pacific Neolithic period, and which remained in use well into the time of the "alien impact", with some remnants continuing up to the present.

2 The Physical and Cultural Background

The Pacific Ocean covers one third of the earth's surface. Around its periphery are land masses but vast areas of the ocean basin are devoid of any land (Fig. 1). Only between 30° north and 30° south in the Central and South Pacific are there island groups such as the archipelagos of the Marquesas, Tuamotu, Tokelau, Fiji, and Solomons; in the North Central Pacific there is the Hawaiian Chain, while in the West Central area around the equator there are the small islands of Micronesia. Isolated inhabited islands very distant from their nearest neighbours include Clipperton (500 nm), Easter (over 1,000 nm), Johnston (470 nm), Marcus (980 nm), and Norfolk (400 nm) (Fig. 1).

The planetary wind systems predominate in the vast sea area between 30° north and 30° south where they blow as the north east and south east trades. In the middle latitudes there are westerlys and also regions of hurricanes and doldrums (Figs. 2

---

[1] Professor of Maritime Studies, University of Wales, Institute of Science and Technology, Cardiff, U.K.

Localization and Orientation in Biology and Engineering
ed. by Varjú/Schnitzler
© Springer Verlag Berlin Heidelberg 1984

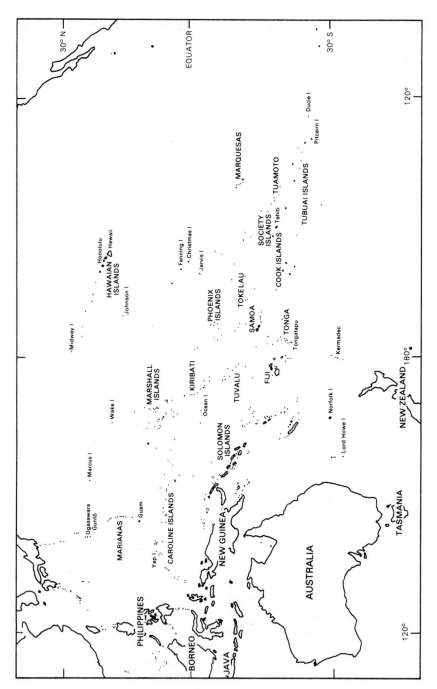

Fig. 1.

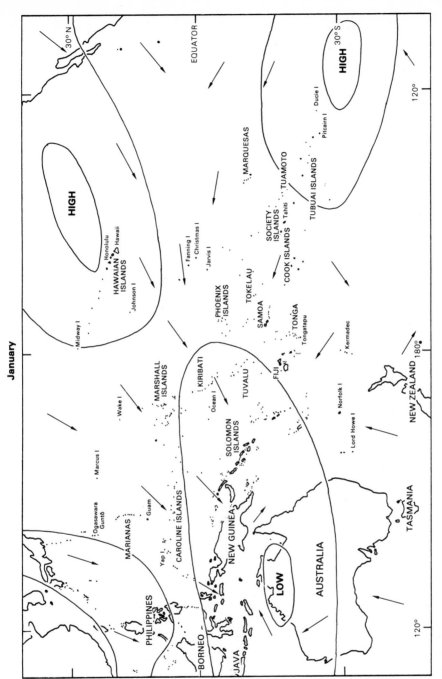

Fig. 2.

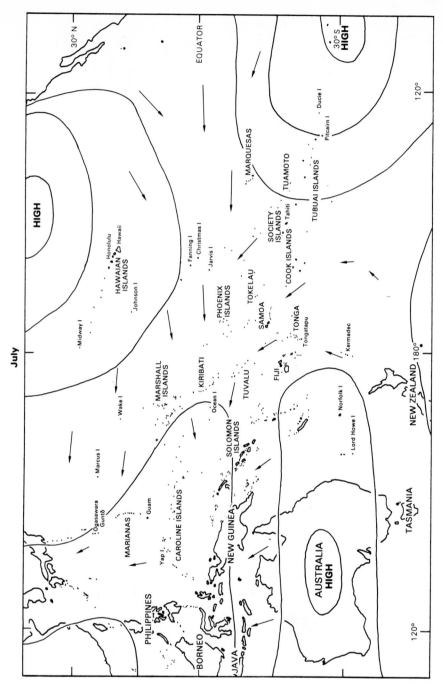

Fig. 3.

and 3). Storms occur in the tropical Pacific, but seasonally they are fairly predictable; so that frequently the seas and winds provide favourable sailing conditions. The sea currents in turn are produced by the surface water moving under the force and direction of the winds, and are also predictable in general direction although there are complex counter currents especially in the doldrum region (Fig. 4).

The island world of the Pacific peoples is often divided into three enthnographic regions: Micronesia, Melanesia, and Polynesia. Micronesia represents the small islands lying about 10° on each side of the equator in the West Central Pacific including the Marshall Islands, Gilbert Islands (now Karibati), and the Caroline Islands. These are inhabited by people of brown skin, light build and slightly mongoloid features. To the south is Melanesia, the black islands inhabited by an oceanic Negroid group of people, including those in the Solomons, New Hebrides, New Caledonia, and Fiji. The Polynesians inhabit the many islands to the south east and east of Fiji, and north to Hawaii. The people are somewhat Caucasian in type and include the inhabitants of Samoa, Tonga, Cook Islands, New Zealand, Easter Islands, Hawaii, and also Tuvalu which lies between Micronesia and Melanesia.

The islands inhabited by these people fall into several categories. There are the high continental islands such as Fiji and New Guinea; the high oceanic islands such as Easter Island and the main chain of islands of Hawaii; and the low coral islands and atolls as in Karibati. There are many variations within these categories in the same archipelago. The combinations of wind systems, island topography, geology, and soils, together with distance from nearest neighbours, have presented contrasting home environments.

The low coral islands of Micronesia are particularly difficult environments, many of them semi arid, allowing cultivation of only a rough root crop by digging deep pits to reach damp coral sand. The inhabitants live in villages where they are primarily sustained by the coconut crop, reef products, and fish. They use the raw material of the coconut and the pandanus tree for construction purposes.

The Polynesians occupy a region with a combination of low and high islands. The high volcanic islands have a far more favourable environment than the coral atolls. In many places there was normally little difficulty in supplying food, and the economy was based on gardening and fishing, including the growth of yams, sweet potatoes, breadfruit, coconut, and the gathering of shell fish on reef flats, and lagoon and ocean fishing. Bark cloth, derived from the paper mulberry tree, was produced and used throughout the islands. The Polynesians of the continental islands of New Zealand also grew sweet potatoes and similar crops in the north, but towards the south were primarily hunters, gatherers, and fishers, and adopted furs, skins, and bird feathers instead of bark cloth.

In Melanesia the high islands and continental islands were well supplied with food, timber, and often animal resources, while the low islands depended upon the ubiquitous coconut, reef products and sea resources. Often differences in resource supplies were equalised by trading between islands, and between inland and coastal people.

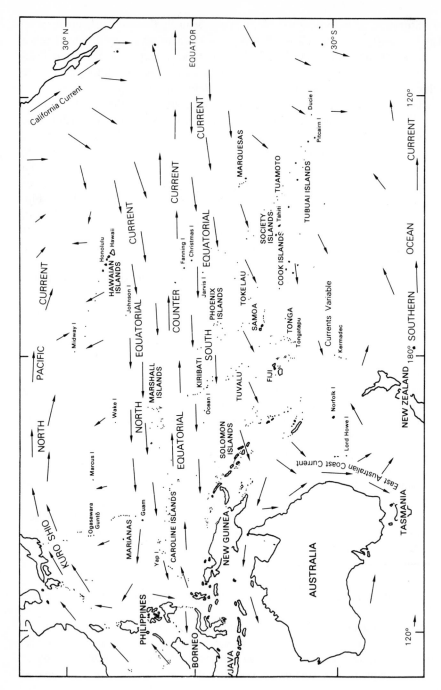

Fig. 4.

3 Origins of Pacific Islanders

The early European navigators tried to explain the human occupation of these isolated, scattered and very diverse islands. Many of the explorers concluded that the inhabitants must have been forced away while on local trading or fishing trips from some as yet undiscovered continent lying not too far to the south. It was only in the 18th century that Captain James Cook suggested that the Pacific islands might have been deliberately colonised (Parsonson 1962, p. 15). Later when he rescued survivors from a canoe at sea, he adopted the contrary opinion that the island world of the Pacific was probably colonised by accident over many years by this type of survivor. The then Hydrographer of the British Navy, Andrew Dalrymple, supported this firmly and also believed that the inhabitants of these isolated islands originated in a southern continent. Dumont d'Urville, on the other hand, postulated that they were the remnants of the population of a vast land area which had slowly submerged beneath the sea, leaving small groups to occupy the remaining peaks above sea level (Parsonson 1962).

Archeological evidence is now providing knowledge which will enable the origins and patterns of migration of the Pacific peoples to be more accurately pieced together. The initial migratory pattern appears to have been a movement of Negroid hunters and gatherers from the Asian equatorial region. They came to occupy parts of South East Asia and New Guinea possibly as early as 2,300 B.C. Further movements took place during the South East Asian Neolithic age with the building of sea craft using more advanced Neolithic tools. The voyagers carried with them coconuts, yams, bananas, pigs, dogs, and fowls, all of which appear to be of South East Asian origin (Vayda 1968), moving from Asia to Micronesia around 3,000 B.C. Another stream, the so-called Lapita people (from their characteristic pottery), entered Fiji around 1,300 B.C. and reached Tonga and Samoa around 1,100 B.C. From these areas the culturally distinctive Polynesians emerged and moved out to other islands. Tahiti appears to have been colonised by 500 A.D., New Zealand 600 A.D., Hawaii 750 A.D., and by 1,000 A.D. all of the islands of the Pacific which were habitable, appear to have been occupied. There is evidence of the colonisation of Easter Island the Marquesas by 400 A.D., a distance of 2,400 miles (Emory 1974, p. 739). The distances involved in Pacific migrations included: Tonga to Samoa 480 nm, Cook Islands to New Zealand 1,680 nm, and Marquesas to Hawaii 2,070 nm. Such voyages, to many small remote islands, represent one of the greatest maritime achievements of mankind. There is also evidence that after voyages of exploration and initial colonisation, trade routes were established with previous homelands as well as with nearer areas.

4 Inter-Island Trade and Voyages

The type of trade that was conducted between islands persisted into modern times, and, in fact partly still exists. The motives for trade were partly for quasi-economic exchange of complimentary resources, of stone, shell, timber, and foodstuffs, many noncommercial, ceremonial exchanges, the circulation of articles of religious significance, the alleviation of disasters such as hurricanes and droughts through evoking reciprocal obligations, the collection of birds' eggs on uninhabited islands, and the buildings of ships where timber was plentiful. There were also accidental voyages, forced migrations, and deliberate castaways, and voyages undertaken for purposes of raiding and war as well as for life crisis and religious events. Islanders voyaged from the low coral Tuamotu Islands to the high Society Islands to obtain stone for axes; from Yapp Island to Palau Island for calcite money discs; from the Trobriand Islands to Muria Island for stone axe heads; from Tonga to Fiji for sandelwood; from Tahiti to Samoa for coconut oil, and from and to many other regions. Much of the actual patterns and channels of trade were guided by kinship networks (Couper A.D. 1974).

5 The Ships

The ships of the Pacific islanders varied in type according to area. In the low coral islands they were constructed of coconut and pandanus timber, garvel built (in the style of almost all Indo-Pacific craft), with frames inserted after the shell construction and fitted outriggers. In some areas of Melanesia and Polynesia multi-hull vessels were built, several exceedingly large, such as the Ndrui in Fiji which had a length of over 100 ft and could carry up to 200 people, the Tongan trading craft of 80–100 ft and the Hawaiian double canoe which was also over 100 ft in length. Captain Cook's ship the Endeavour, by contrast, had a length of 106 ft.

6 Geographic Knowledge and Navigational Skills

There is only fragmentary evidence, derived from the records of early explorers, of ancient voyages made in the Pacific. It is clear from these records that indigenous navigators had a store of spatial information memorized and could point to the direction of other islands and provide some indication of the length of voyage. The Tahitian Chief Tupaia named about 70 islands for Captain Cook and claimed to have visited many of them. A map of this distribution of islands was drawn for Cook under the instructions of Tupaia; the original map was on display at the Cook bi-centenary exhibition in London in 1981. The European navigators noted very little of Pacific navigational techniques. There are some accounts of Pacific legends gathered in the 19th century but most of these are of little real value in terms of navigational detail.

It is only in recent times that systematic attempts have been made to determine the nature of the indigenous navigational methods used in the Pacific. By the time this took place, however, charts and compasses had been introduced and adopted in several island regions. The indigenous trading systems which formed much of the basis for major voyages by islanders, had also declined in many places, and some long distance canoe voyages were banned by the European authorities as early as the 1900's. The old navigators lost much of their status and raison d'etre with the regular arrivals of European ships at their islands for cargo and passengers. The famous Pacific companies such as the Godeffroy Trading and Plantation Company of Hamburg, Burns Philp of Sydney, and On Chong of Hong Kong, changed the network basis of island trade from kinship to cash.

Some island administrators such as Arthur Grimble and H.E. Maude in the Gilbert Islands systematically recorded as much as they could of the remnants of local customs. The records of early castaways and beachcombers on the Pacific islands such as those of William Mariner of Tonga were scrutinised for information. The great anthropologist Malinowski provided details and explanations relating to the voyages undertaken in the Kula Ring in his classic study *Argonauts of the Western Pacific* (1922). Later researchers added many pieces of evidence for types of voyaging and techniques which were still in use, and they accompanied island navigators on such voyages. Most notable among these researchers were Firth (1931), Goodenough (1953), Gatty (1958), Gladwin (1970), and particularly David Lewis (1972).

7 The Navigational System Reconstructed

The outline of the navigational system which is presented here is drawn from several sources and from personal observations and research in the Pacific. It is not intended to imply that there is conclusive evidence for the operation of a complete system such as is depicted in any part of the Pacific. The various techniques, signs and procedures are all, however, known to have been used by some Pacific navigators and collectively may represent most of the levels of information and sequence of techniques which an indigenous navigator would use to determine his course and position.

7.1 Preparation for the Voyage

Goodenough (1953) reports the existence of local star calendars in the Caroline Islands which were used to predict seasonal winds, currents, rains, and overcast periods. The navigators employed these to determine the most favourable time of year for voyaging, i.e. when the sea, wind, and sky were most suitable. They would not doubt wait for a day in this period when winds and currents appeared to be appropriate in force and direction to take them to a particular destination. There is evidence in various parts of the Pacific of the use of wind compasses in this connection. In Fiji, for example, the wind compass has four cardinal points and winds are named between these according to the island areas from which they blow.

7.2 Departure

With all of the above factors being favourable, the voyaging canoes loaded with coconuts, dried fish, pandanus paste and other non-perishable foods, along with water in gourds and bamboo canes would take up their position beyond the reef. They would orientate themselves in relation to directive stones or transit marks

1. SETTING THE COURSE

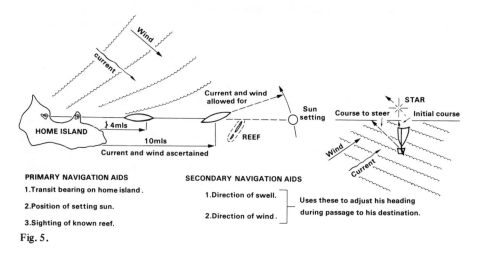

PRIMARY NAVIGATION AIDS
1.Transit bearing on home island.

2.Position of setting sun.

3.Sighting of known reef.

SECONDARY NAVIGATION AIDS
1.Direction of swell.

2.Direction of wind.

Uses these to adjust his heading
during passage to his destination.

Fig. 5.

ashore (Fig. 5). In Arorae Island in Karibati, for example, there is a group of stones said to have been used to direct early canoe voyages to neighbouring islands (Hilder 1962).

The vessels would depart just before sunset along the transit lines and adjust their orientation by the setting sun in the direction of their destination. They would also when on course check the direction of wind, waves, and swell impact on the canoe hulls and sails as secondary course indicators, applicable in the event of overcast conditions. As they sailed clear of the island they would determine the set and drift of the currently by comparing their course steered, with the course made good by observing transit marks. This would provide an indication of the direction and rate of current which they would then compensate for by adjusting sails and canoe trim accordingly.

Soon after sunset the navigators would pick up a guiding star a little above the horizon. This would be a star with a bearing towards the destination and a course would be set in relation to it, or an alternative star to port or starboard, to allow for the set and drift (Fig. 5).

7.3 The Sidereal Compass

The navigators would continue using stars to assist in direction for the remainder of the voyage. The sky for this purpose represents a vast compass which Goodenough (1953) has called the Sidereal Compass. The fixed points of this are the Pole Star (magnitude 2) which would only be visible with certainty at about 10° north, but the pointers to the Pole Star would be visible to voyagers below the equator, possibly as far as latitude 20° south on clear nights. The opposite Pole is marked by the Southern Cross, which when upright, indicates almost true south and is visible to voyagers to about latitude 20° north. Orion's Belt, Altair, and other stars of low declination in turn, rise close east, and set close west, and this bearing does not change appreciably

Fig. 6.

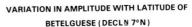

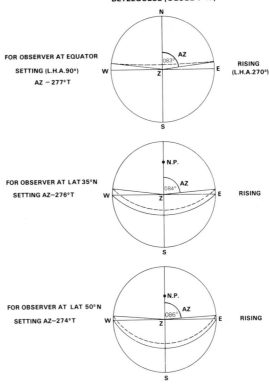

with changes in latitude. For example, Betelguese (declination $7°24'$ N) bears $083°$ on rising to an observer on the equator, and in latitude $35°$ north bears $084°$ (Fig. 6).

Four cardinal points were thus, readily available in most regions. In the Caroline Islands, at least, the navigators used more, for their Sidereal Compass covered 32 star points providing specific directions to a range of islands, and related to different stars at different times of the year (Fig. 7).

A star rises 4 min earlier each day so that a rising star would become a setting star 6 months later. Consequently navigators had to recognize the sky as a whole, the relationship of one star to another, and the way the sky changed with time and latitude. The navigators would recognize the so-called companion stars, that is; those with wide differences in sidereal time which would be rising and setting on complimentary bearings, and so could look either ahead, or aft, in order to check their direction.

The star courses to various destinations were memorised as were the stars themselves. Grimble (1931) noted that in his time some 50 stars were still known by names to the Gilbertese. These guiding stars would be used until the altitude reached $15°$ to $25°$, the bearing then beginning to change somewhat faster (unless that is, if it had a declination corresponding the latitude of the observer). A high star is difficult to steer by, and would therefore be replaced by another star rising with a similar amplitude.

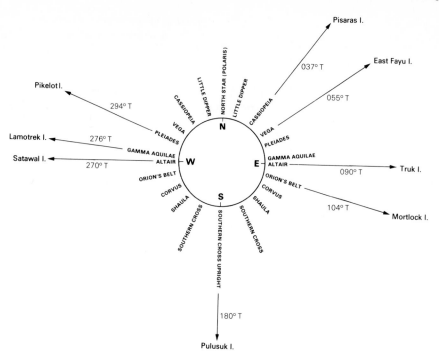

Fig. 7.

There are in fact many bright stars with a wide spread of sideral hour angles suitable for steering in the tropical sky.

During the day the rising and setting sun would provide an indication of direction (although as the sun rises above the horizon, its rate of change of bearing is too great to provide a reliable point for steering and when high in the sky it cannot be used for this purpose (Fig. 8). The meridian passage of the sun may provide an approximate indicator of north or south, but as a steering guide it is far inferior to that of the stars. Daytime courses could also be related to the secondary direction indicators of wind and swell.

The course may have been planned en route to cross some reef areas known to the navigators thereby providing a position check. The reef would also confirm the direction of current and some indication of its rate, which could not be determined in the open sea and adjustments would again be made (Fig. 5).

8 Position Finding

In the absence of a reef crossing or identification of a distant island, position finding, as distinct from latitude determination, was by dead-reckoning. This represented difficulties in the absence of charts to plot the course and distance run. A very important

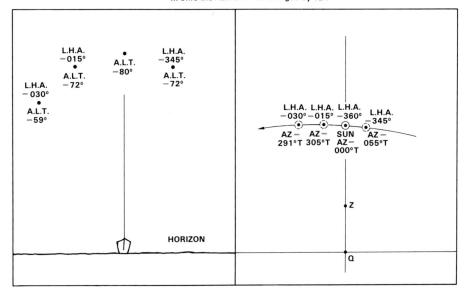

Fig. 8.

method used to produce a mental map and DR position was the ETAK System of the Caroline Islands which has been described by Gladwin (1970). The method (Fig. 9) divides the voyage from origin to destination, into stages by star bearings of an ETAK Island. An island is chosen lying between origin and destination, but distant from the courseline. This island may not be visible, but the navigator knows the various star bearings it will lie under at each stage of the voyage. When the voyage commences the navigator notes the star under which the ETAK Island lies and after he has proceeded at an estimated speed the ETAK Island will have moved and will lie under another star; meanwhile the island of destination will have moved closer. The conceptual framework of this system is that of a canoe lying stationery in the ocean and the ETAK Island moving past it, while the destination island moves towards it, against the background of fixed stars. The Caroline DR System allows the navigator to orientate and position himself in relation to his home island, the island of destination, a reference island, and the stars.

Also known to the Pacific Islanders was the zenith star method of determining latitude. Stars in their paths across the sky will pass directly above all places having a latitude equal to their declination. The meridian passage of the star will be directly apparent to an observer in that latitude. Sirius (the brightest star in the sky, magnitude − 1.6) with a declination of almost $17°$ south is overhead at Vanaulavu in Fiji. Altair (mag. 0.9) with a declination of $8.5°$ north is overhead in the Carolines, and Arcturus (mag. 0.2) with a declination of $19\,^3/_4°$ north is overhead in parts of Hawaii ($23°$ north to $19°$ north). By observing the maximum meridian altitude of the known zenith star of an island, the navigator would know that he was in the latitude of that island.

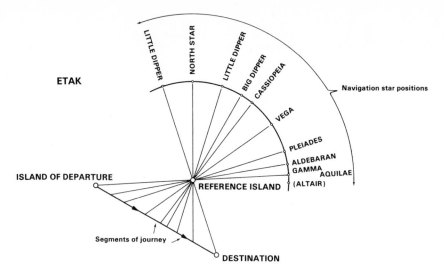

Fig. 9.

However he would have no way of knowing astronomically whether he lay east or west of the island. Because of this inability to determine longitude it appeared to be the practice when bound for an island to steer a course well upwind of the destination. When the zenith star was overhead the navigator would know he was in the required latitude, the course would then be altered and the vessel would run downwind

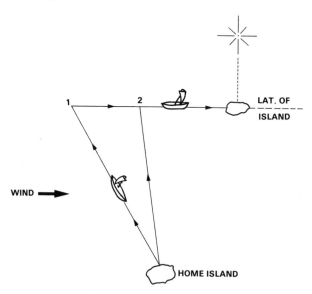

1.Helmsman chooses star well to port of his desired course.

2.Allowance for wind and current.

Fig. 10.

(Fig. 10). There would be no problem of tacking and the navigator would note all the signs which would tell him when he was in the vicinity of land.

9 Signs Approaching a Landfall

Normally, a vessel would make a landfall in the vicinity of a group of islands, consequently the navigator would not normally be searching for a specific small island but for the outlying reefs, surf and other evidence that would indicate that he was close to his destination. The target he was aiming for would thereby be greatly expanded, particularly approaching archipelagos, so that the chance of missing a destination when he turned downwind was greatly reduced. The crossing of a reef may have given an early indication of position in relation to the archipelago. Soundings may have been taken over the reef to determine depth, or the navigator might have traced the edge of the reef and from its distinctive water colour and shape would have been able to distinguish it from others he knew thereby fixing his position in relation to the archipelago.

He would also have been looking for many other signs as he ran downwind (Fig. 11). Birds which feed at sea and nest on land, such as the terns, noddies, and boobies, would indicate the direction of an island to be approximately 25 miles during their early morning and evening flights. The distortion of the swell in the vicinity of islands would also be a very important indicator (Fig. 12). In the Marshalls, sticks, and shell charts representing swell and islands were used for teaching, thus testifying to the extent to which the swell pattern was emphasised by indigenous navigators in determining the direction and proximity of land. The effect of high islands to windward, might also be signalled by changes in wind force and direction and by wave behaviour.

The characteristic shapes and movements of clouds would be observed (Fig. 13), cumulus clouds tend to tower thousands of feet above islands and often remain stationery, or show an inclination away from the island under the influence of upper winds. Slightly closer to an atoll the discolouration of the underpart of the cloud due to reflection from the lagoon would be noted.

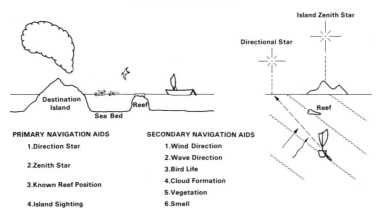

3. MAKING A LANDFALL

Fig. 11.

PRIMARY NAVIGATION AIDS
1. Direction Star
2. Zenith Star
3. Known Reef Position
4. Island Sighting

SECONDARY NAVIGATION AIDS
1. Wind Direction
2. Wave Direction
3. Bird Life
4. Cloud Formation
5. Vegetation
6. Smell

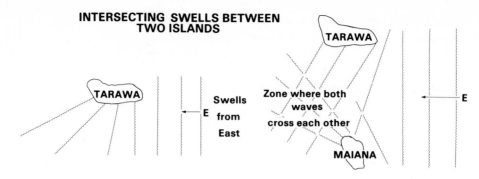

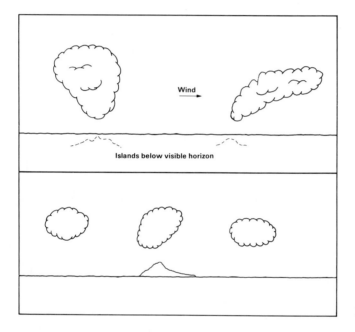

Fig. 12. After Lewis (1972)

Fig. 13.

Other characteristics such as smell of vegetation, driftwood, and sea temperature, would all indicate the proximity and direction of the land to the experienced navigator. Dr. Lewis provides an account of phosphorous flashes observed at sea on voyages with Gilbertese navigators. They called this Te lapa and it is reported to be visible below the surface of the sea some 80 miles or so from land (Lewis 1972).

10 Learning the System

A number of observers have recorded some of the methods of teaching astronomy and navigation in the Pacific. In the Gilbert Islands the Maneaba (meeting house) was used for this purpose during Arthur Grimble's administration. The ridge pole of Maneaba was considered to be the meridian, the various constellations were allocated places on the thatch, while the eaves of the Maneaba were considered to be the horizon. Gladwin (1970) describes the formal instruction in the Carolines which demanded that great amounts of factual information be commided to memory.

11 Conclusion

The remarkable mental orientation feats of the Pacific Islanders must be one of the greatest achievements in navigation by man, considering they had no compass, chart, or, as far as can be determined, any instruments. Long before Columbus sailed, and the great European voyages of Dicovery took place, they had reached every part of the vast Pacific Ocean and had engaged in inter island trade. Their navigational system depended on maintaining a constant mental traverse and framework of references in relation to their home island, en route signs, and the star paths.

References

Couper AD (1974) Islanders at sea: change and the maritime economies of the pacific. In: Brooksfield HC (ed) The pacific in transition. Edward Arnold, London

Denning GM (1962) The geographical knowledge of the polynesians and the nature of inter-island contact. In: Golson J (ed) Polynesian navigation. The Polynesian Society of New Zealand, p 102

Emory KP (1974) The coming of the polynesians. National Geographic, vol 146/6, December, p 739

Firth R (1931) A native voyage to tikopea. Oceania, vol 2/2

Firth R (1936) We the tikopea. London

Gatty H (1958) Nature is your guide. London, Collins

Golson J (1962) Polynesian navigation. The Polynesian Society of New Zealand

Goodenough WH (1953) Nature astronomy in the Central Carolines. University Museum, Pennsylvania

Gladwin (1970) East is a big bird: navigation and logic in puluwat atoll. Harvard University Press

Grimble A (1931) Gilbertese astronomy and stronomical observations. Journal of the Polynesian Society, vol 40

Hilder B (1962) Primitive navigation in the pacific. In: Golson J (ed) Polynesian navigation. The Polynesian Society of New Zealand

Lewis D (1972) We the navigators. Australian National University Press, Canberra

Makemson MW ( ) The morning star rises. University Press

Malinowksi B (1922) Argonauts of the western pacific

Parsonson GS (1962) The settlement of oceania: an examination of the accidental voyage theory. In: Golson J (ed) Polynesian navigation. The Polynesian Society of New Zealand

Vayda AP (1968) Peoples and cultures of the pacific. The American Museum of Natural History, New York

Interplanetary Navigation: An Overview

J. F. JORDAN[1] and L. J. WOOD[2]

1 Introduction

Navigation, for interplanetary space flight, is defined as the process of determining the position and predicted flight path of a spacecraft and correcting that flight path to achieve the mission objectives. The space navigation process involves a sequence of actions: the acquisition of radio tracking and optical measurements, the determination of a best statistically estimated orbit based on the measurements, and the computation of the trajectory-correcting spacecraft maneuvers.

At the end of a second decade of interplanetary exploration, we can now look back on a history rich in challenging navigation applications. In the United States' NASA program alone, there have been numerous flights to the inner planets of the solar system, including orbiters about Mars and Venus and landers upon Mars. There have also been flights to the outer planets, most recently the exacting Voyager flights to Jupiter and Saturn. In the 1980's we can look forward to the continuation of the Voyager 2 mission to Uranus and Neptune and to NASA's Galileo mission, which will tour Jupiter's major satellites and probe its atmosphere; to NASA's Venus Radar Mapper Mission, and to flights to Halley's Comet and Comet Giacobini-Zinner sponsored by the European Space Agency, the Soviet Union, Japan, and NASA. Requirements on navigation have increased dramatically during the past 20 years, as have the achieved navigational accuracies. The interplanetary missions of the 1980's impose further demands upon the navigation process and require advances in the technology of interplanetary navigation.

In this paper, we begin with an examination of the Voyager navigation system, the foundation upon which future interplanetary navigation systems are likely to be based. We discuss the demonstrated performance of this system, emphasizing the November 1980 encounter of Voyager 1 with Saturn. We then discuss the advances in navigation technology which are required for the Galileo mission, the most demanding of the missions to be launched in the 1980's from the standpoint of navigation.

---

[1] Technical Manager, Navigation Systems Section, Jet Propulsion Laboratory, California Institute of Technology, Pasadena, Calif. USA
[2] Technical Group Supervisor, Navigation Systems Section

Localization and Orientation in Biology and Engineering
ed. by Varjú/Schnitzler
© Springer Verlag Berlin Heidelberg 1984

2 The Voyager Navigation System

The Voyager navigation system consists of elements of the Deep Space Network (DSN), elements of the spacecraft, ground-based computational facilities and software, and various support functions (see Fig. 1). The DSN provides various radio metric data types for orbit determination purposes (Renzetti et al. 1982). For Voyager, these data types include two-way coherent Doppler, range and two-station near-simultaneous range. The Doppler data yield information about the sum of the geocentric range rate of the spacecraft and the projection of the geocentric motion of the tracking station onto the station/spacecraft line of sight. Thus, the Doppler signature is approximately a sinusoid with a 24-h period, superimposed upon the geocentric range rate of the spacecraft (see Fig. 2). The phase of this sinusoid is linearly related to the right ascension angle of the spacecraft. The amplitude is proportional to the cosine of the declination angle of the spacecraft. Consequently, the declination of the spacecraft is difficult to determine accurately from conventional Doppler data when the declination is small (Melbourne 1976, Melbourne and Curkendall 1978, Jordan 1981).

The declinations of both the Voyager spacecraft were close to zero at the times of their encounters with Saturn. As a consequence, a new radio metric data type, two-station near-simultaneous ranging, was developed for these encounters. If the two receiving stations have different latitudes, this interferometric data type has no zero-declination singularity (see Fig. 2). Conventional Doppler data are fundamental-

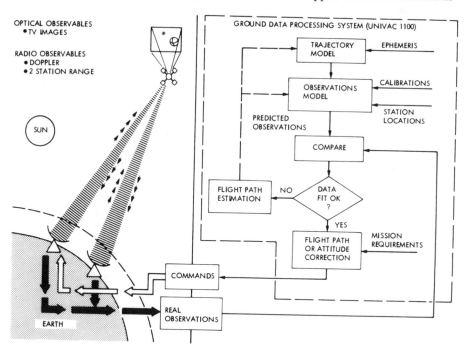

Fig. 1. Voyager navigation system

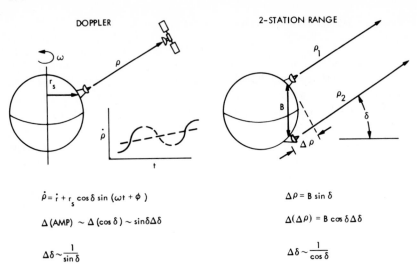

Fig. 2. Doppler and two-station range data as navigation measurements

ly more accurate than long baseline two-station near-simultaneous ranging data, at declination angles greater in magnitude than about $6°$. However, the reverse is true for smaller declination angles (Melbourne and Curkendall 1977, Curkendall 1978, Jordan 1981).

Information is uplinked to the Voyager spacecraft at a single carrier frequency (S-band). Two downlink carrier frequencies, both coherent with the uplink, are available, in the S- and X-bands. Deviations of the phase velocity and group velocity of the telecommunication signal from the vacuum speed of light are proportional to the integrated electron density along the signal path and inversely proportional to the square of the frequency. Thus, the dispersive effects of charged particles along the signal path may be calibrated using the dual-frequency downlink (Royden et al. 1980).

The portions of the spacecraft which may be considered part of the navigation system include the transponder, which coherently relays the uplink Doppler carrier back to Earth; the imaging subsystem, which provides data useful for orbit determination; and the propulsion subsystem, which provides the means of effecting trajectory changes. Optical navigation data consist of images of a planet or a planetary satellite against a background of catalogued stars. Such images are recorded using the narrow angle (1,500 mm focal length) optics in the spacecraft imaging subsystem, in conjunction with a scanning electron beam vidicon photodetector. The onboard optical observations are target-relative, rather than Earth-relative, as in the case of the radio metric data, a considerable advantage given that the target body ephemeris is not known precisely and given that the spacecraft is much closer to the target than to the Earth when the final flight path correction maneuvers are performed. Line and pixel spacings on the Voyager spacecraft vidicons are 0.015 mm, so that 1/2 pixel resolution produces an angular resolution of 5 μrad, for the narrow angle optics. The ultimate position accuracy obtainable with optical data is also limited by one's ability to determine the gravitational center of the target body from the limb and terminator information in the optical images. Currently, our centerfinding capability is within 1% of target body radius.

The accuracy of the current Jet Propulsion Laboratory (JPL) navigation system can be summarized in a very simplified way as follows. The accuracy of radio navigation is characterized by the uncertainty in the direction to the spacecraft as viewed from the Earth. The ability of Doppler data to determine a spacecraft's declination is inversely related to the sine of the declination. Today's Doppler system enables a spacecraft in cruise at high absolute declination, say $23°$, to be located within a total angular uncertainty of $1/4$ μrad. The two-station range measurement provides a 1 μrad backup at low declinations. Optical data locate the spacecraft relative to the target to an angular accuracy of 5 μrad. Earth-based radio navigation and its less accurate but target-relative counterpart, optical navigation, thus form complementary measurement sources, which provide a powerful sensory system to produce high-precision orbit estimates.

Orbit determination and maneuver computations are performed on the ground in the Voyager mission. The orbit determination process is basically one of adjusting estimates of the trajectory initial conditions and selected model parameters until the differences between the actual observational data and the predicted observational data are minimized in a weighted least-squares sense. A flight system ephemeris is generated for the period of interest by numerically integrating the spacecraft equations of motion starting from an assumed initial state (position and velocity), with some assumed set of values for the model parameters. Partial derivatives of the spacecraft state with respect to the initial state and the model parameters are obtained concurrently by numerical integration. Partial derivatives of observed quantities with respect to the initial state and the model parameters can then be obtained. Predicted values of the observed quantities can be deduced from the flight system ephemeris. Data residuals (differences between observed values and predicted values) can then be suitably processed to update the estimates of the initial state and the model parameters. These orbit determination computations are carried out using the JPL Orbit Determination Program (ODP), in conjunction with several smaller, specialized software modules (Moyer 1971, Ekelund 1980, Jordan 1981). Various support systems are also needed for the orbit determination process, to supply accurate planetary and satellite ephemerides, tracking station locations, universal time and polar motion calculations and transmission media calibrations (Jordan 1981). The estimated trajectory serves as the basis for computation, within the Maneuver Operations Program, of the velocity corrections required to meet the mission targeting objectives.

3 Voyager Navigation at Saturn

Two Voyager spacecraft were launched in the summer of 1977. Voyager 1 arrived at Jupiter in March 1979 and passed within 350,000 km of the planet, while being targeted to a close (20,000 km) flyby of the satellite Io. Voyager 2 encountered Jupiter in July 1979, passing within 730,000 km of the planet and encountering the satellite Ganymede at a distance of 62,000 km. (All distances are measured from the centers of the various bodies.)

Voyager 1 flew by Saturn on November 12, 1980. The flight path was targeted such that the spacecraft flew within 7,000 km of Titan, 18 h before closest approach to Saturn. After the close encounter with Titan, the flight path passed directly behind the satellite, as viewed from the Earth, which allowed the atmosphere of Titan to affect the radio signal, and thus provide data for atmospheric studies. The spacecraft then passed close by (185,000 km), and then behind, Saturn and its rings, so that similar effects on the radio signal due to the planetary atmosphere and ring system could result. The spacecraft then continued on out of the Saturnian system, passing through the E ring and making imaging passes by several Saturnian satellites, the closest approach being of Rhea (73,000 km).

Voyager 2 encountered the Saturnian system on August 25, 1981. The spacecraft passed by Saturn at a distance of 161,000 km and then passed within 100,000 km of the Saturnian satellites Enceladus and Tethys. The targeting of the flight path of Voyager 2 at Saturn enables an approach to Uranus in January of 1986 and Neptune in August of 1989.

The highly successful navigation of Voyagers 1 and 2 relative to Jupiter has been discussed by Campbell et al. (1980), Gray and Van Allen (1980), and Jordan (1981). At Saturn, Voyager 1 was required to perform a diametric Earth occultation with Titan, within a tolerance of 265 km, and then, after closest approach to Saturn, pass through a 5,000-km-wide corridor in the E ring, where it was believed that particles would have been swept away by the satellite Dione, and the chances of a disastrous impact consequently minimized. The close encounter with Titan created a difficult navigational challenge for controlling the instrument pointing for observations of the Saturnian satellites on the outbound leg of the Saturnian system encounter. Uncertainties in the dispersed flight path after the Titan and Saturn flybys made it mandatory to quickly and accurately redetermine the trajectory after the Titan closest approach.

Voyager 2 has been navigated to Jupiter and Saturn with the primary concern being retention of adequate propellant to reach Uranus. During a planetary swingby, an error in either the approach trajectory or the mass of the planet leads to an error in the outbound direction of travel, which must be corrected with a propulsive maneuver. This maneuver reduces the remaining propellant available for both attitude control and subsequent flight path corrections. Thus, for Voyager 2, as for Voyager 1, a premium was placed on the accuracy of delivery of the spacecraft on its incoming trajectory at both Jupiter and Saturn.

Optical navigation was required for the first time on Voyager to meet a deep space mission's objectives. The radio Doppler system was used as a baseline cruise system, but optical measurements were employed over the final few months before each encounter and served as the most accurate navigation measurements for the planetary encounters. The Doppler system served as an initialization and backup to optical navigation during the Jupiter approaches. The Doppler system, augmented with the two-station range measurement system, was adopted as a backup at Saturn. Optical navigation has been performed in the Voyager mission using TV images of the satellites of Jupiter and Saturn, not the planets themselves. The smaller sizes of the satellites and their sharp (rather than diffuse) surfaces make them more attractive as optical navigation targets than the central planets. In addition, many of the encounter target conditions were relative to the satellites themselves, so that knowledge of the flight path of the target satellite was required to the same precision as the spacecraft flight path.

Special satellite ephemeris propagation software systems for the Galilean and Saturnian satellite systems were developed, and an extensive pre-encounter ephemeris generation activity, including the acquisition of many new astrometric plate observations, was undertaken. The ephemeris propagation software used in the generation of the ephemerides was also used in the flight navigation system. The satellite ephemerides were corrected as the spacecraft orbit was determined from the optical navigation measurements.

The Voyager 1 encounter with Saturn was, from the standpoint of navigation, the most complex planetary encounter yet experienced. First, the Saturn satellite ephemeris propagator, used for the preflight ephemeris generation, was not accurate enough for the reduction of optical measurements of Titan. Therefore, the pre-flight Titan ephemeris theory parameters had to be transformed to a cartesian state vector, and the subsequent path of Titan numerically integrated. The conversion of both the ephemeris parameters and their uncertainties to cartesian coordinates caused difficulties in the orbit determination process. Navigating Voyager 1 to its 7,000 km close approach and occultation of Titan was tedious, with many orbit solutions processed and reviewed before a final, successful one was chosen. Voyager 1 achieved its flyby of Titan by means of a 1.9 m/s maneuver performed 33 days before encounter and a 1.5 m/s cleanup maneuver 5 days out. The flight path was accurate to 330 km. There was a strong science desire that the occultation be as nearly diametrical as possible. The error in the flight path was only 37 km in this key direction; thus, a nearly perfect diametrical occultation with Titan was achieved.

During the Titan encounter, Doppler data, which were then greatly influenced by Titan and hence very sensitive to the flyby distance and Titan's mass, were processed to quickly determine the flyby trajectory to high precision. This action led to accurate instrument pointing adjustments for the outbound imaging of the satellites Mimas, Enceladus, Dione, and Rhea.

The delivery to Titan was also sufficiently accurate that as the spacecraft passed through the E ring, on its escape from the Saturnian system, it passed through a gap thought to be more nearly devoid of particles than other regions. This section of the E ring may have been swept free by Dione. Whether the region is actually safer than other regions is, of course, still open to speculation. However, the spacecraft did survive the E ring passage and continued its science data acquisition sequences.

The Voyager 2 navigation requirements at Saturn were neither as stringent nor as complex as those of Voyager 1. The closest satellite encounter was with Enceladus at 87,000 km. There was, however, a high premium placed on navigating the incoming trajectory as accurately as possible, so that the post-Saturn correction maneuvers would be small.

The Voyager navigation experience has brought to reliable maturity the optical navigation process. The two-station range backup system, although a wise investment in overall mission reliability, was never actually required. Optical navigation can now take its place alongside radio navigation as a baseline system which can be totally relied upon for future missions.

The navigation of the Voyager 1 and 2 encounters with Saturn has been described in greater detail by Campbell et al. (1982), Cesarone (1982), Jacobson et al. (1982), and Taylor et al. (1982). Plans for navigating Voyager 2 at Uranus and Neptune have been described by Gray et al. (1982) and Van Allen et al. (1982).

4 The Galileo Mission

The Galileo mission represents perhaps an even greater navigational challenge than Voyager. In 1986, a Space Shuttle/Centaur launch will send the dual-spin Galileo spacecraft to Jupiter. Attached to the spinning portion of the vehicle will be an atmospheric entry probe, and on the despun portion will be an articulated platform with science instruments, including an imaging system will a charge-coupled device photodetector.

The heliocentric cruise orbital plane will be broken in midflight by a velocity correction of roughly 200 m/s. The spacecraft will then be targeted to meet the flight path entry requirement for a descent into the Jovian atmosphere. Then, 150 days before encounter with Jupiter, the entry probe will be released. The remaining spacecraft will be deflected a few days later by a 57 m/s burn to fly over the probe as it descends into the Jovian atmosphere, and act as a relay link for data transmitted by the probe. The current mission design calls for the spacecraft to fly within 1,000 km of Io, for purposes of gravity retardation, on the way to a 270,000 km closest approach to Jupiter, where a 670 m/s orbit insertion maneuver will be performed. The spacecraft then becomes a long period (200-day) orbiter of Jupiter. At its first apoapsis passage, a 350 m/s burn will be performed, to raise the periapsis radius to about 820,000 km, where the spacecraft is less susceptible to radiation damage. The spacecraft will then be targeted to eleven or more close encounters with the Galilean satellites Ganymede, Callisto, and Europa.

The primary navigational challenges for the Galileo mission are: (1) to deliver the atmospheric probe to the atmosphere of Jupiter and to deliver the orbiter to a flight path which will enable it to relay the probe data to Earth, and (2) to deliver the orbiter to a bound orbit around Jupiter and guide it through a tour of the Galilean satellites.

The probe must be released 150 days from Jupiter, so that the orbiter deflection maneuver is small; yet the probe's atmospheric entry angle must be controlled to 1.6°. The orbiter must then be maneuvered over the probe so that data can be relayed to Earth for at least 1 h after the probe reaches the 0.1 bar atmospheric pressure level. This requires that the probe-orbiter relative geometry be controlled to within a Jupiter-centered angular uncertainty of 2°. The probe trajectory in the Jovian atmosphere must be reconstructed to great accuracy to enable the precise correlation of the probe-acquired temperature, pressure, and acceleration data with the descent altitude. The atmospheric entry angle must be determined to 0.05° (1 σ) to make this possible.

The Galilean satellite tour must be navigated with extreme precision. For each close satellite encounter, the science instrument fields-of-view must be controlled accurately enough that the desired images are obtained. Most important, however, is the requirement that the entire tour, 11 encounters in 20 months, be navigated within a 205 m/s ΔV propellant allocation. These stringent requirements imply that flight paths must be controlled to an accuracy of roughly 30 km (1 σ) at the Galilean satellites.

Several developments will permit changes in the Voyager navigation system, which will enable it to meet the stringent Galileo requirements. The first development is the use of Differential Very Long Baseline Interferometry (ΔVLBI) as a new radio metric data type, to improve Earth-based navigation accuracy. The second is the ground automation of the optical navigation processing system, to decrease the processing time during satellite approaches and therefore allow more accurate optical navigation during the Galilean tour. Finally, the imaging system will use a charge-coupled device, rather than a scanning electron beam vidicon, photodetector.

ΔVLBI, as a navigation measurement technique, involves the simultaneous tracking of first a spacecraft, then a nearby extragalactic natural radio source (EGRS), from two widely separated ground stations. Through the ground correlation processing of the spacecraft data, the difference between the times the two stations receive the same signal can be computed to extreme precision, about 1 ns. If the same correlation process is applied to EGRS data, a similar time delay difference is obtained. Both computed time delay differences are sensitive to station location errors, transmission media errors, and station timekeeping errors, but they are sensitive in the same way. If the respective time delay differences are differenced again, the resulting doubly differenced delay is relatively insensitive to the major error sources. Whereas the individual time delays provide information which can yield an angular position fix to about 0.25 μrad, the double differencing technique provides a spacecraft direction angle fix accurate to 0.05 μrad, or better, relative to the observed EGRS (Melbourne and Curkendall 1977, Curkendall 1978). This measurement of time delays assumes that the signal transmitted by the spacecraft has a relatively wide bandwidth. If the spacecraft signal is only narrow-band, the ΔVLBI technique can only measure the differenced time delay rate of change, which is proportional to the rate of change of the spacecraft geocentric direction.

Figure 3 illustrates how the wideband ΔVLBI system produces time delays. The Galileo spacecraft transponder is being equipped with 19 MHz side tones at X-band and 4 MHz side tones at S-band, to allow the correlation process to extract the time delays.

Wideband ΔVLBI is thus a downlink doubly differenced range system, which requires only short, 10-min passes of data, and is accurate to 0.05 μrad. Its accuracy characteristics are similar to the Voyager system's differenced two-station, two-way range, in that there is no appreciable degradation at low geocentric declinations; but it is, of course, some 20 times more accurate.

For the Galileo mission, the processing of optical measurements during target approaches, which has been performed in the ODP system along with the processing of radio metric data for the Voyager mission, will be performed in a dedicated minicomputer. The Automated Optical Navigation System (AON) will provide a ground-based automated navigation data processor for Galileo, which will produce approach navigation products faster than the traditional system. The AON will also serve as a ground prototype for onboard navigation systems which may be needed in the planetary exploration program in future years.

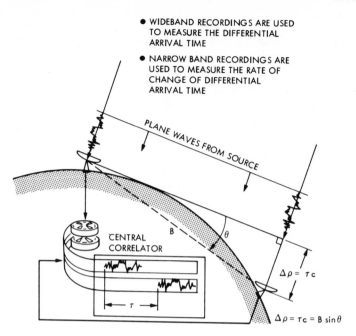

Fig. 3. VLBI as a navigation measurement

As an initial condition for target approach, AON requires a radio data determined orbit, to be computed in the ODP and transferred to the AON software. Then, as TV pictures of the target against the star background are received on Earth during the approach, the AON system will automatically extract angular information from the pictures, compute an updated estimated flight path with each newly received and validated picture, and compute the trajectory correction parameters for command to the spacecraft. Figure 4 depicts the sequence. Klumpp et al. (1980) have discussed this process in more detail.

Vidicon TV cameras, as have been used in Voyager and previous interplanetary missions, experience deflections in the scanning electron beam due to electric and magnetic disturbances. If left uncalibrated, beam bending destroys the metric accuracy in the angular measurements by tens of microradians. The calibration process requires analysis on a picture-by-picture basis, which prohibits automation. However, Galileo will carry a charge-coupled device (CCD) as a photodetector. A CCD is not subject to these electric and magnetic disturbances, and should allow automation of the angular measurement extraction process. In addition, a CCD-based system offers improved dynamic range and sensitivity relative to a vidicon-based system.

Orbit determination for the cruise and atmospheric probe delivery and trajectory reconstruction portions of the Galileo mission will be based on Doppler and wide-band ΔVLBI measurements. Doppler data alone are sufficient to enable probe delivery to an entry angle accuracy of $1.5°$ $(3\,\sigma)$. The time of probe entry can be predicted to an accuracy of 90 s $(3\,\sigma)$, and the time of flight of the orbiter spacecraft can be estimated to an accuracy of 15 s $(3\,\sigma)$ near the time of probe entry, which allows the

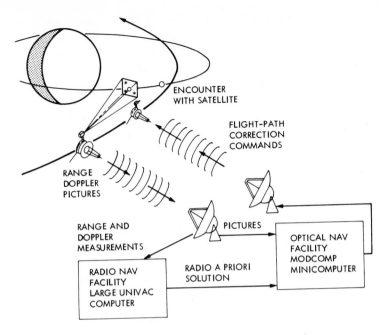

Fig. 4. Ground-automated optical navigation system

orbiter-probe line-of-sight vector at the time of entry to be controlled to an accuracy of $2°$ (3σ). Doppler data alone allow the estimation of probe entry angle only to $0.25°$ (1σ), whereas $0.05°$ (1σ) is desired. ΔVLBI data may allow this latter requirement to be met.

The Io flyby during the final approach to Jupiter will be navigated optically. The spacecraft will be delivered to Io with an uncertainty of about 80 km (1σ). This flyby enables a reduction of about 150 m/s in the Jupiter orbit insertion maneuver. The Jupiter orbit insertion maneuver can be planned based upon the anticipated Io flyby trajectory; and the perijove raise maneuver can be planned using a subsequent Doppler-determined orbit.

The tour of the Galilean satellites will rely heavily on optical navigation, with wideband ΔVLBI measurements used as a backup. Orbit correction during the tour will be a complex process. After each close satellite encounter, a maneuver to correct the spacecraft orbital period will be performed, principally to time the encounter with the next satellite. At apojove, a second maneuver will be performed to adjust the flight path to the correct encounter aim-point. Then, optical measurements become all important as the spacecraft approaches the satellite to be encountered. On previous orbital passes, optical measurements will have been acquired both when the satellite was at the same orbital position as at encounter and $180°$ from this position. These measurements will help to accurately locate the satellite for the encounter. The final maneuver, based on the optically determined orbit, will be processed in the AON system and performed three days before the encounter. This maneuver will not target to the approaching encounter, but to the subsequent satellite encounter in the sequence, thus extending the tour. Of course, as optical navigation is performed on each successive encounter, the ephemeris of the encountered satellite will be improved, to aid subsequent tour encounters. Most of the satellite flybys will be controlled to the limit of optical resolution, 20 to 30 km ($1\,\sigma$). It is expected that the tour can be completed within the allocated fuel budget.

Nominal trajectories for the Galileo mission have been described by D'Amario and Byrnes (1983) and Diehl et al. (1983). Navigation of the Galileo mission has been described in greater detail by Miller et al. (1983).

5 Summary

The history of deep space navigation over the past 20 years may be characterized by two major stages: (1) the development and maturation of radio doppler navigation in the 1960's and 1970's, culminating in the navigation of Mariner 9 and Vikings 1 and 2 to Mars, and (2) the development and maturation of optical navigation, which, with the Voyager successes, takes its place beside radio navigation as a proven reliable technology.

The near future appears to bring additional changes to deep space navigation. First is the development of ΔVLBI as a navigation measurement source, and along with it, the concept of high precision quasar-relative navigation. ΔVLBI will be used primarily in the 1980's for precision reconstruction of orbits, but could eventually replace Doppler as the primary radio navigation measurement, if the planetary ephemerides are improved during the next decade and computed in a quasar-relative coordinate frame. Next is the increasing autonomy in ground processing, with its potential for faster and more reliable data processing. Another change lies in the increasing complexity of maneuver planning. The methods associated with optimal maneuver location and second-body targeting algorithms, first exercised for the Voyager mission, will reach critical importance as missions like Galileo become more complex, and yet remain, as always, propellant-constrained.

Acknowledgements. The research described in this paper was carried out by the Jet Propulsion Laboratory, California Institute of Technology, under contract with the National Aeronautics and Space Administration (NASA).

References

Campbell JK et al. (1980) Voyager 1 and Voyager 2 Jupiter encounter orbit determination. Paper 80-0241, AIAA 18th Aerospace sciences meeting, Pasadena, California
Campbell JK et al. (1982) Voyager I and Voyager II Saturn encounter orbit determination. Paper 82-0419, AIAA 20th Aerospace sciences meeting, Orlando, Florida
Cesarone RJ (1982) Voyager 1 Saturn targeting strategy. J spacecraft and rockets, vol 19, no 1, pp 72–79
Curkendall DW (1978) Radio metric technology for deep space navigation: A development overview. Paper 78–1395, AIAA/AAS Astrodynamics conf, Palo Alto, California

D'Amario LA, Byrnes DV (1983) Interplanetary trajectory design for the Galileo mission. Paper 83-0099, AIAA 21st Aerospace sciences meeting, Reno, Nevada

Diehl RE, Kaplan DI, Penzo PA (1983) Satellite tour design for the Galileo mission. Paper 83-0101, AIAA 21st Aerospace sciences meeting, Reno, Nevada

Ekelund JE (1980) The JPL orbit determination software system. In: Penzo P et al. (ed) Advances in the astronautical sciences: Astrodynamics 1979, vol 40, part I. Univelt, San Diego, pp 79–88

Gray DL, Van Allen RE (1980) Voyager Jupiter maneuver targeting strategies. Paper 80-1772, AIAA Guidance and control conf, Danvers, Massachusetts

Gray DL, Cesarone RJ, Van Allen RE (1982) Voyager 2 Uranus and Neptune targeting. Paper 82-1476, AIAA/AAS Astrodynamics conf, San Diego, California

Jacobson RA, Campbell JK, Synnott SP (1982) Satellite ephemerides for Voyager Saturn encounter. Paper 82-1472, AIAA/AAS Astrodynamics conf, San Diego, California

Jordan JF (1981) Deep space navigation systems and operations. In: Proceedings of an international symposium on spacecraft flight dynamics. European Space Agency, Paris, SP-160, pp 135–150

Klumpp AR et al. (1980) Automated optical navigation with application to Galileo. Paper 80-1651, AIAA/AAS Astrodynamics conf, Danvers, Massachusetts

Melbourne WG (1976) Navigation between the planets. Scientific American, vol 234, June, pp 58–74

Melbourne WG, Curkendall DW (1977) Radio metric direction finding: A new approach to deep space navigation. AAS/AIAA Astrodynamics conf, Jackson Hole, Wyoming

Miller LJ, Miller JK, Kirhofer WE (1983) Navigation of the Galileo mission. Paper 83-0102, AIAA 21st Aerospace sciences meeting, Reno, Nevada

Moyer TD (1971) Mathematical formulation of the double-precision orbit determination program (DPODP). Technical report 32-1527, Jet Propulsion Laboratory, Pasadena, California

Renzetti NA et al. (1982) The deep space network – an instrument for radio navigation of deep space probes. Jet Propulsion Laboratory publication 82-102, Pasadena, California

Royden HN, Lam VW, Green DW (1980) Effects of the charged particle environment on Voyager navigation at Jupiter and Saturn. Paper 80-1650, AIAA/AAS Astrodynamics conf, Danvers, Massachusetts

Taylor TH et al. (1982) Performance of differenced range data types in Voyager navigation. Paper 82-1473, AIAA/AAS Astrodynamics Conf, San Diego, California

Van Allen RE, Cesarone RJ, Gray DL (1982) Voyager 2 navigation to Uranus and Neptune. Paper 82-1474, AIAA/AAS Astrodynamics conf, San Diego, California

Image-Supported Navigation

R.-D. THERBURG[1]

1 Introduction

If a flight navigation system has to operate independently of ground stations or satellites, it has to be updated by acquisition of ground data. Terrain features will be discernible under most weather conditions by a passive sensor with a suitable spectral response (Moore 1974). The visual field, specially from low altitudes, is very limited and rarely contains unique structures. Thus it is difficult to determine a position from a single image. Flying at low altitudes, perspective distortions will be generated in the region of ground elevations or buildings, complicating aerial correlation from airframes with maps carried on board (Rehfeld 1977). However roads and other line-shaped picture elements are undistorted exactly when the optical axis of a vertical looking sensor crosses them. These patterns, undistorted even if viewed from low altitudes, are detected by our sensor system which correlate the image with a rotating slit. As this sampling pattern is matched to line-shaped image patterns, traffic routes are detected in a highly structured background, in spite of weak contrast.

2 Contour-Sensor

Our electro-optical preprocessor, called contour-sensor, images the surface of the earth underneath the aircraft onto a wedge-shaped slit. The slit rotates about its vertex and when it crosses a contour as in Fig. 1, the difference between two samples of the sensor signal marks the gradient of intensity. The moment of crossing yields the orientation of the contour (Mühlenfeld and Therburg 1981). In this way the contour-sensor performs a two-dimensional correlation with a wedge-shaped slit and computes the tangential component of the correlation gradient during rotation of the slit. The slit eliminates image defects by averaging many pixles, accidental or irregular image-structures are thus suppressed in this procedure (Therburg and Mühlenfeld 1979). Consequently, traffic routes will be detected even if they are covered by trees or shades partially and buildings and area-shaped structures such as corn-fields and forests may be suppressed. Figure 2 shows an image of the lab-version of the contour-sensor with a mechanical rotating slit.

---

[1]   Institut für Elektrische Informationstechnik, Leibnizstraße 28, 3392 Clausthal-Zellerfeld, FRG

Localization and Orientation in Biology and Engineering
ed. by Varjú/Schnitzler
© Springer Verlag Berlin Heidelberg 1984

Fig. 1. A scanning star is shifted across the image searching for contours

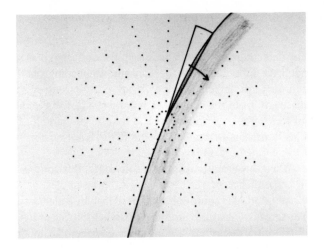

Fig. 2. Lab version of the contour-sensor with a mechanical rotating slit

Without any mechanical rotation the correlation can be processed by a star-shaped scanning pattern on the target of an image-dissector camera. There are no fundamental differences between a contour-sensor viewing at the original scene, or at a display of images taken by radar-, microwave-, or infrared technology and made visible for the pilot.

For navigation updating data of detected traffic routes are fed to a micro-computer to be compared with stored map data. All measures are processed with a technique similar to Kalman-filtering. Data acquisition and precision of position update are suitable for midcourse and terminal guidance of aircrafts and missiles.

3 Contour Detection

For detecting contours in an image, the optical axis as rotation-centre of the slit does not have to point onto the contour. Depending on the strength of image defects, contours are detectable even if the rotation-centre has a distance of up to half the slit length from the contour (Therburg 1982). Even for high flight velocities of mach 3 no traffic routes were omitted as long as the samples followed each other regularly. This is already guaranteed by the FORTRAN-version of the detection-software with a cycle-time of less than 20 ms.

When the aircraft crosses a traffic route, this contour can be detected several times in consequence of the frequency of the samples, as the measures follow a normal distribution. The contour itself is described by a plane gaussian probability distribution. The spread of measures produces a parallel-shifting of the contour only. The probability distribution describing a contour is gaussian-like in the orthogonal, and uniformly distributed in the direction of the contour. Position r_i, contour direction φ_i, and variance s_i^2 in the orthogonal direction are to be ascertained from a sequence of measures as contour parameters. With the definition of a base-vector for the orthogonal-direction,

$$w_i = (\sin \varphi_i, -\cos \varphi_i)^T$$

it is possible to formulate a density function (Therburg 1982) of the probability distribution of a detected contour K_i.

$$K_i = (r_i, \varphi_i, s_i^2)$$

$$f_i(r) = \frac{1}{\sqrt{2\pi}\, s_i} \exp\left\{-\frac{1}{2}(r - r_i)^T \frac{w_i w_i^T}{s_i^2}(r - r_i)\right\}.$$

To demonstrate the detectivity of the contour-sensor we arranged in aerial photographs simulated flight-paths in regions with highly structured background. Figures 3 and 4 illustrate photographs of this kind. The contour-sensor detects in all images only the traffic routes, which are searched for and this shows us same essential properties of the contour-sensor and the utilized algorithms:

- Area-shaped terrain structures like forests, regions of agriculture, but also shapes of clouds can be suppressed.
- Traffic routes and creeks are detected.

Hence, the contour detection and the position-fixing is appreciable

- independently of alterations of illumination by clouds and rain,
- independently of seasonally alterations of vegetation,
- independently of alterations of agriculture,
- independently of snow, if the streets to be detected have been used previously by vehicles.

Figs. 3 and 4. The contour-sensor detects in aerial photographs only the traffic routes, which are sought

4 Navigation Updating

Our method of terrain sensing guidance permits a high accuracy with only small demands on the inertial navigation system (INS) (Heilbron 1980), for unmanned aircrafts as well. Figure 5 shows the combined acting of the systems' components.

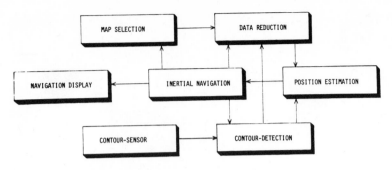

Fig. 5. Collective acting of the system's components

For position fixing, data of detected traffic routes are fed to a micro-computer to be compared with stored map data. As the contour-sensor extracts roads and rivers only, the digital reference map carried on board has to contain these patterns only. In this way, the map data may have been acquired by a ground based sensor of the same type from maps or aerial photographs. This yields a considerable reduction of stored information compared with complete digitalised images or profiles from TERCOM. By way of example, 400 Bytes are needed for a square kilometer of a country-like region of Germany.

For in-flight navigation the actual reference map is selected by the INS-position. Proceeding from a known startposition the inevitable drift error (Britting 1971) of the INS causes a displacement between the true and the INS-position. In time the position variance increases uniformly. Thus it can be modelled by a diffusion process (Krebs 1980):

$$\sigma_i^2 = g \left[(\sigma_{i-1}^2), (t_i - t_{i-1}) \right] .$$

When the contour-sensor detects a traffic route K_i (Sect. 3) all map data for the directions $\varphi_i - 1$, φ_i, $\varphi_i + 1$ are selected from the actual reference. This takes the quantization error into account. The quantity of contours K_{ij}, probably crossed by the aircraft, can be reduced with a plausibility-check. The true position has to lie in the 3σ circle around the INS-position with a probability of 99% (Therburg 1982):

$$K_{ij} = (r_j, \varphi_j, s_j^2) ; \quad j = 1, 2, \ldots$$

with

$$\varphi_j \in (\varphi_i - 1, \varphi_i, \varphi_i + 1)$$

$$|r_j - r_i| \leqslant 3 \sigma_i$$

and

r_i: = Position of the detected contour
r_j: = Position of a mapped contour .

If none of the contours K_{ij} meet this demand, the measures are dismissed. Otherwise, we obtain a line as hypothesis for the true position in relation to the INS-position. As contours K_{ij} are more improbable on the edge of the confidential-sphere, we diffuse the appropriate density functions f_{ij} (Krebs 1980). Parameter is the distance to the position of the detected contour:

$$s_j^2 = g\,(s_i^2, |r_j - r_i|)\ .$$

This approach weights all possible contours and minimizes mistakes of the contour-sensor by aimed disturbances.

In time the position information carried by the position lines will become more uncertain, due to the continuing drift (Fig. 6). We adjust this with a diffusion of the density functions. Crossing the next traffic route the contour-sensor delivers a further contour K_{i+1}. If map data exists under the above conditions, we obtain new position lines $K_{i+1,j}$ with the apertaining density functions $f_{i+1,j}$. All former position lines K_{ij} and density functions f_{ij} can be actualized. After translation and diffusion, an a-posteriori joint distribution conditioned by the measures can be constructed. Taking the statistical nature of all concerned processes into account, the joint distribution arises by multiplication of the actual and the actualised density functions (Winkler 1977):

$$f_{i,i+1}\,(r) = c\,\underset{j_i,j_{i+1}}{\pi}\ f_{i,j_i}'\,(r)\,f_{i+1,j_{i+1}}\,(r)$$

$$\int f_{i,i+1}\,(r)\,dr = 1$$

with

$$K_{i,j_i}' = (r_j', \varphi_j, s_j'^2)$$

and

$$r_j' = r_j + (r_{i+1} - r_i) \qquad \text{(Translation)}$$

$$s_j'^2 = g\,(s_j^2, t_{i+1} - t_i) \qquad \text{(Diffusion)}$$

In this way a joint distribution for the true position can be constructed, which is conditioned by all measures:

$$f\,(r/K_{uv}) = k\,\underset{u,v}{\pi}\ f_{u,v}\,(r)\ ;\quad \int f\,(r/K_{uv})\,dr = 1\ .$$

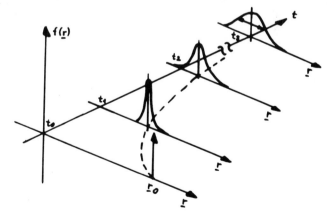

Fig. 6. In the course of time the position information will become more uncertain, due to the drift error of the INS

The local maximum of this joint distribution, fixed by the gradient to be vanished and the hessian matrix to be negative-definite, yields an estimation of the true position. The analyzation of Fishers' information matrix shows that the Cramer-Rao-inequation turns into an equation for this estimation (Kröschel 1973). Beyond that, the error-covariance is minimal and we obtain an optimal estimation in the sense of the filter-theory.

The difference between the estimated position $\hat{r}$ and the actual contour position yields the position offset of the INS:

$$\Delta r = r_{i+1} - \hat{r} \ .$$

This displacement and the confidential-sphere of the estimation, computed from the error-covariance matrix, can be displayed for the pilot, and is also to be used to update the INS. In the proposed description, the estimator needs a dynamically growing memory for the position lines, which is hardly suitable for practical missions. However, we succeeded in describing the whole history, meaning, the joint distribution conditioned by all measures, by a state-vector with only five linearly independent components. This state-vector can be actualized with each new position line, recursively. In addition, position lines of the contour-sensor point position data with noted co-variance can be included in the computation.

5 Conclusion

In accordance to the recursive formulation of the filter we could implement the tests on a micro computer PDP 11/23 with 32 k words memory.

- Besides the software for detection and filtering there are all data of the actual reference map in the memory.
- To demonstrate the high accuracy of the implemented system we observed many simulated flights and during these tests, we realized simulated flight velocities up to 800 m/s.
- The drift error of the INS was supposed to be 1% of the covered distance. All position-estimations computed in a time of 250 ms for a position fixing were compared with the true position.

We examined in all tests an updating error less than 2 m!

References

Britting KR (1971) Inertial navigation systems analysis. John Wiley & Sons
Heilbron H (1980) Ein Rückblick auf die Entstehung der Trägheitsnavigation. Ortung und Navigation 1/80
Krebs V (1980) Nichtlineare Filterung. Oldenbourg
Kröschel K (1973) Statistische Nachrichtentheorie. Springer

Moore RP (1974) Microwave radiometric all-weather imaging and piloting techniques. AGARD report Nr. 8 of AGARD-CPP/148, May 1974, Stuttgart

Mühlenfeld E, Therburg R.-D. (1981) Erkennung und Vermessung stark gestörter linienhafter Bildstrukturen. Technisches Messen 48, Heft 5

Rehfeld N (1977) Möglichkeiten und Grenzen der Flächenkorrelation zum Navigations-Updating, ermittelt mit dem inkohärent-optischen Korrelator. IITB/Bericht A 1029 (Nr. 9215)

Therburg R.-D. (1982) Rekursive Filterung signalabhängig gestörter Meß- und Navigationsdaten. Dissertation

Therburg, R.-D., Mühlenfeld, E. (1979) Extraktion von Konturen aus verrauschten Bildern. Inform. Fachberichte 20. Springer

Winkler G (1977) Stochastische Systeme. Akademischer Verlag Wiesbaden

Range Measuring Techniques in Aviation

W. SKUPIN[1]

1 Introduction

In aviation distance or range determination plays an important part in the task of navigation. The most common use is measuring distance to characteristic points of the flight path, such as distance to touch down during the landing phase, or distance to an air-track waypoint during an en route operation. However, distance determination is also a tool for fixing a position on the basis of multiple ranging and is a necessary function of collision avoidance techniques. Based on these operational requirements, different range measuring techniques have been developed and systems have been established, each dedicated to a specific task. All these techniques and systems can be classified into two categories of range determination methods. The principle methods will be presented here and characteristic performance problems on an established system will be discussed.

2 Principle Range Measuring Techniques

The basic principle of all range measuring systems is the determination of signal travelling time between source and drain. With the known propagation velocity – which for radio systems usually corresponds to the velocity of light – the desired distance can be calculated as follows:

Distance = Propagation Velocity · Travelling Time .

Travelling times can be measured in two different ways:

a duplex travelling time measurement;
b simplex travelling time measurement.

The basic function of these two methods is shown in Fig. 1. For duplex travelling time measurement an interrogator transmits a signal burst which triggers a response signal by a responder unit located at a distance d. On its round trip the signal is delayed by twice the travelling time T_T. From the measured time difference T_M (Interrogation-Reply) the distance d can be determined as given in Fig. 1. Response

---

[1] SEL Stuttgart FRG

Localization and Orientation in Biology and Engineering
ed. by Varjú/Schnitzler
© Springer Verlag Berlin Heidelberg 1984

Fig. 1. Concepts for range measuring techniques

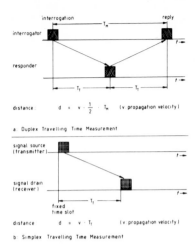

a. Duplex Travelling Time Measurement

b. Simplex Travelling Time Measurement

to the interrogation signal can be achieved by passive devices, i.e. reflectors, as well as by active responders, so called transponders (*trans*mitting res*ponders*). A typical example of a duplex system with passive reflectors is Radar (or Sonar for submarine applications). Radars are widely used in aviation, for instance for air traffic control and surveillance, or as airborne weather sensors. However, Radar is a complex subject with a great variety of applications which are extensively covered in this congress by other authors, and therefore this presentation will focus on a duplex system with active responders. This is the so-called *D*istance *M*easuring *E*quipment (DME), which is widely used in aviation and will be described in Section 3.

The principle function of a simplex travelling time measurement is shown in Fig. 1b. The signal source, i.e. the transmitter located at the reference point, and the signal drain, represented by the user receiver, are synchronized to a common time frame. At a fixed time slot within this frame the signal source transmits a characteristic signal, which is received by the signal drain delayed by the travelling time T_T. Based on the common time frame, this travelling time can be measured easily and yields the desired distance. This method requires identical time reference at signal source and signal drain to be sufficiently accurate, thus demanding utilization of highly stable time standards by both transmitter and receiver. This requirement has prevented extensive use of this method for a long time.

Although trends can be observed to deploy this technique for modern systems, none of these systems is fully operational yet. So this method will not be discussed further here.

3 Distance Measuring Equipment (DME)

DME is an active duplex travelling time measuring system. It is used as range function for en route navigation and landing systems in aviation. The basic system function is shown in Fig. 2. The airborne interrogator transmits randomly distributed pulse pairs,

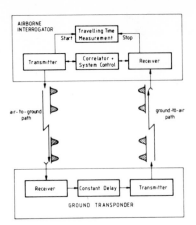

Fig. 2. DME system function

which are received by the ground transponder. Deployment of pulse pairs allows simple discrimination between wanted signals and noise. Because of the inevitable delay times in the transponder circuitry, all delay times are summarized and supplemented to a constant delay time. Following this delay the ground transponder re-transmits reply pulse pairs, which are received by the airborne interrogator. As the ground transponder is usually serving a large number of aircraft at the same time the airborne interrogator has to detect its pertinent replies out of many pulse pairs transmitted by the transponder. This is achieved by correlating replies received with transmitted interrogations. As DME is a co-operative system requiring a signal exchange in both directions, it is susceptible to saturation effects. Interrogations can be garbled by pulse coincidence from different sources and will be suppressed, if they occur during a reply transmission period of the transponder. This problem could be solved by synchronizing all interrogators to a common time frame. However, this would limit the number of subscribers, whereas the random system operation causes an undesirable – but eventually acceptable – increase in reply efficiency reduction, when the (nominal) system capacity is exceeded.

There are different error sources which deteriorate the performance of DME. These sources and their impact on the system function are listed in Table 1. Instrumentation error can widely be overcome by a careful system design, e.g. choice of a suitable power budget and adaptive calibration procedures. Multipath arises from diffuse and specular reflections from the ground, buildings, vehicles, and other large structures. As multipath is unavoidable, its impact on system performance has to be minimized by the choice of an appropriate signal format. Because of their longer propagation paths, multipath signals always arrive later than the direct signal, so the direct signal will not be contaminated by multipath at the very beginning of a pulse. Hence pulse shapes with fast rising leading edges are preferred, where the reference point for the time-of-arrival (TOA) measurement is being reached prior to the incidence of multipath signals. However, fast rising pulse edges give rise to spreading of the signal spectrum. Care needs therefore to be taken not to violate spectral restrictions. Figure 3 shows a comparison of TOA errors for two different pulse shapes. Pulse shape 1 is a conventional bell-shaped pulse, referred to as a Gaussian pulse. It has a 2.5 μs rise

Table 1. DME error sources

1. *Instrumentation*		
Drift of equipment delay times	→	travelling time measurement error
Pulse shape distortion by filters	→	TOA error
Receiver noise	→	TOA error
2. *Multipath*		
Pulse shape distortion by echoes	→	TOA error
Signal fading	→	pulse drop out
3. *Traffic Loading*		
Pulse shape distortion due to pulse coincidence	→	TOA error
Pulse suppression by pulse coincidence	→	pulse drop out
System guard and shut down times	→	pulse drop out
Dynamic response of airborne data filter	→	trailing error
TOA: time of arrival (Measurement)		

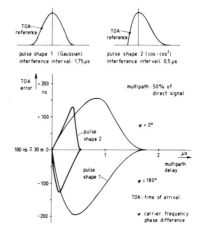

Fig. 3. Multipath TOA errors for characteristic pulse shapes

time and the TOA reference at 50% of amplitude. Pulse shape 2 has recently been internationally agreed upon as a new DME signal format for special applications. The signal is characterized by a cos and a $\cos^2$ slope. Rise time is 1.6 μs and TOA reference is 25% amplitude. The TOA error originating from interference between the direct signal and a single multipath echo with 50% amplitude of the direct signal is plotted as a function of the multipath delay time. The error curves indicate the TOA error for both extremes of signal phaseshifts direct – multipath of 0° and 180°.

TOA errors for all other phase differences are confined within the plotted curves. It can be seen clearly that pulse shape 2 offers enhanced multipath resistance due to the reduced interference interval. The influence of multipath on system performance can be further reduced by suitable signal processing. Because of their partly random nature, multipath errors can be minimized by utilization of outlier rejection circuits and data smoothing filters.

As already stated, DME is a co-operative system and therefore is susceptible to saturation effects, which degrade system performance. Traffic loading has two main effects on the system performance. Firstly, coincidence of pulses at the transponder receiver causes interference (garble), which result in either TOA error or pulse drop out in severe cases. The second, more substantial effect, however, is given by the guard periods in the ground transponder initiated by decoded interrogations and receiver shut down times during transmission. As a result of this an increasing traffic loading reduces reply efficiency (reply-to-interrogation ratio). Figure 4 shows that the system capacity is determined by the minimum reply efficiency that can be tolerated. The numerical example shown in Fig. 4 indicates that a reply efficiency in the order of 50% will limit the number of interrogators served by a common transponder to 450. Obviously a reduction in reply efficiency degrades system performance. This problem can be solved partially by introducing appropriate signal processing methods. An adaptive tracking filter (tracking gate) operates as an outlier rejection circuit and data smoothing is achieved by means of a predictor filter. Figure 5 shows test data obtained from a (simulated) flight test with 50% reply efficiency. The flight path is a ground transponder overshoot at 600 ft height with a velocity of 150 Knots. The aircraft velocity relative to the ground transponder is plotted in the center frame of Fig. 5. In the bottom frame the reply efficiency is shown as a histogram. The main result of this test, however, is contained in the top frame, where the range error and the tracking gate limits are plotted. It can be seen clearly that the range error is close to zero where the flight profile offers quasi-constant dynamics. Due to the virtual acceleration in the overflight region the error performance is deteriorated. This trailing error originates from the considerably reduced up-date rate for the predictor filter, which therefore hardly copes with the given dynamics. However, during the entire flight test the interrogator under test keeps track and provides continuous range information. This test shows, that system operation can be performed safely even at low reply efficiencies, but attention has to be paid to the dynamic behavior of the tracking filter.

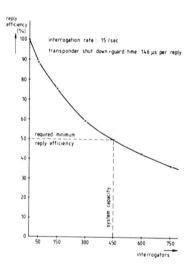

Fig. 4. DME reply efficiency depending on traffic

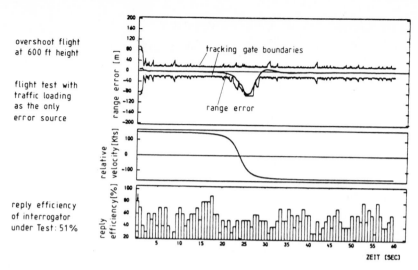

Fig. 5. Transponder overshoot flight at 50% reply efficiency

4 Conclusions

In general it can be stated that DME has turned out to be a reliable range measuring system, which offers accuracy to between 30 m (advanced equipment) and 200 m (standard equipment). Because of this accuracy and the overall system performance, a new utilization of DME has been suggested recently. This is position fixing by multi-DME, i.e. intersection of different DME signals. This method is well suited to en route navigation and its utilization will be extended in the future.

The Geometrical Characteristics
of the Navigation Satellite Systems

M. SZYMOŃSKI[1]

1 Introduction

From the view of the practical use of Navigation Satellite Systems, it is interesting to explain how the geometry of transit of the artificial Earth satellite in the observers radiocommunication area (service area) influences the correct determination of position. The geometry is directly dependent on the orbit parameters of the navigation satellites, and also on the observers location against the orbit surface.

One of the choices of the orbit parameters and the number of satellites in the system is connected with discrete observations. For this reason satellite systems can be divided into two essential groups:

1. the system securing periodical observation of navigation satellites with a projected discretion,
2. the system securring the continuous observation of one or several satellites simultaneously.

Taking into consideration the altitude ranges of navigation satellites, the systems can be subvided as shown in Table 1.

The American System TRANSIT/NNSS belongs to a system with satellites in low altitude orbits. This system is used generally on merchant ships all over the world, for geodetic and hydrographic purposes.

The Soviet System, called CYKADA – started with satellite KOSMOS-1000 on March 31, 1978, reported by Pravda on April 2, 1978 – TASS Agency – should also be mentioned.

The NAVSTAR/GPS, developed in the USA is a system in medium altitude orbits and belongs to the group with geostationary satellites in high altitude synchronous

Table 1. Characteristics of navigation satellite altitude ranges

Altitude range	Altitude (km)	Period
Low	900– 2,700	100–150 min
Medium	13,000–20,000	8–12 h
Synchronous	22,000–48,000	24 h

[1]   Institute of Navigation, Gdynia, Poland

Localization and Orientation in Biology and Engineering
ed. by Varjú/Schnitzler
© Springer Verlag Berlin Heidelberg 1984

orbits, although it is being used exclusively for sea radiocommunication purposes. The possibility of the future use of the geostationary satellites for navigation purposes cannot be excluded.

2 The Geometry of the Navigation Satellite-Transit

The value of the Navigation Satellite Systems is that they provide global coverage with great accuracy. The geometrical relation of the observer and the navigation satellite, determines the conditions of receiving the radio signals from the satellite, and consequently, the accuracy of position at determination.

Figure 1 shows the general case of investigation of transit of navigation satellite by the observer being located on the Earth surface at the point P_0.

To analyze the geometry of transit of navigation satellites it is convenient to use the simpliflied model, and assume that satellites are moving around circular orbits, in the centre of which the globe is situated. The point P_0 is dismissed from the orbit plane. The distance between the observers position and the orbit plane is determined by the circular arc $P_0 L'$ or the arc Φ of the great circle.

The geocentric angle a determines the radiocommunication area (service area) of the observer.

Figure 2a shows the changes of the geocentric angle a, limiting the service area, the time t of satellite transit in the service area, and the elevation angle h in the culmination as a function of angle Φ.

In the case of location of the observer in the orbit plane ($\Phi = 0°$) of satellite of TRANSIT system the service area is determined by angle $2\,a = 64°$, which embraces the surface of the globe limited by a circle with radius equal to about 1,900 nm. This is not much compared with the service area for the satellite of NAVSTAR System, where $2\,a = 114.8°$, which corresponds with the surface limited by the circle with radius 4,300 nm. It should be stressed that the satellite in low altitude orbit is observed in the service area not longer than about 18 min, whilst the satellite in medium altitude orbit may be observed for 5 h.

It should be mentioned that the geostationary satellite, placed on the altitude to 25,000 km covers with its activity the area with the radius of about 4,700 nm, i.e. from the equator to the latitudes ±78°.

In a given moment, the satellite in the low altitude orbit covers a much smaller service area than the geostationary satellite, but passing fast around the globe in 24 h secures practically the navigation on the whole Earth surface. The service area passes according to the movement of the artificial Earth satellite. During one period of a navigation satellite passing around the globe in low altitude orbit, the service area will change for angle $\Delta a = 26°$, on the equator. For satellites in medium altitude orbit, it is equal to 180°.

The covering of the globe by the system is a very important parameter characterizing the geometrical quality of the Navigation Satellite System. The covering depends on the number of satellite orbits, and on the number of satellites in one orbit.

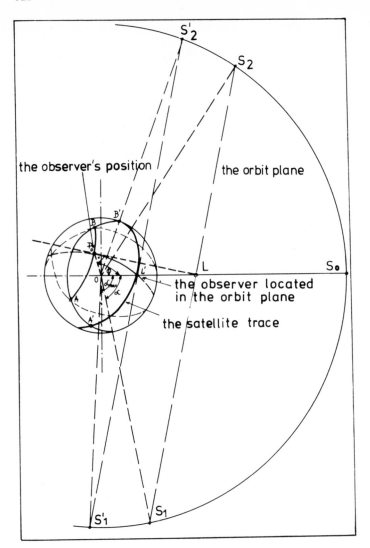

Fig. 1. Transit of navigation satellite by observer located on the Earth surface

It appears that one satellite of the TRANSIT system gives the covering in 58% on the equator, while in the case of the NAVSTAR system the same value of covering will be possible with three navigation satellites placed one by one in three orbit planes. In both cases the covering is similar, but in the first, the navigation satellite appears periodically over a chosen point of the globe. In the second case, one satellite can be seen continuously by the observer. The angle Δa and the covering of the globe by the system are changeable for different latitudes, as shown in Fig. 2b.

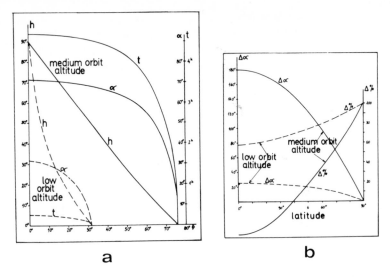

Fig. 2a,b. a Changes of geocentric angle a; **b** Changes at different latitudes

3 Conclusions

Using the satellites in low altitude orbits, fast changes of measured parameters are achieved. The use of the measurement method of the Doppler shift of the frequency is very useful here. The observer may obtain a great number of the navigation parameters during the satellite transit in the service area. Obviously, the increase of the number of measurements allows greater effect in the minimum mean-squared error sense.

The transmitting radio station on board of a navigational satellite and the receiving equipment of the observer are rather simple, and the requirements of the stability of reference oscillators, are met by classical cristal oscilators. The main limitations of this realisation of Navigation Satellite System are:

1. discretion of observation which can be diminished by increasing the number of satellites;
2. results of accuracy of position determination depend on parameters of the observers movement.

The system with satellites in medium altitude orbits seems to be the most universal, because of constantly increasing requirements in navigation, because of the possibility of practically continuous observation of the position, and because measurements results are independent of the knowledge of parameters of the observers movement.

The simultaneous observation of several satellites in a given region of the globe is assured by a reasonable number of navigation satellites. For example, in the NAVSTAR system 18 satellites moving subsynchronously six by three orbit planes assure the realisation of at least three simultaneous measurements against three satellites which are in the service area. Twenty-four satellites in this system assure the simultaneous observation of at least five navigation satellites. The range method in this case seems to be the most useful, because the measurements of Doppler frequency shift are limited by a small (in comparision with satellites in low altitude orbit) relative velocity between the satellite and the receiver, and the long delay of satellite transit in the service area. In the most simple case of position determination in two-dimensional application, with correction of the user clock bias, one can state that it can be solved by simultaneous performance of at least three range measurements.

A suitable geometry of navigation satellite locations at the moment of observation assures the highest accuracy of position determination, and continuous measurements diminishes the possibility of great error.

The system with geostationary satellites, however, does not assure the covering of subpolar regions, and near the equator the accuracy of position determination is too small. Thus, the system of geostationary satellites can be used for navigation purposes only with cooperation with satellites in orbits of a different type.

Elements of Bird Navigation

K. SCHMIDT-KOENIG[1]

1 Compass and Distance or Vector Navigation

Migratory birds may cover many thousands of kilometers. An explanation of how they find their destinations can be provided only for the first autumnal migration of the young-of-the-year, and then only in part. According to the results of many displacement experiments with banded birds, particularly those of Perdeck (1958) with starlings sketched in Fig. 1, young birds departing in the fall possess two sets of data: information concerning (a) direction and (b) the distance to fly. A system employing such data is referred to as "vector navigation". A vector is characterized by direction and length = distance, hence the name. The directional component of the system has been demonstrated to include three compasses: a sun compass, a star compass, and a magnetic compass. These compasses not only use different environmental cues, they also operate on different principles.

1.1 The Sun Compass

The operation of the sun compass is well established and generally accepted. It was discovered by Kramer (1950), its mode of operation was further analyzed by Hoffmann (1954) and by Schmidt-Koenig (1958, 1961, 1972). The birds allow for the sun's apparent azimuthal movement. Altitude, the other sun variable, is ignored. The chronometer or "internal clock", which is a component of the animal's circadian biorhythm, interacts with the azimuth position. Problems remain concerning the accuracy and ontogenetic development of this system. The accuracy of the sun compass itself has been shown to be on the order of magnitude of several degrees (McDonald 1972). Ontogenetically in young birds it appears to be aligned to the magnetic compass (Wiltschko et al. 1976, Wiltschko 1980). Contrary to the initial idea that the sun compass is innate, details seem to have to be learned (see refs above).

1.2 The Star Compass

For star compass orientation, birds learn the spatial arrangement of constellations as such, independent of rotation. Thus, in contrast to what one is inclined to extrapolate

[1] Abt. f. Verhaltensphysiologie, Universität Tübingen and Dept. of Zoology, Duke University, Durham N.C., USA

Localization and Orientation in Biology and Engineering
ed. by Varjú/Schnitzler
© Springer Verlag Berlin Heidelberg 1984

Fig. 1. Displacement experiment of Perdeck with starlings and the demonstration of vector navigation and of true navigation. Starlings migrating from their breeding range to the wintering range (*differently shaded*) were caught near The Hague and displaced to Switzerland by airplane. Young starlings continued parallel to the ancestral direction of migration (*straight black arrow*) and established a new winter area (*dotted*). Adult starlings flew in the direction of the ancestral wintering range (*straight white arrow*). (After Perdeck 1958)

from knowledge of the sun compass, a clock is not required. Several investigations have shown that under experiment conditions, e.g. in a planetarium, completely arbitrary and unnatural patterns of light-dots are learned as readily as actual astronomical constellations (Wallraff 1969, Wiltschko and Wiltschko 1976). We have no knowledge of the accuracy of the star compass. The question of how star compasses are established ontogenetically is also still open. According to Emlen's (1972) results with indigo buntings, young birds watch the nocturnal sky to discover which part rotates least. In the northern hemisphere, this is the area around Polaris which is then used as reference. Wiltschko and Wiltschko's (1975a,b) findings on European robins and on old world warblers indicate that the magnetic compass is the basic reference system to which the star compass is aligned. It can be re-aligned, e.g. during a southward migration, under latitudinally changing starry skies.

1.3 The Magnetic Compass

Providing convincing evidence for a bird's capacity to use the geomagnetic field for directional orientation, proved the longest and thorniest task. In a tedious experimental series that extended over many years, Merkel and Fromme (1958), Fromme (1961) and then Merkel and Wiltschko (1965) confirmed that migratorily restless European robins were able to maintain their ancestral migratory direction, roughly south in fall, north in spring, even when enclosed in cages in sealed rooms. The crucial experiment unequivocally establishing the role of the magnetic field was carried out by Wiltschko (1968, 1973). The birds predictably followed the rotation of an artificial magnetic field produced by Helmholtz coils. An example is given in Fig. 2. Some details of how the geomagnetic field may be used by European robins were found by Wiltschko and Wiltschko (1972). The birds magnetic compass does not use the polarity of the field, as does our man-made magnetic compass. It seems to evaluate the inclination of the field. More precisely, the birds seem to take the smaller angle between field lines and the vertical as "pole-ward". This compass cannot discriminate North Pole from South Pole. How the magnetic field is sensed and measured by the bird remains an open question.

The major weakness of this magnetic compass model, especially for trans-equatorial migrants, is that it does not function in the horizontal field at the magnetic equator. Kiepenheuer's (in press) modification of the model would overcome this difficulty.

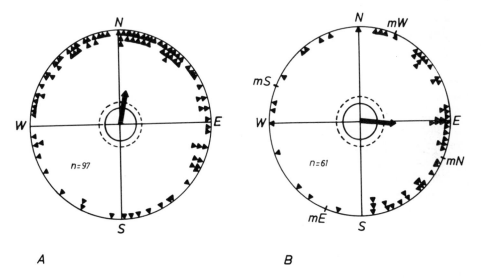

Fig. 2A,B. Spring orientation of European robins (**A**) in the normal geomagnetic field and (**B**) upon experimental rotation of magnetic North (*mN*) to east-south-east. In these experiments migratorily restless birds were kept in octogonal cages excluding external visual cues. The directional component of migratory restlessness was recorded by microswitches attached to radial pearches. Each *black triangle* represents the mean direction of one bird night, *n* indicates the total number. The *centrifugal arrow* represents the mean vector of each summary. The *inner concentric circle* (*solid*) represents the 5%, the *second circle* (*dashed*) the 1% critical vector length (Raleigh test) for uniformity. (Wiltschko 1973)

According to this model the magnetic compass is a probe that has a fixed spatial orientation, specific to each species and population. This arrangement would enable the migrating bird to fly in such a way so as not to cross the lines of force of the geomagnetic field. Kiepenheuer calculated migratory routes, predicted by his model, for a number of species and found them to agree with the routes documented by recoveries of banded birds.

1.4 The Distance

While many details of the compasses remain to be investigated, the mechanisms documented so far suffice to explain the directional component of bird orientation including, of course, vector navigation. As to the distance component of vector navigation, experimental evidence has been accumulated, mostly by Gwinner (1968, 1969, 1972, 1974) and by Berthold et al. (1972), Berthold (1973, 1978) on some species such as European warblers of the genera *Phylloscopus* and *Sylvia*. In laboratory experiments, they recorded the temporal pattern of migratory restlessness of related species with differently distant winter ranges and differently long migratory routes, respectively. The time of migratory restlessness and the lengths of migration routes turned out to be proportional. A long-distance migrant (e.g. wintering in southern Africa) was proportionally more active or active for a longer period than a medium-distance migrant (e.g. wintering in central Africa), and the medium-distance migrant was proportionally more active, etc., than was a short-distance migrant (e.g. wintering in northern Africa). These results support the view that an endogenous temporal program may determine the duration (or amount) of migratory activity, and that this is equivalent to information on distance. How this mechanism works physiologically, how the birds measure activity, energy expenditure or distance is entirely unknown. The accuracy of such an indirect system certainly cannot be expected to be overwhelmingly great. It is, however, well-documented by banding recoveries that young birds are less accurate than experienced, older individuals. In principle, a system as outlined above, using information on the direction, plus information on the distance to travel, would work, and would explain one aspect of bird migration: how a young bird reaches the winter range during its first autumnal migratory season. In subsequent migrations, vector navigation recedes in favor of true navigation. This capacity is discussed in the following paragraph.

2 True Navigation

Compasses indicate directions only. If a bird uses one or two or even three of its compasses, it can select and maintain directions, alter its heading, and so on, but it cannot identify the direction of a goal. Even vector navigation as outlined above is a system of limited scope. It is insufficient to explain what could be called "true navigation" i.e. the well-documented ability of wild birds to return to their nest sites or wintering grounds as shown by Perdeck for starlings (Fig. 1) or of domestic homing pigeons to return to their lofts upon experimental displacement. A large proportion of experimental efforts to solve the mystery of bird navigation has centered on the homing pigeon. It is a fairly convenient work-horse and possibly at least partially adequate as a model of a migratory bird. Although domestic birds with non-migratory ancestry, pigeons and migratory birds have in common at least some basic orientational mechanisms such as the sun and the magnetic compass. Pigeons navigate in the true sense of the word. If displaced to a location where they have not been before they home in a nonrandom fashion. Which sensory capacities could be used by pigeons for navigation?

2.1 Some Sensory Capacities

In addition to the ability to use the geomagnetic field as already discussed, in recent years a number of surprising sensory capacities have been discovered in pigeons (Kreithen and Keeton 1974a,b; Kreithen 1978). They see ultraviolet light and the plane of polarization pretty much as is known for some insects, e.g. bees. Pigeons are sensitive to changes in ambient pressure on the order of magnitude of 1 mbar, which is equivalent to changes of altitude of about 10 m. Pigeons hear infrasound down to 0.06 Hz. These capacities were all discovered in laboratory experiments in artificial environments. Whether or not these capacities are used under natural conditions is as yet unknown. At least some of these capacities could be used for navigation. The suggested role of olfaction in navigation will be discussed below (Sect. 2.3).

For many years the key role for navigation was attributed to the eye, the apparently dominant sensory organ of a bird. The sun arc navigation hypothesis of Matthews (1953, 1955) is an example of a navigation hypothesis requiring rather keen visual capacities. Many experiments have since demonstrated that the sun is in fact not used for navigation although it is used for compass purposes (cf. Keeton 1974, Schmidt-Koenig 1979). Our experiments in which we drastically reduced vision with frosted lenses attached to the eyes, showed that vision, particularly image vision, is not required at all for successful homing, excepting the final approach to the loft (Schmidt-Koenig and Schlichte 1972, Schmidt-Koenig and Walcott 1978). The emphasis thus shifted to non-visual means for navigation.

2.2 Inertial Navigation

One theoretical possibility would be inertial navigation, as outlined by J.S. Barlow (1964, 1966). To date, the relevant experimental evidence is clearly negative and most experts agree that inertial navigation has a very low probability of actually being important in bird navigation. The final crucial experiment to disprove this hypothesis has as yet to be carried out.

2.3 Olfactory Navigation

Presently, a focus of attention is the hypothesis of olfactory navigation, as advanced, and later modified, by Papi et al. (1972, 1973). At first glance it seems to be a very simple idea. Pigeons and possibly other birds as well, create an olfactory map (centered on their home) in which each area is composed of a characteristic pattern of volatile, presumably organic, odorants. In an early phase of their lives the birds learn to associate odors with the directions from which winds carry the odors. During displacement, they keep track of the direction of displacement by recording the various odors through which they are being transported, and they reverse and integrate the directions to obtain the home direction. The sun or the magnetic compass would be used to fly home. The implications of the hypothesis are, however, not as simple as they seem to be at first glance. Moreover, there is as yet no direct evidence from laboratory experiments that pigeons can perceive and discriminate subtle environmental odors, as required by the hypothesis. Nevertheless, the results of a large number of homing experiments carried out by Papi and his collaborators in Italy seem to support this hypothesis. Repetitions of the original experiments or new experimental approaches elsewhere, however, yielded less clear-cut results, or even contradictory evidence.

If one considers the two main criteria customarily used in homing experiments, (a) homing performance and (b) initial orientation, separately, one fairly consistent result emerges. Experimental elimination of olfactory input by e.g. bisecting the olfactory nerves, inserting plastic tubes into the nasal passage or plugging the nostrils, reduced homing performance, including homing success (e.g. Papi et al. 1974, Benvenuti et al. 1973, Papi et al. 1971, 1978, Hartwick et al. 1977, Keeton et al. 1977, Hermayer and Keeton 1979). Often, however, initial orientation remained unchanged, although the interference with olfaction would necessarily have to influence initial orientation if olfaction were the basic navigational element. There is considerable laboratory evidence that experimental interference with the olfactory system has distinct effects on non-olfactory functions, such as learning and discrimination, including orientational tasks, and also motivational aspects of behavior (Lindley 1930, Wenzel and Salzman 1968, Phillips 1969, Marks et al. 1971, Alberts and Friedman 1972, Binder 1979). Such behavioral elements are undoubtedly also involved in homing without necessarily affecting navigation as such.

Crucial experiments on the role of olfaction in navigation would have to produce predictable deflections of initial orientation. There is one experiment that predictably and consistently shifts initial orientation: the deflector aviary experiment (Baldaccini et al. 1975). Young pigeons were raised in aviaries equipped with vanes that deflected winds, thereby potentially producing an association of odorous substances with false directions (Fig. 3). In repetitions elsewhere (Kiepenheuer 1978, Waldvogel et al. 1978, Waldvogel and Phillips 1982, Phillips and Waldvogel 1982) deflected initial orientation was recorded as well as in the original experiment as shown in Fig. 3. In experiments with modified methodology, however, deflections persisted in birds made temporarily anosmic with local anaesthesia (Kiepenheuer 1981) or for the entire time spent in the aviary by bilateral bisection of the olfactory nerve (Kiepenheuer, to be published). Deflection was also observed in a "whirlwind" deflector aviary constructed in such a way that the birds could not associate specific odors with wind directions (Kiepenheuer 1981a). Thus, the original interpretation of the deflector aviary experiment as support for the olfactory navigation hypothesis cannot be maintained. In addition to inconsistencies in repetitions of experiments, especially those performed outside Italy, there are some experiments in which pigeons

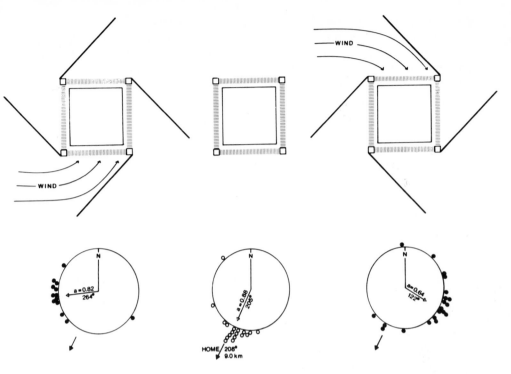

Fig. 3. The deflector aviary experiment with homing pigeons of Baldaccini et al. (1975). *Upper row* aviaries with deflectors and wind chanelling lamellae and control aviary (*center*) with lamellae but without deflectors. Initial orientation as recorded in the original experiment is given in the circular diagrams below each aviary. Each *dot* (*black* for experimentals, *white* for controls) at the *periphery of each circle* represents the vanishing point of one pigeon. The sample mean vector is given in the center of each circle with direction and length

homed, although they had had no olfactory information whatsoever. There were yet others in which they had had no olfactory information from the direction of release. One such example is the fan experiment by Papi et al. (1974). Pigeons were raised in aviaries shielded with plastic and bamboo, which permitted air to enter only diffusely. The treatment by fans was such that the birds were never able to associate wind directions and environmental odors, as required by the hypothesis. Nevertheless, many birds homed.

Two similar studies were described by Wallraff (1979) and by Ioalè (1982). Both raised pigeons in partially shielded aviaries so that winds could not enter from certain directions. Experimental birds did, however, home, even from those directions from which they had never experienced wind and olfactory information. These results indicate that if olfaction is involved as a navigational element in homing at all, it is certainly neither the only, nor an indispensable component as repeatedly claimed by Papi (e.g. 1982) and by Wallraff (e.g. 1981) respectively.

Recently, attempts have been made to identify the odorous substances potentially involved in homing (Wallraff and Foà 1981). During transport to the release site, experimental pigeons were supplied with environmental air filtered through a charcoal filter. In contrast to the results and interpretations of preceding experiments (e.g. Wallraff 1980) based on theoretical considerations (Wallraff 1974), the initial orientation of experimental birds that had breathed charcoal-filtered air was inferior to that of control birds. The charcoal filter seemed to have removed navigationally relevant information. Such apparently conflicting results and other inconsistencies need clarification.

2.4 The Geomagnetic Field

Last though not least there is considerable (although still insufficient) evidence that the geomagnetic field may, in addition to being used for compass purposes, also play a role for navigation. Magnets or small Helmholtz coils attached to pigeons have been shown to affect initial orientation (Keeton 1972, 1974, Walcott and Green 1974). This may be explained by a disturbing effect on the compass upon release. Disorientation observed in pigeons upon release in geomagnetic anomalies (Walcott 1978, Kiepenheuer 1981b) is less readily explained by assuming an interaction with the compass. Finally, pigeons have been displaced in experimentally manipulated magnetic fields (Kiepenheuer 1978, Wiltschko et al. 1978, Benvenuti et al. 1982). For example a northward displacement was magnetically simulated while they were actually taken to a release site south of the loft or pigeons were transported in iron containers that strongly reduced the geomagnetic field. Initial orientation was usually disturbed, indicating that the birds pay attention to magnetic information during displacement. Two examples of particularly striking results are given in Fig. 4.

For a maximally effective system additional capacities such as those for measuring and processing intervals of time, would be required. The ancestors of our homing pigeons flew to feeding places from their nests. Perhaps they evolved a system that was adapted to their flight speed, and that is incapable of coping with the variable and usually higher speeds of transport by car. At all events, the role of the geomagnetic field in navigation by birds is by no means fully understood.

3 Conclusion

The sun compass, the star compass, and the magnetic compass are integral constituents of bird navigation. Their existence is well established. Although many functional details remain to be clarified, the existence of these compasses suffices to explain directional orientation as revealed by radar studies of bird migration or as shown in laboratory experiments or in homing of pigeons.

In vector navigation, compass orientation and information on the distance to travel are integrated into a navigation system of limited scope.

Some researchers favor as the basis for a true navigation system sensors that respond to environmental odors. This is looked upon with scepticism by others. There is some evidence that the geomagnetic field is also utilized in true navigation.

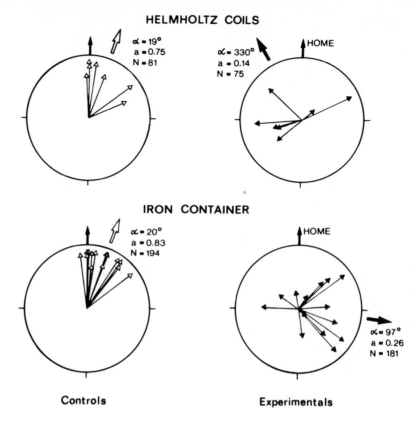

Fig. 4. Two examples of obviously altered initial orientation of homing pigeons as a result of magnetic manipulations during displacement by Helmholtz coils (*upper diagrams*) or iron containers (*lower diagrams*). *Arrows originating in the center* of the circle mean vectors of roughly 15 (6 to 19) vanishing points. *Black* or *white arrow off the periphery* the over all mean vector per diagram with direction, length *a*, and sample size *N*. *Black arrow at 0°* home. (Data are taken from the original Fig. 1 in Benvenuti et al. 1982, uncorrected for some graphical errors we found)

Even if all of these elements were the only ones operative, we still would not understand how they are integrated into a system permitting the navigational performances documented for birds. Many more details, more extensive discussions, as well as references in addition to those listed below, may be found e.g. in Keeton (1974), Schmidt-Koenig and Keeton (1978), Schmidt-Koenig (1979, 1980), Gauthreaux (1980), Papi and Wallraff (1982).

Acknowledgement. I am grateful to Drs. J. Kiepenheuer and P.H. Klopfer for many helpful discussions and for improving the English.

References

Alberts JR, Friedman MI (1972) Olfactory bulb removal but not anosmia increases emotionality and mouse killing. Nature 238:454–455

Baldaccini NE, Benvenuti S, Fiaschi V, Papi F (1975) Pigeon navigation: Effects of wind deflection on home cage and homing behavior. J Comp Physiol 99:177–186

Barlow JS (1964) Inertial navigation as a basis for animal navigation. J Theoret Biol 6:76–117

Barlow JS (1966) Inertial navigation in relation to animal navigation. J Inst Navigation 19:302–316

Benvenuti S, Baldaccini NE, Ioale P (1982) Pigeon homing: Effect of altered magnetic field during displacement on initial orientation. In: Papi F, Wallraff HG (eds) Avian Navigation. Springer, Berlin Heidelberg New York, pp 140–148

Benvenuti S, Fiaschi V, Fiore L, Papi F (1973) Homing performances of inexperienced and directionally trained pigeons subjected to olfactory nerve section. J Comp Physiol 83:81–92

Berthold P (1973) Relationships between migratory restlessness and migration distance in six Sylvia species. Ibis 155:594–599

Berthold P (1978) The conception of the endogenous control of migration in warblers. In: Schmidt-Koenig K, Keeton W (eds) Animal migration, navigation, and homing. Springer, Berlin Heidelberg New York, p 462

Berthold P, Gwinner E, Klein H, Westrich P (1972) Beziehungen zwischen Zugunruhe und Zugablauf bei Garten- und Mönchsgrasmücke (Sylvia borin und S. atricapilla). Z Tierpsychol 30: 26–35

Binder E (1979) Das optische Lernvermögen der Brieftauben (Columba livia) in Abhängigkeit von Störungen des olfaktorischen Systems. Wiss Arbeit zur Zulassung an Gymnasien, Tübingen

Emlen ST (1972) The ontogenetic development of orientation capabilities. In: Galler SR, Schmidt-Koenig K, Jacobs GJ, Belleville RE (eds) Animal orientation and navigation. NASA SP-262, US Govt Printing Office, Washington DC, pp 191–210

Fromme HG (1961) Untersuchungen über das Orientierungsvermögen nächtlich ziehender Kleinvögel, Erithacus rubecula, Sylvia communis. Z Tierpsychol 18:205–220

Gauthreaux SA (1980) Animal migration, orientation, and navigation. Academic, London New York

Gwinner E (1968) Circannuale Periodik als Grundlage des jahreszeitlichen Funktionswandels bei Zugvögeln. Untersuchungen am Fitis (Phylloscopus trochilus) und am Waldlaubsänger (P. sibilatrix). J Ornithol 109:70–95

Gwinner E (1969) Untersuchungen zur Jahresperiodik von Laubsängern. J Ornithol 110:1–21

Gwinner E (1972) Endogenous timing factors in bird migration. In: Galler SR, Schmidt-Koenig K, Jacobs GJ, Belleville RE (eds) Animal orientation and navigation. NASA SP-262, US Govt Printing Office, Washington DC, pp 321–338

Gwinner E (1974) Endogenous temporal control of migratory restlessness in warblers. Naturwissenschaften 61:405–406

Hartwick RF, Foa A, Papi F (1977) The effect of olfactory deprivation by nasal tubes upon homing behavior in pigeons. Behav Ecol Sociobiol 2:81–89

Hermayer KL, Keeton WT (1979) Homing behavior of pigeons subjected to bilateral olfactory nerve section. Monitore Zool Ital 13:303–313

Hoffmann K (1954) Versuche zu der im Richtungsfinden der Vögel enthaltenen Zeitschätzung. Z. Tierpsychol 11:453–475

Ioale P (1982) Pigeon homing: Effects of differential shielding of home cages. In: Papi F, Wallraff HG (eds) Avian navigation. Springer, Berlin Heidelberg New York, pp 170–178

Keeton WT (1972) Effects of magnets on pigeon homing. In: Galler SR, Schmidt-Koenig K, Jacobs J, Belleville RE (eds) Animal orientation and navigation. NASA SP-262, US Govt Printing Office, Washington DC, pp 579–594

Keeton WT (1974) The orientational and navigational basis of homing in birds. In: Adv Study Behav 5, Academic, London New York, pp 47–132

Keeton WT, Kreithen ML, Hermayer KL (1977) Orientation by pigeons deprived of olfaction by nasal tubes. J Comp Physiol 114:289–299

Kiepenheuer J (1978) Inversion of the magnetic field during transport: its influence on the homing behavior of pigeons. In: Schmidt-Koenig K, Keeton WT (eds) Animal migration, navigation, and homing. Springer, Berlin Heidelberg New York, pp 135–142

Kiepenheuer J (1978) Pigeon homing: a repetition of the deflector loft experiment. Behav Ecol Sociobiol 3:393–395

Kiepenheuer J (1981) A preliminary evaluation of the factors involved or not involved in the deflector loft effect. In: Papi F, Wallraff HG (eds) Avian navigation. Springer, Berlin Heidelberg New York, pp 203–210

Kiepenheuer J (1981b) The effect of magnetic anomalies on the homing behavior of pigeons: an attempt to analyse the possible factors involved. In: Papi F, Wallraff HG (eds) Avian navigation. Springer, Berlin Heidelberg New York, pp 120–128

Kiepenheuer J (to be published) The magnetic compass mechanism of birds and its possible association with the shifting course direction of migration. Behav Ecol Sociobiol

Kiepenheuer J (to be published) Homing behavior of anosmic pigeons raised in a deflector loft.

Kramer G (1950) Weitere Analyse der Faktoren, welche die Zugaktivität des gekäfigten Vogels orientieren. Naturwissenschaften 37:377–378

Kreithen ML (1978) Sensory mechanisms for animal orientation – can any new ones be discovered? In: Schmidt-Koenig K, Keeton WT (eds) Animal migration, navigation, and homing. Springer, Berlin Heidelberg New York, pp 25–34

Kreithen ML, Keeton WT (1974a) Detection of changes in atmospheric pressure by the homing pigeon, Columbia livia. J Comp Physiol 89:73–82

Kreithen ML, Keeton WT (1974b) Detection of polarized light by the homing pigeon, Columbia livia. J Comp Physiol 89:83–92

Lindley SB (1930) The maze learning abilities of anosmic and blind anosmic rats. J genetic Psychol developm comp and clinical Psychol 37:245–265

Marks HE, Remley NR, Seago JD, Hastings DW (1971) Effects of bilateral lesions of the olfactory bulbs of rats on measures of learning and motivation. Physiol Behav 7:1–6

Matthews GVT (1953) Sun navigation in homing pigeons. J Exp Biol 30:243–267

Matthews GVT (1955) Bird navigation. Cambridge University Press, Cambridge

McDonald DL (1972) Some aspects of the use of visual cues in directional training of homing pigeons. In: Galler SR, Schmidt-Koenig K, Jacobs GJ, Belleville RE (eds) Animal orientation and navigation. NASA SP-262, US Govt Printing Office Washington DC, pp 293–304

Merkel FW, Fromme HG (1958) Untersuchungen über das Orientierungsvermögen nächtlich ziehender Rotkehlchen, Erithacus rubecula. Naturwissenschaften 45:499–500

Merkel FW, Wiltschko W (1965) Magnetismus und Richtungsfinden zugunruhiger Rotkehlchen, Erithacus rubecula. Vogelwarte 23:71–77

Papi F (1982) Olfaction and homing in pigeons: ten years of experiments. In: Papi F, Wallraff HG (eds) Avian navigation. Springer, Berlin Heidelberg New York, pp 149–159

Papi F, Wallraff HG (eds) (1982) Avian navigation. Springer, Berlin Heidelberg New York

Papi F, Fiore L, Fiaschi V, Benvenuti S (1971) The influence of olfactory nerve section on the homing capacity of carrier pigeons. Monit Zool Ital 5:265–267

Papi F, Fiore L, Fiaschi V, Benvenuti S (1972) Olfaction and homing in pigeons. Monit Zool Ital 6:85–95

Papi F, Fiore L, Fiaschi V, Benvenuti S (1973) An experiment for testing the hypothesis of olfactory navigation of homing pigeons. J Comp Physiol 83:93–102

Papi F, Keeton WT, Brown AI, Benvenuti S (1978) Do American and Italian pigeons relay on different homing mechanisms? J Comp Physiol 128:303–317

Papi F, Ioale P, Fiaschi V, Benvenuti S, Baldaccini NE (1974) Olfactory navigation of pigeons: the effect of treatment with odorous air currents. J Comp Physiol 94:187–193

Perdeck AC (1958) Two types of orientation in migrating starlings, Sturnus vulgaris L, and chaffinches, Fringilla coelebs L, as revealed by displacement experiments. Ardea 46:1–37

Phillips DS (1969) Effects of olfactory bulb ablation on visual discrimination. Physiol Behav 5:13–15

Phillips JB, Waldvogel JA (1982) Deflected light cues generate the short-term deflector-loft effect. In: Papi F, Wallraff HG (eds) Avian navigation. Springer, Berlin Heidelberg New York, pp 190 to 202

Schmidt-Koenig K (1958) Experimentelle Einflußnahme auf die 24-Stunden-Periodik bei Brieftauben und deren Auswirkungen unter besonderer Berücksichtigung des Heimfindevermögens. Z Tierpsychol 15:301–331

Schmidt-Koenig K (1961) Die Sonne als Kompass im Heim-Orientierungssystem der Brieftauben. Z. Tierpsychol 68:221–244

Schmidt-Koenig K (1972) New Experiments on the effect of clock shifts on homing in pigeons. In: Galler SR, Schmidt-Koenig K, Jacobs J, Belleville RE (eds) Animal orientation and navigation. NASA SP-262, US Govt Printing Office Washington DC, pp 275–285

Schmidt-Koenig K (1979) Avian orientation and navigation. Academic, London

Schmidt-Koenig K (1980) Das Rätsel des Vogelzugs. Hoffmann und Campe, Hamburg

Schmidt-Koenig K, Keeton WT (eds) (1978) Animal migration, navigation, and homing. Springer, Berlin Heidelberg New York

Schmidt-Koenig K, Schlichte HJ (1972) Homing in pigeons with impaired vision. Proc Nat Acad Sci USA 69:2446–2447

Schmidt-Koenig K, Walcott C (1978) Tracks of pigeons homing with frosted lenses. Anim Behav 26:480–486

Walcott C (1978) Anomalies in the earth's magnetic field increase the scatter of pigeon's vanishing bearings. In: Schmidt-Koenig K, Keeton WT (eds) Animal migration, navigation, and homing. Springer, Berlin Heidelberg New York, pp 143–151

Walcott C, Green RP (1974) Orientation of homing pigeons altered by a change in the direction of an applied magnetic field. Science 184:180–182

Waldvogel JA, Benvenuti S, Keeton WT, Papi F (1978) Homing pigeon orientation influenced by deflected winds at home loft. J Comp Physiol 128:297–301

Waldvogel JA, Phillips JB (1982) Pigeon homing: new experiments involving permanent-resistent deflector-loft birds. In: Papi F, Wallraff HG (eds) Avian navigation. Springer, Berlin Heidelberg New York, pp 179–189

Wallraff HG (1969) Über das Orientierungsvermögen von Vögeln unter natürlichen und künstlichen Sternmustern. Dressurversuche mit Stockenten. Verh Dt Zool Ges Innsbruck 1968: 348–357

Wallraff HG (1974) Das Navigationssystem der Vögel. Oldenbourg, München

Wallraff HG (1979) Goal-oriented and compass-oriented movements of displaced pigeons after confinement in differentially shielded aviaries. Behav Ecol Sociobiol 5:201–225

Wallraff HG (1980) Does pigeon homing depend on stimuli perceived during displacement? Experiments in Germany. J Comp Physiol 139:193–201

Wallraff HG (1981) The olfactory component of pigeon navigation: steps of analysis. J Comp Physiol 143:411–422

Wallraff HG, Foa A (1981) Pigeon navigation: Charcoal filter removes relevant information from environmental air. Behav Ecol Sociobiol 9:67–77

Wenzel BM, Salzman A (1968) Olfactory bulb ablation or nerve section and behavior of pigeons in nonolfactory learning. Exp Neurologz 22:472–479

Wiltschko R (1980) The development of the sun compass in young homing pigeons. Congr Int Ornith 17:599–603

Wiltschko R, Wiltschko W, Keeton WT (1978) Effect of outward journey in an altered magnetic field in young homing pigeons. In: Schmidt-Koenig K, Keeton WT (eds) Animal migration, navigation, and homing. Springer, Berlin Heidelberg New York, pp 152–161

Wiltschko W (1968) Über den Einfluß statischer Magnetfelder auf die Zugorientierung der Rotkehlchen, Erithacus rubecula. Z Tierpsychol 25:537–558

Wiltschko W (1973) Kompaßsysteme in der Orientierung von Zugvögeln. Akad Wiss Lit Mainz, Reihe Inf Org II. Steiner, Wiesbaden

Wiltschko W, Wiltschko R (1972) The magnetic compass of European robins. Science 176:62–64

Wiltschko W, Wiltschko R (1975a) The interaction of stars and magnetic field in the orientation system of night migrating birds. I. Autumn experiments with European warblers (Gen. Sylvia). Z. Tierpsychol 37:337–355

Wiltschko W, Wiltschko R (1975b) The interaction of stars and magnetic field in the orientation system of night migrating birds. II. Spring experiments with European robins, Erithacus rubecula. Z Tierpsychol 39:265–282

Wiltschko W, Wiltschko R (1976) Interrelation of magnetic compass and star orientation in night migrating birds. J Comp Physiol 109:91–99

Wiltschko W, Wiltschko R, Keeton WT (1976) Effects of a permanent clock-shift on the orientation of young homing pigeons. Behav Ecol Sociobiol 1:229–243

The Development of the Navigational System in Young Homing Pigeons

R. WILTSCHKO and W. WILTSCHKO[1]

1 Introduction

Bird navigation is generally looked upon as a two-step process described by Kramer's map and compass model (1953). In step 1, the bird determines its home direction as a compass direction; in step 2, it uses a compass to localize this direction in space (cf. Wallraff 1974, Wiltschko W. and Wiltschko R. 1982). For both these steps, alternative mechanisms are available to the birds so that they can determine their home direction (1) by information obtained en route during the outward journey; or (2) by local information collected at the starting point of the return flight (in experiments: at the release site). In the first case, a simple strategy called route reversal is applied: the bird records the compass direction of the outward journey, integrating detours if necessary, and reverses this direction to obtain the homeward course (cf. Wiltschko et al. 1978). When using local information, we must assume that the birds make use of some kind of environmental gradients which form the grid of a "map". Knowing the directions in which these gradients increase, a bird can extrapolate their values at unfamiliar locations, and so obtain its home direction by comparing the data of a given site with those of home (for details, see Wallraff 1974, Wiltschko W and Wiltschko R. 1982).

Likewise, a bird has two possibilities of localizing the home direction once it is determined: (1) it may use a magnetic compass which functions as an inclination compass, distinguishing pole-ward and equator-ward by the inclination of the axial course of the field lines (for detail, see Wiltschko and Wiltschko 1972); or (2) it may use a sun compass which derives directions from the sun's azimuth, compensating for its apparent movement in the course of the day by means of the internal clock (cf. Schmidt-Koenig 1961, Wiltschko 1980).

2 The Navigational Strategy of Very Young, Inexperienced Pigeons

Experimental evidence indicates that old, experienced pigeons prefer to determine their home direction using local information, and they make use of their sun com-

[1] Fachbereich Biologie der Universität, Zoologie, Siesmayerstraße 70, 6000 Frankfurt a.M., FRG

Localization and Orientation in Biology and Engineering
ed. by Varjú/Schnitzler
© Springer Verlag Berlin Heidelberg 1984

pass whenever the sun is visible. This is not true, however, for very young, inexperienced pigeons: when they become able to fly, their most important orientation mechanism appears to be the magnetic compass.

Figure 1 demonstrates the essential role of magnetic information in the orientation of such birds. Transportation in a distorted magnetic field meant denying them the possibility of obtaining meaningful directional information en route and caused disorientation (Fig. 1a,b; cf. Wiltschko W. and Wiltschko R. 1981); disorientation also occurred when the magnetic compass was disturbed upon release by fixing small bar magnets to the pigeon's back (Fig. 1c,d; comp. Keeton 1972). These findings suggest that the very young, inexperienced birds' first strategy is route reversal, using the magnetic field for determining the direction of the outward journey and for localizing the home direction. This is not surprising, since the magnetic compass is available to the birds right from the beginning. Perceiving the magnetic field per se provides them with a means of distinguishing directions in the horizontal plane, just as perceiving gravity provides a reference in the vertical. For the simple strategy described above, this ability is sufficient; no other knowledge of the directional relationship of environmental factors is required.

As soon as the young birds begin to gather flying experience, either spontaneously by venturing away from their loft, or forced upon them by training releases, they begin to calibrate a variety of other orientation cues found in their environment. The basic homing strategy based on the magnetic compass alone is then rapidly replaced by learned mechanisms which are easier to use and/or offer more safety. In particular, the birds switch from information collected en route during the outward journey to local information collected at the release site, and for localizing directions, the magnetic compass is replaced by the sun compass as long as the sun is visible.

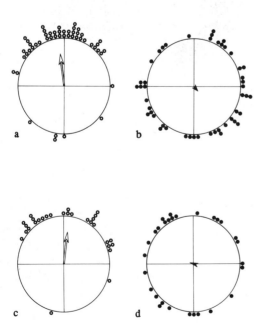

Fig. 1a–d. Very young, inexperienced pigeons need undisturbed magnetic information during the outward journey as well as upon release. (All birds were released under sun.) **a** Normally transported controls. **b** Birds transported in a distorted magnetic field. (Data from W. Wiltschko and R. Wiltschko 1981). **c** Control birds carrying brass bars. **d** Birds carrying small magnets. (Data from Keeton 1972). The home direction is set equal to 0°. *Symbols at periphery circle* vanishing bearings of individual birds; *arrows* mean vector in relation to the radius of the circle = 1

In several series of experiments we have tried to learn more about the transition of the birds' orientation system, and here we present a short summary of the results.

3 Establishing the Sun Compass

Sun compass orientation is based on the birds' "knowledge" of the relationship between the sun's azimuth, time of day, and geographic direction, a relationship which varies with geographic latitude and the season. The following series of experiments clearly indicates that this knowledge is based on experience:

A group of young pigeons was raised in a permanently shifted photo-period which began 6 h after sunrise and ended 6 h after sunset, so that they experienced an abnormal relationship between the sun's azimuth, geographic direction and time. In critical tests, they oriented like their controls, as long as they lived in that shifted photoperiod. When their internal clock was reset to normal time (i.e. advanced by 6 h), they showed a characteristic deviation from the mean of their controls which corresponded to the reaction of normal birds which had undergone a 6 h forward phase shift. These findings show that the experimental birds had established a "false" sun compass which was adapted to the experimental situation (W. Wiltschko et al. 1976): for them, the sun in the south had been the morning sun etc.

The next experimental series concerned the nature of the learning process. We raised a group of pigeons that was allowed to see only the descending part of the sun's arc in the afternoon. Tested in the early morning, these birds were well-oriented. Critical tests involving shifting their internal clock and fixing magnets on their back, however, revealed that they did not use their sun compass at that time of the day, but relied on their magnetic compass instead. Their knowledge of sun only in the afternoon had not been sufficient to establish the sun compass for the entire day. Obviously the birds must observe every portion of the sun's arc to associate it with time and geographic direction (R. Wiltschko et al. 1981).

This led to the question of how the birds calibrate the sun's movements. Here the magnetic field appeared to provide an obvious reference system. To test this hypothesis, we raised a group of young pigeons that was allowed to see the sun only in an abnormal directional relationship to the magnetic field. They lived in a small loft surrounded by Helmholtz coils which turned magnetic north clockwise to geographic east. They could observe the sun from a roof-top aviary, and they were released for exercise and for training flights only when the sky was totally overcast. On their first flight under sun, the experimentals showed a clockwise deviation from their controls, indicating that the artificially altered magnetic field had affected their orientation under sun. The deflection was smaller than expected, however, and later experiments suggest that the interrelation of the pigeons' sun compass and magnetic compass may be more complex than a simple calibration process (W. Wiltschko et al. 1983).

To find out at what age the learning process establishing the sun compass takes place, we performed a series of clock-shift experiments testing young pigeons of various ages. In inexperienced pigeons, the reaction to clock-shifting appeared to

depend on their age. When the birds were less than 11 weeks old, they mostly departed
in the home direction (Fig. 2a,b): only a few individuals showed a deviation in the
expected direction; at an age of 12 weeks and older, all birds showed the characteristic
deviation (Fig. 2c,d). Young birds that had had a few previous training flights, however, began to use the sun compass much earlier (Fig. 2e,f). They showed the typical
deviation, indicating sun compass orientation already at an age of only 8–10 weeks
(R. Wiltschko and W. Wiltschko 1981). These findings point out that flying experience,
rather than absolute age determines when the sun compass is developed. For un-
trained pigeons, increasing age meant an increasing number of spontaneous flights
around their loft. They gathered experience by "training themselves", which obviously had the same effect as training releases. By confronting the young birds with the
need to orientate, the learning processes of establishing the sun compass were started.

Once the sun compass is established, it is used whenever the sun is visible, and the
magnetic compass serves only as a back-up system for overcast days. We do not know
why the sun compass is preferred, but it must offer some advantages to the birds, as
otherwise it would be hardly possible to understand why evolution favored the formation of such a complex mechanism.

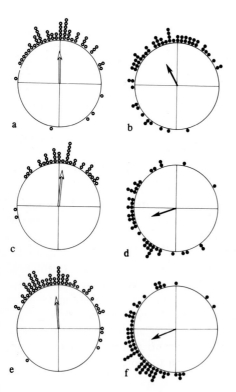

Fig. 2a–f. The development of the sun
compass depends on the young pigeons'
flying experience. Inexperienced pigeons hardly responded to shifting their
internal clock when they were younger
than 11 weeks (**a, b**), whereas they did
when they were older (**c,d**). Young pigeons that had completed a few training
flights already showed the typical deviation indicating the use of the sun compass when they were 8–10 weeks old
(**e, f**) (symbols as in Fig. 1); *open symbols* controls; *solid symbols* pigeons
whose internal clock had been reset 6 h
fast. (Data from R. Wiltschko and W.
Wiltschko 1981)

4 Establishing the Navigational "Map"

Much less is known about the birds' change in strategy from using route reversal to navigation based on local factors. The physical nature of these factors in the navigational "map" is still unclear (but see Papi 1976, Walcott 1980, Wallraff 1980 and others): we tend to assume that they are some kind of environmental gradients (cf. Wallraff 1974). To derive their home direction, the birds must be familiar with their course and their variability in space. Young pigeons appear to acquire this knowledge on their spontaneous flights at the loft and on their training flights (W. Wiltschko and R. Wiltschko 1982); for some factors, a direct measurement of the gradient direction at the home loft may also be possible (Wallraff 1974).

So the navigational "map" of birds is assumed to be a system which has to be learned, yet the details of the learning processes are still rather unclear. An analysis of the vanishing bearings of inexperienced homing pigeons, though, seemed to suggest that it takes place at an age of about 10-12 weeks, when the young birds begin to gather flying experience by extending their spontaneous flights (W. Wiltschko and R. Wiltschko 1982). Direct evidence that outward journey information was no longer of primary importance, however, was still lacking, and so we began a series of experiments in which we deprived birds of meaningful directional information during displacement. This was done either by transporting them in a distorted magnetic field (R. Wiltschko et al. 1978) or by transporting them in total darkness (W. Wiltschko and R. Wiltschko 1981).

In very young, inexperienced pigeons, such treatment had a dramatic effect: those experimentals were totally disoriented (Fig. 3b), while the normally transported controls were very well home-oriented (with the exception of two releases, see below) as must be expected when information collected during the outward journey is used (Fig. 3a). The behavior of the control birds showing larger deviations from home and that of the inexperienced pigeons older than 12 weeks seems to indicate the next step in the development. Deprivation of outward journey information ceases to have a large effect (Fig. 3d), and at the same time the controls begin to be less well home-oriented (Fig. 3c). The young pigeons have started to use local information, but they frequently interpret it incorrectly (resulting in deviations from home) because of their still limited experience, or their interpretations disagree (resulting in short vector lengths) because the experience within a group gathered on spontaneous flights varies. Yet a small effect of denying them outward journey information is still found: the controls test homeward-oriented, the experimentals do not.

Experienced pigeons that have completed numerous training flights do not seem to be noticeably affected by transportation without meaningful outward journey information (Fig. 3e,f). The birds now rely on local information, and on their training flights, their "map" has become a more or less realistic picture of the true spatial distributions of the factors involved. Experienced pigeons are generally well oriented; deviations from the home direction (so-called release site biases) are still observed, but they are seldom extreme (see W. Wiltschko and R. Wiltschko 1982, Grüter et al. 1982).

Young pigeons, therefore, as soon as they have gathered some flying experience begin to use local information, and information collected during the outward journey

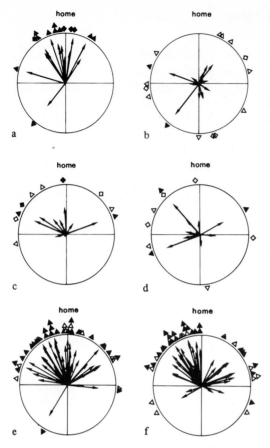

Fig. 3a–f. As young pigeons grow older and more experienced, they cease to use information collected during the outward journey. Deprivation of meaningful outward journey information caused disorientation in very young, inexperienced pigeons (**a, b**), whereas it had much less effect in inexperienced birds older than 12 weeks (**c, d**). The treatment did not seem to have any effect on experienced birds (**e, f**). The home direction is set equal to 0°. *Arrows* represent the mean vectors of the individual releases in relation to the radius of the circle = 1. *Left diagrams* controls; *right diagrams* experimentals. *Symbols at the periphery* mark the mean directions (*solid* mean direction is significant with p < 0.05; otherwise *open*): *diamonds* experimentals transported in a distorted magnetic field; *triangles* transported in total darkness. (See text; data from Wiltschko in prep.)

gradually becomes less important. That this change in orientation strategy, like the development of the sun compass, depends on experience rather than age is supported by the observation that extremely young pigeons, after a few training flights, frequently show a similar release site bias as older, inexperienced birds (cf. R. Wiltschko 1983). The reason for the preferential use of local information is obvious: the navigational "map" allows a bird to redetermine its home direction as often as it feels necessary, and by this incorrect courses may be corrected, which is not possible if outward journey information alone is used. Hence navigation based on local factors offers more safety.

5 Conclusions

During the first months, therefore, the navigation system of a young pigeon undergoes a great change: the original mechanism, route reversal by the magnetic compass, is replaced by more complex mechanisms, like the sun compass and the navigational "map". During this development, the magnetic compass serves as a directional reference calibrating the environmental factors on which these learned mechanisms, are based. Establishing them by experience makes the birds' orientation system very flexible and offers important advantages. A bird can make use of highly complex, multi-factorial mechanisms which cannot be genetically encoded because of their complexity and local variability, and these mechanisms can be perfectly adapted to the situation in which the individual bird must orient, e.g. the sun compass is tuned to the sun's arc at that specific geographic latitude, and the navigational "map" incorporates those factors which are the most useful in a given region. The "map" may be very different, however, in the various regions of the world.

Acknowledgements. Our work is supported by the Deutsche Forschungsgemeinschaft in the program SFB 45.

References

Grüter M, Wiltschko R, Wiltschko W (1982) Distribution of release-site biases around Frankfurt a.M. In: Avian navigation. Springer, Berlin Heidelberg New York, p 222

Keeton WT (1972) Effects of magnets on pigeon homing. In: Animal orientation and navigation, NASA SP-262. U.S. Gov. Print. Off., Washington D.C., p 579

Papi F (1976) The olfactory navigation system of the homing pigeon. Verh Deutsch Zool Ges Hamburg, p 184

Schmidt-Koenig K (1961) Die Sonne als Kompaß im Heim-Orientierungssystem der Brieftauben. Z Tierpsychol 18:221–244

Walcott C (1980) Magnetic orientation in homing pigeons. IEEE Trans on Magnetics, Mag-16: 1008–1013

Wallraff HG (1974) Das Navigationssystem der Vögel. Schriftenreihe „Kybernetik". R Oldenbourg Verlag, München Wien, p 136

Wallraff HG (1980) Olfaction and homing in pigeons: nerve-section experiments, critique, hypotheses. J Comp Physiol Psychol 139:209–224

Wiltschko R (1980) Die Sonnenorientierung der Vögel. I. Die Rolle der Sonne im Orientierungssystem und die Funktionsweise des Sonnenkompaß. J Ornithol 121:121–143

Wiltschko R (to be published) The ontogeny of orientation in young pigeons. Comp Biochem Physiol A

Wiltschko R, Nohr D, Wiltschko W (1981) Pigeons with a deficient sun compass use the magnetic compass. Science 214:343–345

Wiltschko R, Wiltschko W (1981) The development of sun compass orientation in young homing pigeons. Behav Ecol Sociobiol 9:135–141

Wiltschko R, Wiltschko W, Keeton WT (1978) Effect of outward journey in an altered magnetic field on the orientation of young homing pigeons. In: Animal migration, navigation, and homing. Springer, Berlin Heidelberg New York, p 152

Wiltschko W, Wiltschko R (1972) Magnetic compass of European robins. Science 176:62–64

Wiltschko W, Wiltschko R (1981) Disorientation of inexperienced young pigeons after transportation in total darkness. Nature 291:433–434

Wiltschko W, Wiltschko R (1982) The role of outward journey information in the orientation of homing pigeons. In: Avian navigation. Springer, Berlin Heidelberg New York, p 239

Wiltschko W, Wiltschko R, Keeton WT (1976) Effects of a "permanent" clock shift on the orientation of young homing pigeons. Behav Ecol Sociobiol 1:229–243

Wiltschko W, Wiltschko R, Keeton WT, Madden R (1983) Growing up in an altered magnetic field affects the initial orientation of young homing pigeons. Behav Ecol Sociobiol 12:135–142

The Homing Behavior of Pigeons Raised in a Reversed Magnetic Field

J. KIEPENHEUER[1]

1 Introduction

Homing pigeons use a magnetic compass to maintain their home direction (Keeton 1971, Walcott and Green 1974). But how do they establish their relative position with respect to home? A magnetic compass or a related mechanism may possibly be used in determining the home vector from the sequence of movements and turns during transport. In pigeons, manipulation of the magnetic field during transport results in a bias of initial orientation (Kiepenheuer 1978, Wiltschko et al. 1978). This may be caused, at least to some extent, by the inability to correctly perceive information on turns during displacement (Kiepenheuer 1980). When simulating transportation in a direction opposite to actual movement by reversing the magnetic field, inconsistencies inevitably arise, either because the static artificial field is not perfect, or because it does not copy the dynamic parameter, the movement through the field. Some of these problems may be overcome if the birds are raised in a reversed field and then displaced in the normal field.

2 Methods

Fledglings (4–5 weeks) were allotted to two cubic aviaries (1.5 m) which stood in the open, slightly shielded by our main loft to the east. One aviary (experimental birds, E) was surrounded by a box coil (Rubens 1945) oriented N–S, powered by a constant DC current device. The artificial field was twice as strong as and opposite to the natural horizontal field component. The resulting field was opposite to the natural field in all respects. The control aviary (C) was 2 m from the E aviary. The birds remained inside the cages for 4–5 months. They were then transported (Fig. 1) inside covered crates. Series A birds were not subjected to any exercise prior to the critical releases. Series B birds had exercise flights of 300–1,000 m and were 2 months old before confinement. Pigeons were released one at a time E and C alternating between, with the sun visible. Vanishing bearings and vanishing time intervals were measured and the return of the pigeons to the loft was recorded. For statistics see Batschelet (1982); in some cases the Fisher exact probability test was used.

[1] Abt. f. Verhaltensphysiologie, Inst. für Biologie III, Universität Tübingen, FRG

Localization and Orientation in Biology and Engineering
ed. by Varjú/Schnitzler
© Springer Verlag Berlin Heidelberg 1984

Fig. 1 a,b. a Series A: Vanishing bearings of inexperienced birds. *Dashed line* points towards home (● 1982, ○ 1980, ▲ 1979, $\beta = 54°$, d = 11.5 km; △ 1979, $\beta = 54°$, d = 16.3 km). The mean vector and a 95% confidence range (*dotted*) of the angle are given; **b** Series B: Vanishing bearings of pigeons which have had some exercise flights before confinement (○ 1980, ●, △ 1981, $\beta = 146°$, d = 11.5 km; ▲ 1981, $\beta = 62°$, d = 12.2 km; □ 1981, $\beta = 54°$, d = 16.3 km)

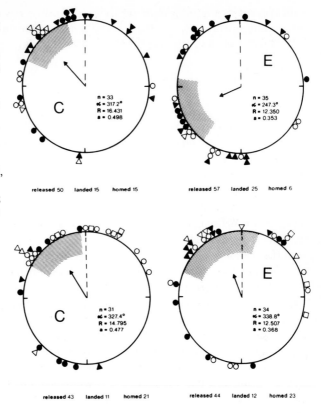

n = 33
α = 317.2°
R = 16.431
a = 0.498

released 50   landed 15   homed 15

E
n = 35
α = 247.3°
R = 12.350
a = 0.353

released 57   landed 25   homed 6

n = 31
α = 327.4°
R = 14.795
a = 0.477

released 43   landed 11   homed 21

E
n = 34
α = 338.8°
R = 12.507
a = 0.368

released 44   landed 12   homed 23

3 Results

All birds showed difficulties in flying and many landed almost immediately. Most, however, took off decidedly after some hesitating wing beats. E pigeons of series A landed more often than the controls (p = 0.005); in series B no difference was observed (p = 0.48, see Fig. 1). Small differences in vanishing intervals were of no significance.

In spite of their total confinement during the previous months some birds of series A and many more of series B homed (see Fig. 1). In series A, more C than E birds homed (p = 0.011). In series B there was no difference between C and E birds (p = 0.50). In general, homing success was better in series B than in series A (p < 0.001).

Figure 1a shows the vanishing bearings of E and C birds of series A. The mean direction of the C birds ($a = 317° ± 26°$) is very close to the normal vanishing direction of our main loft birds, with a significant bias (p < 0.01) to the left of home. Orientation is significant (p < 0.001). Both distributions are approximately unimodal and symmetric around the mean. E birds are more scattered than C birds but they are still significantly oriented (p = 0.013), although not in the home direction (V test, u = -1.15). The number of bearings in the homeward and those in the opposite semicircle are significantly different (p = 0.006). In series B, both E and C birds are significantly oriented (p < 0.01, p < 0.001) in the same direction as in the C birds of series A with only a slightly larger scatter in the E birds (Fig. 1b).

4 Discussion

The behavior of pigeons brought up in a reversed field indicates that they were in a conflict as to which direction to fly, or that the information available (e.g. from the sund and the magnetic compass) was conflicting, because they were opposed to each other. According to Wiltschko (1980, 1983) the pigeon's sun compass is aligned to the magnetic compass, within the time the birds were confined to the aviary.

Under sunny conditions older pigeons use only the sun compass to determine their headings, see Wiltschko (1980). The navigational process from which the direction of home must be derived is, however, at least partly based on information obtained from the magnetic field during transport (Kiepenheuer 1978, Wiltschko et al. 1978). What does that mean for an E bird in the present experiment? When carried out-side the artificial field, it first experiences a turn of 180° with respect to the magnetic field; then it is subject to a sequence of movements and turns in the natural magnetic field till it reaches the release site. Following the information from the magnetic compass it should head towards home. Under sun the bird will, however, utilize the sun compass which now is opposed to the magnetic compass. The E birds should, therefore, fly off in a direction opposite home. On the other hand, the navigational system should tell the bird that this direction is wrong and this should result in the conflict situation actually observed. Final dominance of the correct navigational information over the sun compass would then account for the fact that some E birds finally did return. We would expect different results if we assume a mechanism relying, for example, on olfactory cues rather than on the magnetic field: E birds should fly in the home direction, since under sunny conditions neither navigation nor the compass would be influenced by the reversed field in the home cage. We may therefore conclude that information derived from the magnetic field en route is used to determine the position of the release site, relative to home. In the case of series B we must assume that by the time of confinement (the pigeons being about 2 months old), the sun compass was already fully aligned to the magnetic compass.

References

Batschelet E (1982) Circular statistics in biology. Academic Press, London New York
Keeton WT (1971) Magnets interfere with pigeon homing. Proc Natl Acad Sci USA 68:102–106
Kiepenheuer J (1978) Pigeon navigation and magnetic field. Naturwissenschaften 65:113–114
Kiepenheuer J (1980) The importance of outward journey information in the process of pigeon homing. Congr Int Ornith 17 Berlin, pp 593–598
Rubens SM (1945) Cube surface coil for producing a uniform magnetic field. Rev Sci Instrum 16/9:243–245

Walcott CH, Green RP (1974) Orientation of homing pigeons altered by change in the direction of the applied magnetic field. Science 184:180–182
Wiltschko R (1980) The development of the sun compass in young pigeons. Congr Int Ornith 17 Berlin, pp 599–603
Wiltschko R (to be published 1983) Behav Ecol Sociobiol
Wiltschko R, Wiltschko W (1978) Evidence for the use of outward journey information in homing pigeons. Naturwissenschaften 65:112–113

The Magnetic Compass as an Independent Alternative to the Sun Compass in the Homing Performance of Young Pigeons

D. NOHR[1]

Since it was demonstrated that homing pigeons use the sun as a compass for their orientation whenever it is visible (Schmidt-Koenig 1961), many tests have been carried out and in 1976 a series of releases showed that sun compass orientation is learned rather than innate (Wiltschko et al. 1976). This learning process was studied and a first series of tests with young pigeons that had only seen the descending part of the sun's arc, demonstrated that these pigeons did not use their sun compass when they were released in the early morning, i.e. they did not react to a 6-h shifting of their internal clock but were homeward orientated, while 6-h-shifted control birds, that knew the complete arc of the sun, showed a deflection in their mean vanishing direction from the home direction (R. Wiltschko and W. Wiltschko 1980).

The next step was to find out what kind of information these experimentals used for correct orientation. One possibility was the magnetic field of the earth and if this were correct small bar magnets fixed on the pigeons back should cause disorientation, as is known in experienced birds under overcast skies (Keeton 1972).

For our tests a group of pigeons was raised in a light-tight room under a timer-controlled natural photoperiod. The birds were only allowed to go into their aviary and to see the sun in the afternoon. The control birds were raised in an identical room but they were allowed to go into their aviary and to see the sun the whole day long. Both groups had training flights of up to 30 km in the cardinal compass directions, the control birds at different times of the day, the test birds only in the afternoon. When the program of test releases was started, the birds were about 4 months old.

The critical tests took place under clear skies early in the morning at different release sites. Four groups were tested: control, experimentals, and a half of each group wearing small bar magnets (2.6 × 0.6 × 0.3 cm, 4 g). These magnets were fixed on the pigeons backs with branding cement, the north pole pointing to the birds' head. During release the pigeons were tossed singly and alternatively from all four groups to ensure a minimum of difference in any change of environmental conditions for the different treatments. The pigeons were watched by two observers with 10 × 40 binoculars until they vanished from sight, the last bearing being recorded to the nearest 5°. The vanishing times were recorded by stop-watch. The mean vector of each test was calculated by vector analysis, and the different tests combined by setting the home direction to 360°. The computer work was carried out at the Hochschulrechenzentrum of the university, it using the Rayleigh test for significant direc-

---

[1] Fachbereich Biologie der Universität, 6000 Frankfurt/Main, FRG

Localization and Orientation in Biology and Engineering
ed. by Varjú/Schnitzler
© Springer Verlag Berlin Heidelberg 1984

tional tendencies and the Mardia Watson Wheeler test for differences in the scatter of the two samples (Batschelet 1965, 1972). The data are shown in Table 1, the results are shown in Fig. 1.

The control birds (Fig. 1a) were well homeward orientated, as were the experimentals (Fig. 1b) which had never been outside their loft before at this time of day. The small bar magnets did not impair the orientation of the control, which were used to flying at this time of day (Fig. 1c) but the test-birds with magnets were disorientated (Fig. 1d). Their departure bearings were significantly more scattered than those of the other three groups [$p < 0.001$ for both control groups, $p < 0.01$ for the test birds without magnets, (Mardia test)].

Table 1. Statistical data of four groups of pigeons

Group	N (number of birds released)	Bearings	Mean vector Direction	Length	P (Rayleigh test)
Control	74	57	356°	0.70	< 0.001
Control with magnets	61	57	353°	0.74	< 0.001
Experimental	72	54	353°	0.54	< 0.001
Experimental with magnets	62	54	13°	0.18	> 0.05 [a]

[a] Not significant

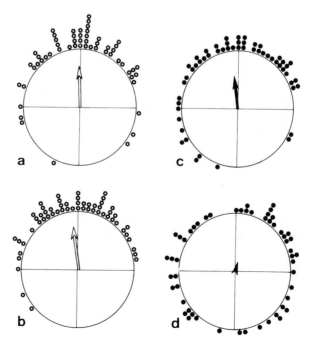

Fig. 1a–d. Results of the tests with magnets in the morning. a controls, n = 57, $a = 356°$, r = 0.70, (p < 0.001); b controls with magnets, n = 57, $a = 353°$, r = 0.74, (p < 0.001); c experimentals, n = 54, $a = 353°$, r = 0.54, (p < 0.01); d experimentals with magnets, n = 54, $a = 13°$, r = 0.18, (p > 0.05). The home direction is equal to "north" or 360°

We came to the conclusion that both control groups used their sun compass for orientation, while the experimentals that had never seen the sun in the morning were unable to do so. It seems clear that they used information about the earths magnetic field and their magnetic compass, as the small bar magnets, which represent the only difference between the two groups of experimental pigeons, caused disorientation similar to that seen in experienced pigeons under overcast skies (Keeton 1972).

Thus we found out that the magnetic compass is a functionating orientation system for young pigeons, whether or not their use of the sun compass has been established, and that the magnetic field of the earth is the first source of compass information for homing pigeons before they learn sun compass orientation.

Acknowledgements. The study was supported by the Deutsche Forschungsgemeinschaft.

References

Batschelet E (1965) Statistical methods for the analysis of problems in animal orientation and certain biological rhythms. Am Inst Biol Sci, Washington D.C.

Batschelet E (1972) Recent statistical methods for orientation data. In: Galler et al. (eds) Animal orientation and navigation NASA SP-262, U.S. Gov Print Off, Washington D.C., pp 61–93

Keeton WT (1972) Effects of magnets on pigeon homing. In: Galler et al. (eds) Animal orientation and navigation. NASA SP 262, U.S. Gov Print Off, Washington D.C., pp 579–594

Schmidt-Koenig K (1961) Die Sonne als Kompaß im Heim-Orientierungssystem der Brieftauben. Z Tierpsychol 18:221

Wiltschko W et al. (1976) Effects of a "permanent" clock shift on the orientation of young homing pigeons. Behav Ecol Sociobiol 1:229

Wiltschko R, Wiltschko W (1980) The process of learning sun compass orientation in young homing pigeons. Naturwissenschaften 67:512

Typical Migratory Birds: Endogenously Programmed Animals

P. BERTHOLD[1]

1 Introduction

Typical migratory bird species, i.e. intercontinental long-distance migrants, are characterized by exceedingly precise aviation in a temporal sense despite the fact that they often frequent areas with weak environmental information on seasons. For centuries "endogenous" guiding components for migrants have been postulated, but more comprehensive information about the endogenous basis of control mechanisms of migration in birds has only recently been collected and will be outlined here.

2 Circannual Rhythms

As has been shown in 12 avian species, endogenous physiological oscillations can regularly bring migratory birds into a migratory disposition (above all fat deposition providing fuel for migration, expressed by body weight gain) and into a certain stage of actual migration (exhibition of migratory "restlessness" in capative individuals, measurable in registration cages, Fig. 1) although the birds are kept in constant experimental conditions. Despite the elimination of environmental seasonality in these experimental birds, the physiological rhythms still persist for up to 10 years. Thus, they must be considered truly endogenous, i.e., self-sustaining and of lifelong efficacy.

The most prominent characteristic of these internal rhythms is the regular deviation from the calendar year. As a result they are called circa-annual, in short circannual rhythms. This deviation from 12 months proves that the observed rhythms are internal biological clocks and that they are not driven by some environmental annual rhythms. On the other hand, their deviation requires that environmental factors adjust the approximately annual rhythms to the biological seasons. Photoperiod has been proven to be an important synchronizing *Zeitgeber* (for reference, see, e.g., Berthold 1980 and Gwinner 1981). With support by the Deutsche Forschungsgemeinschaft.

[1] Max-Planck-Institut für Verhaltensphysiologie, Vogelwarte Radolfzell, Schloß, 7760 Radolfzell, FRG

Localization and Orientation in Biology and Engineering
ed. by Varjú/Schnitzler
© Springer Verlag Berlin Heidelberg 1984

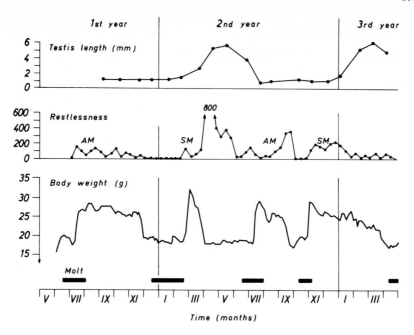

Fig. 1. Circannual rhythms of four annual events in an individual Garden Warbler *Sylvia borin* kept in constant experimental conditions (daily light–dark ratio 10:14 h) during a long-term experiment from juvenile development onward. *AM, SM* autumn and spring migratory period. (Berthold et al. 1971)

3 Innate and Inheritable Programs for the Course of Migration

3.1 Migratory Disposition and Activity

The study of circannual rhythms (Sect. 2) has increasingly supported the view that both the onset and termination of migratory events and the extent and pattern of these processes are internally programmed. In the Garden Warbler it was found that neither suppressing temporarily migratory restlessness or disposition by interposed periods of complete nocturnal darkening or by simulated rainfall, nor preventing the birds from fattening, had any effect on the subsequent course of the patterns of restlessness and migratory disposition. Thus their entire course appears to be endogenously controlled. A "sliding set-point" seems permanently to determine the required seasonal values.

These results support the view that extent and patterns of migratory disposition and restlessness in typical migratory birds are innate and inheritable like morphological features, colourations, etc. This was in fact demonstrated by a cross-breeding experiment (Fig. 2): F_1-hybrids from two Blackcap *Sylvia atricapilla* populations with migratory journeys of quite different lengths and accordingly, different extents of migratory restlessness showed an intermediate pattern of restlessness. Thus these migrants seem equipped with genetically fixed time-programs for migration.

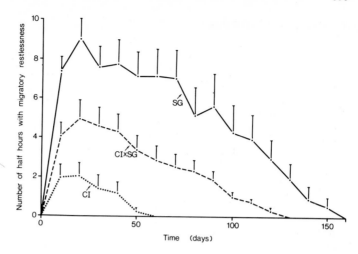

Fig. 2. Time course of nocturnal migratory restlessness of groups of 24–32 Blackcaps from a German (*SG*) and an African (*CI*) population and from their hybrids. Mean values for 10-day periods and standard errors. (Berthold and Querner 1981)

3.2 Patterns of Migratory Activity and Migratory Journey

Inter- and intraspecific comparisons of the amount of migratory restlessness and the distance covered during the first migratory period revealed a strong positive correlation between these two features (e.g., Berthold 1983).

With respect to the patterns of restlessness, the following characteristics were found: in European-African long-distance migrants, normally crossing the Sahara Desert, we regularly found fairly symmetric patterns with high central values. These peaks were in temporal accordance with the time of the Sahara crossing. In shorter-distance migrants, very flat patterns were observed, skewed to the left. Again this low overall restlessness and the peak of activity towards the end of the period of unrest are in full accordance with the course of migration in free-living conspecifics. These and other conformities (for reference, see Berthold 1983) suggest that the patterns of migratory restlessness represent endogenous adaptations to specific migratory journeys.

3.3 Adaptedness of Moult and Morphological Features

In typical migratory birds, wings, as a rule, are comparably longer and more pointed, and moult is of shorter duration and more telescoped with other annual events than in less migratory forms. Recently, we have demonstrated the immediate genetic basis of these adaptations to migration: cross-breeding Blackcaps from different populations with various migratory performance and thus according differentiations in the time course of juvenile moult as well as in wing length and body weight, clearly demonstrated inheritance of these characters in F_1-hybrids (Berthold and Querner 1982a).

354

3.4 Directional Preferences

There is evidence in some bird species that the choice of migratory direction is endogenously programmed (for review, see Berthold 1980). Gwinner and Wiltschko (1978) showed that even a population-specific shift of the directional preference during migration can be internally fixed. The Garden Warbler is known to leave its central European breeding grounds and fly initially in a southwesterly direction towards the Iberian peninsula. Since all the winter quarters in Africa are situated south of this intermediate goal, a shift of the migratory direction has to occur roughly in the area of Iberia to northern Africa. Indeed, a similar alteration of the directional preference was spontaneously performed by experimental Warblers at the appropriate time, although the birds had been kept in constant conditions in S. Germany (Fig. 3). The most plausible interpretation of this result is that the observed shift of the migratory direction is innate and probably closely linked to the genetically controlled course of migratory activity, and thus to the underlying circannual system (Sects. 2, 3.1).

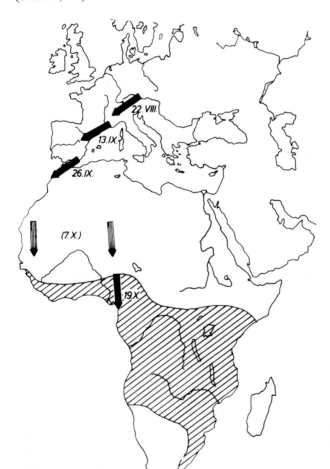

Fig. 3. Directional preferences (*arrows*) of experimental Garden Warblers, investigated in S. Germany, during those periods at which free-living conspecifics pass through certain banding stations (average passage dates at corresponding places along their migratory route from central Europe to the African winter quarters (*hatched area*). (Gwinner and Wiltschko 1978)

4 Genetic Determination of Migrants and Nonmigrants

Most common among migratory bird species is the partial migratory habit: part of the population is migratory (and benefits from the more suitable conditions of the chosen wintering areas), part is nonmigratory (and temporarily suffers from hazardous winter conditions). Fitness is alternately higher in both fractions, and hence evolution in many ways favours this strategy of diphenism. With respect to the control of this diphenism, two controversial hypotheses were established: a "genetic" hypothesis and a "behavioural-constitutional" hypothesis.

In the Blackcap we have found evidence supporting the "genetic" hypothesis. We first observed that the expression of migratory restlessness in experimental conditions in birds of different populations was a fairly good reflection of their fully or partially migratory habits. Secondly we found a substantial increase in the percentage of birds displaying restlessness in F_1-hybrids, compared to their African parents, when cross-breeding poorly migratory African with fully migratory German Blackcaps (Sect. 3.1). We finally obtained the most convincing results from an experiment of selective breeding with birds from the partially migratory southern French Blackcap population. When selectively breeding migrants with migrants and nonmigrants with nonmigrants, the ratio of the parental phenotype went up by about 8% and 30% respectively in the F_1-generations. Thus the characteristics "migratory" and "nonmigratory" in the Blackcap are to a considerable extent inheritable, and the diphenism is due to some type of dimorphism (Berthold and Querner 1982b).

5 Endogenous Control of Habitat Preferences During Migration

Migrants, often travelling months to reach their winter quarters, regularly interrupt their flights in order to rest and replenish fat reserves. Recent studies (Bairlein 1981) have shown that during stop-over periods, migrants are characterized by clear-cut species-specific habitat segregation due to strong habitat preferences. These preferences appear to a great extent to be endogenously programmed and related to morphological prerequisites of the flying apparatus. Thus passage migrants appear to be equipped with some type of innate habitat-conceptions related to their morphological structure.

6 Concluding Remarks

The evidence at present available on the control of typical migration in birds favours the opinion that it is controlled by a series of endogenous, inheritable programs and that typical migrants are to a great extent programmed "automates". The entire migratory behaviour appears to be related to a fundamental annual biological clock – the circannual rhythms. It must be assumed that inexperienced migrants perform their first migratory journey by flying in innately prescribed migratory directions for as long as their internal time-programs for migratory activity provide. By this they should "automatically" reach their hitherto unknown species- or even population-specific winter quarters (Vector-Navigation-Hypothesis). In partial migration, inheritance of the migratory urge seems to play a major role, and even the choice of a resting area during migration obviously has an essentially endogenous basis.

References

Bairlein F (1981) Ökosystemanalyse der Rastplätze von Zugvögeln. Ökol Vögel 3:7–37

Berthold P (1980) Die endogene Steuerung der Jahresperiodik: Eine kurze Übersicht. Proc XVII Internat Orn Congr Berlin, 1978, pp 473–478

Berthold P (to be published 1984) The endogenous control of bird migration: a survey of experimental evidence. Bird Study 31

Berthold P, Gwinner E, Klein H (1971) Circannuale Periodik bei Grasmücken (Sylvia). Experientia 27:399

Berthold P, Querner U (1981) Genetic basis of migratory behavior in European warblers. Science 212:77–79

Berthold P, Querner U (1982a) Genetic basis of moult, wing length, and body weight in a migratory bird species, Sylvia atricapilla. Experientia 38:801–802

Berthold P, Querner U (1982b) Partial migration in birds: experimental proof of polymorphism as a controlling system. Experientia 38:805

Gwinner E (1981) Circannual systems. In: Aschoff J (ed) Handbook of behavioral neurobiology, vol 4. Plenum Press, New York London, pp 391–410

Gwinner E, Wiltschko W (1978) Endogenously controlled changes in migratory direction of the garden warbler, Sylvia borin. J Comp Physiol Psychol 125:267–273

The Influence of the Earth Magnetic Field to the Migratory Behaviour of Pied Flycatchers (*Ficedula hypoleuca* PALLAS)

W. BECK [1]

1 Introduction

During autumn migration, Pied Flycatchers *(Ficedula hypoleuca)* like many other migrating bird species, have to change their initial migratory direction on reaching Portugal and south Spain from SW to SE to reach their central African wintering area (Zink 1977). While this Zugknick seems to be endogenously controlled in other bird species like the Garden Warbler (*Sylvia borin*, Gwinner and Wiltschko 1978), it was not to observed in Pied Flycatchers. While these birds were tested in the local geomagnetic field (0.46 Gauss, mN 360°, 66° Incl) without visual cues no change in the migratory direction or in orientation from the initial direction to SW after the 10th of October was observed, although Pied Flycatchers are able to orient with respect to the earth magnetic field (Beck and Wiltschko 1981). So it is fell that this bird species needs an external factor to change their migratory direction. The present experiments will prove whether the variation of inclination and force of the earths magnetic field along the migratory path of the Pied Flycatchers serves as this factor.

2 Material and Methods

The tests were carried out during autumn 1981 and 1982. The test birds had never migrated before the tests began. They were housed in rooms provided with an artificial photoperiod, simulating the natural photoperiod of Frankfurt. The birds were divided in two groups. Group I (control) was housed and tested only in the local magnetic field (0.46 Gauss, mN 360°, 66° Incl). Group II (test) was housed and tested in the local magnetic field during the beginning of migration. The inclination and the force of the magnetic field were varied to simulate the migratory path of the Pied Flycatchers (Magnetic Carts 1966). These artificial field conditions corresponded in inclination and force to that of the earth's magnetic field the birds would have registered free migrating at the time of variation of the artificial field. The test rooms were small, closed wooden houses. For tests Emlen funnel cages were used (for evaluation and statistical methods cf. Beck and Wiltschko 1981).

---

[1] Fachbereich Biologie (Zoologie), der Universität, 6000 Frankfurt, FRG

Localization and Orientation in Biology and Engineering
ed. by Varjú/Schnitzler
© Springer Verlag Berlin Heidelberg 1984

3 Results and Discussion

It could be observed that the gradual variation of inclination and force of the earth's magnetic field seems to be one of the external factors Pied Flycatchers *(Ficedula hypoleuca)* need to change their initial migratory direction from SW to SE. Only the birds whose migration was "magnetically simulated" (Group II) show the Zugknick

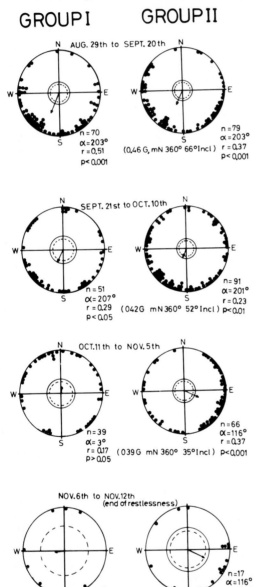

GROUP I   **GROUP II**

Fig. 1. Orientation behaviour of Pied Flycatchers during autumn migration 1981 and 1982.

Group I 16 Individuals housed and tested only in the local geomagnetic field of Frankfurt (0.46 Gauss, mN 360°, 66° Incl).

Group II 16 Individuals housed and tested in a magnetic field simulating the migratory path of Pied Flycatchers magnetically.

The birds of group I showed no oriented behaviour after changing their migratory direction to free migrating. The birds of group II changed their mean direction from SW to SE at the right time. The headings of the individual test nights are given by *dots* at the periphery of a circle with r = 1. The mean direction a of the test period with n tests is shown by the *arrow* from the middle of the circle with the length *r* relative to the radius of the circle. *The two inner circles* give the 5% *(dotted)* and the 1% significance borders of the Rayleigh-test

(Fig. 1). In the absence of this magnetic information the birds (Group I) stopped migration when they had to change their direction to free-migrating. These observations are different from findings with Garden Warblers, *Sylvia borin* (Gwinner and Wiltschko 1978). The different migration distances of these two bird species may explain this. The influences magnetic field perhaps plays an other role for a bird species which has to cross the magnetic equator like Garden Warblers, than in a bird species like Pied Flycatchers which winter north of the equator (Moreau 1972). Additional tests in future will have to be carried out to prove this.

Acknowledgements. This research is supported by the DFG in the program SFB 45, project G1 of Prof. Dr. W. Wiltschko.

References

Beck W, Wiltschko W (1981) Trauerschnäpper *(Ficedula hypoleuca* PALLAS) orientieren sich nichtvisuell mit Hilfe des Magnetfeldes. Die Vogelwarte 31:168–174

Gwinner E, Wiltschko W (1978) Endogenously controlled changes in migratory direction of the Garden Warbler *Sylvia borin.* J Comp Physiol Psychol 125:267–273

Magnetic Carts (1966) U.S. Naval Oceanographic Office (ed) 3rd edn 1700. The magnetic inclination, or dip, 1703. The total intensity of earth's magnetic force, Epoch 1965, 3rd edn. Washington, D.C.

Moreau RE (1972) The paleaarctic-african bird migration systems. Academic Press, London New York

Zink G (1977) Richtungsänderung auf dem Zuge bei europäischen Singvögeln. Die Vogelwarte 29:44–54

Migratory Birds: Path-Finding Using Madar?
(Magnetic Detection and Ranging)

P. N. MBAEYI[1]

This paper is motivated by considerations of the role of magnetic factors in memory processes of neural networks. In this context, the problem of navigational aids of birds offers a fertile ground of investigation as hypotheses exist which suggest that the astonishing navigational capabilities of birds rest on magnetic cues. Here affirmative theoretical evidences are advanced in favour of this likelihood, provided that the magnetic effects are assumed to be exerted indirectly. The presentation is limited to interpreted representation of mathematical modelling results; elaborate quantitative details of the model underlying these results will be contained in a separate paper (Mbaeyi in preparation).

The theoretical observations here are based on the following modelling procedure: The flight of a bird is comparable to the bouyant movement of an appropriately shaped geometrical body in a (non-rarefied) fluid interacting with electromagnetic fields; the fluid is the atmospheric air mass, while the electromagnetic fields are the three-dimensional density distributions of the earth magnetism and the aerial electricity, see Fig. 1.

In this first approximation we observe the problem of bouyancy in magnetohydrodynamical processes; the simplest form of the magnetohydrodynamical equations are assumed (McGraw-Hill Encyclopaedia 1966).

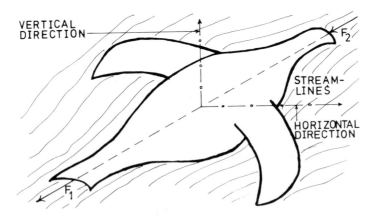

Fig. 1. Flight of bird

---

[1] University of Tübingen, 7400 Tübingen, FRG

Localization and Orientation in Biology and Engineering
ed. by Varjú/Schnitzler
© Springer Verlag Berlin Heidelberg 1984

If a bird slips through the atmosphere absolutely without friction, the forces on it at the nose and tail ends will be approximately equal (Fig. 1). Otherwise boundary layers exist (caused by friction) around the bird's geometrical shape. Directly on the bird's peripheral surface, the first approximation is that the air velocity there is zero. In the equations for magnetohydrodynamics, this means that velocity and all its derivatives vanish. The truncated equations that result lead to the following observations: The peripheral surface of a bird is subject to electrostatic charging and heating which are strongly pressure-dependent; in addition electrostatic charging is dependent on the magnetic field strength at the instantaneous location. Consequently, the following causal chain is obtained: (1) Local magnetic fields determine charge density on the surface of birds; (2) the charge density determines the strength of the contact (frictional) electricity on the surface of birds, and (3) the contact electricity determines the enveloping temperature around a bird (important since birds generally maintain homeothermic temperatures greater than 40 °C) and the pressure around a bird (important for acoustic supplement to the bird's navigational tools, Yodlowski 1977).

As a summary the following causal chain is obtained:

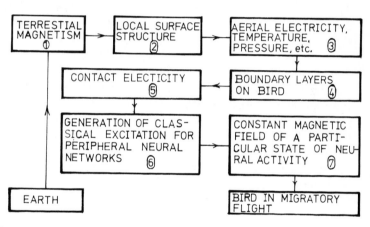

Fig. 2. Causal chain

By assigning memory-dependent behavior to stage 7 of this chain, we obtain a situation which is similar to some types of inertial navigation patterns (Schmidt-Koenig 1979); refer to Warnke (1979), for evidence of electrostatic charging of surfaces of avian species.

The indirect use of magnetic detection and ranging is obtained by interpreting the chain in Fig. 2 inversely, beginning with the bird itself. Through Zugunruhe a bird must put itself into stage 7 (based on endogenic factors, see Berthold, Chap. IV.10, this Volume), there-by prefixing the type of spikes it would preferably desire from peripheral nerves. This can only be achieved by seeking corresponding weather conditions that provide suitable contant electricity; the rest of the chain continues backwards from there. For this purpose, it is essential to presume that Zugunruhe is triggered by certain photo conditions (radiation intensity etc) of the environment; in this connection magnetochemical considerations, (e.g. magnetochemical processes considered by Schulten 1982 and Leask 1977) become relevant. On this basis, the beginning and continuation of migratory navigation is as follows: Birds will transpose themselves internally into totally new states involving higher neural activities (obvious use of Zugunruhe); they then must find optimally corresponding weather situations through a type of matching procedure comparable to terrain matching, as in the case of cruise missile navigation.

If this sort of matching procedure leads to repeatedly the same path finding, year after year, as is the case with migratory birds, then the following theoretical considerations are compelling: The earth's magnetic fields, especially through the influence of the vector $[\vec{t}_B \times \vec{n}_B) \times (\vec{t}_E \times \vec{n}_E)]$, $(\vec{t}_B, \vec{t}_E$, and $\vec{n}_B, \vec{n}_E$ are the tangential and the normal components of the magnetic and electric vectors respectively) assist in creating "Weather Corridors". Granted that this is the case, the resulting question concerns the problem of advance reconnaissance of these weather corridors by migratory birds. To explain this, consider another important aspect of avian behavior – acoustic behaviors. Allow us to assume that through the use of echo soundings, weather corridors facilitate the generation of something like Echo Sign-Posted Air Corridors, (ESPACS), (see Mbaeyi 1983 for corresponding mathematical modelling, and in particular of the nonstationary boundary layer problems essential to this phenomenon). For the maximum exploitation of the navigational aids outlined here, swarms of migratory birds would tend to use flight formations which minimize their energy expenditure, this explanation differing from that of Warnke (1969, pp 70–71). A first alternative is provided by an "Inverted Laminar Funnel" flight formation with densely packed forward end, while the rest of the flock flies in the laminar stream jet generated by this forward group. Another variation is as above, but with a separate forward group organized into an arc and acting as a type of re-connaissance party. In all these formations the resulting laminar streamlines will generate nearly uniform contact electricity throughout the flock; in addition, during periods of still (wind-stagnant) air, the mutual cumulative amplification of wing strokes will generate greater thrust for general speedy flight of the flock.

Conclusions

Magnetic cues affect navigational behavior of migratory birds only through induced indirect ways. This is in agreement with anatomical findings not able to locate magnetically active senders and receivers in avian neural systems, and explains the unsuccessful experiments involving attachments of magnets to avian wings (Schmidt-Koenig 1979). In contrast, magnets located on the skull or spinal cord will disturb confluent fluxes of contact electric fields, creating navigational distortions.

References

Kaufmann L, Williamson SJ (1982) Magnetic location of cortical activity. Ann NY Acad Sci 388:197–213

Leask MJM (1977) Physico-chemical mechanism for magnetic field detection by migratory birds and homing pigeons. Nature 267:144–145

McGraw-Hill (1966) McGraw-Hill encyclopaedia of science and technology, vol 8

Schmidt-Koenig K (1979) Avian orientation and navigation. Academic Press, London New York

Schulten K (1982) Magnetic fields effects in chemistry and biology. Festkoerperprobleme XXII: 61–83

Warnke U (1969) Information transmission by means of electrical biofields. In: Popp FA (ed) Electromagnetic bio-information. Springer, Berlin Heidelberg New York, pp 55–79

Wikswo JP Jr, Freeman JA (1980) Magnetic field of nerve impulse: First measurements. Science 208:53–55

Yodlowski ML (1977) Detection of atmospheric infrasound by homing pigeons. Nature 265: 725–726

Subject Index

Biophysics

Editors: W. Hoppe, W. Lohmann, H. Markl,
H. Ziegler
With contributions by numerous experts
1983. 852 figures. XXIV, 941 pages
ISBN 3-540-12083-1

Contents: The Structure of Cells (Prokaryotes,
Eukaryotes). – The Chemical Structure of Biologically Important Macromolecules. – Structure:
Determination of Biomolecules by Physical
Methods. – Intra-and Intermolecular Interactions. – Mechanisms of Energy Transfer. –
Radiation Biophysics. – Isotope Methods
Applied in Biology. – Energetic and Statistical
Relations. – Enzymes as Biological Catalysts. –
The Biological Function of Nucleic Acids. –
Thermodynamics and Kinetics of Self-assembly.
– Membranes. – Photobiophysics. – Biomechanics. – Neurobiophysics. – Cybernetic. – Evolution. – Appendix. – Subject Index.

Biophysics is the English translation of the completely revised second German edition of this
book. A wealth of information, new sections and
chapters, and extra figures make **Biophysics** an
indispensable text for advanced students and
lecturers of physics, chemistry, biology, and
medicine. Indeed every scientist interested in
biophysical questions will find much in this book
essential for a deeper understanding of modern
biophysical research.

From the reviews: "… Most of these contributions are very informative reviews of 10–20 pages
written for undergraduate students or scientists
who wish to broaden the scope of their knowledge. They provide a convenient source of information for many facets of biophysics and cannot
be found in such a concise way in other textbooks…"
Nature

Springer-Verlag
Berlin
Heidelberg
New York
Tokyo